Applied Mathematical Sciences | Volume 11

Applied Mathematical Sciences

F. John
PARTIAL DIFFERENTIAL EQUATIONS
ISBN 0-387-90021-7

L. Sirovich
TECHNIQUES OF ASYMPTOTIC ANALYSIS
ISBN 0-387-90022-5

J. Hale
FUNCTIONAL DIFFERENTIAL EQUATIONS
ISBN 0-387-90023-3

J. K. Percus
COMBINATORIAL METHODS
ISBN 0-387-90027-6

R. von Mises and K. O. Friedrichs
FLUID DYNAMICS
ISBN 0-387-90028-4

W. Freiberger and U. Grenander
A SHORT COURSE IN COMPUTATIONAL PROBABILITY AND STATISTICS
ISBN 0-387-90029-2

A. C. Pipkin
LECTURES ON VISCOELASTICITY THEORY
ISBN 0-387-90030-6

K. O. Friedrichs
SPECTRAL THEORY OF OPERATORS IN HILBERT SPACE
ISBN 0-387-90076-4

A. H. Stroud
NUMERICAL QUADRATURE AND SOLUTION OF ORDINARY DIFFERENTIAL EQUATIONS
ISBN 0-387-90100-0

W. A. Wolovich

Linear Multivariable Systems

With 28 Illustrations

Springer-Verlag New York · Heidelberg · Berlin
1974

W. A. Wolovich
Division of Engineering
Brown University
Providence, Rhode Island

AMS Classifications
Primary: 93xx, 34Axx, 34C35, C4H05, 08xx
Secondary: 49Exx, 70Gxx; 70Jxx, 12A20, 12Cxx,
12Dxx, 15A12, 15A21, 15A30, 15A54, 15A57

Library of Congress Cataloging in Publication Data

Wolovich, W A 1937-
Linear multivariable systems.

(Applied mathematical sciences, v. 11)
Bibliography: p.
1. System analysis. I. Title. II. Series.
QA1.A647 vol. 11 [QA402] 510'.8s [003] 74-9541

Printed in the United States of America.

ISBN 0-387-90101-9 Springer-Verlag New York · Heidelberg · Berlin
ISBN 3-540-90101-9 Springer-Verlag Berlin · Heidelberg · New York

PREFACE

This text was developed over a three year period of time (1971-1973) from a variety of notes and references used in the presentation of a senior/first year graduate level course in the Division of Engineering at Brown University titled Linear System Theory. The intent of the course was not only to introduce students to the more modern, state-space approach to multivariable control system analysis and design, as opposed to the classical, frequency domain approach, but also to draw analogies between the two approaches whenever and wherever possible. It is therefore felt that the material presented will have broader appeal to practicing engineers than a text devoted exclusively to the state-space approach.

It was assumed that students taking the course had also taken, as a prerequisite, an undergraduate course in classical control theory and also were familiar with certain standard linear algebraic notions as well as the theory of ordinary differential equations, although a substantial effort was expended to make the material as self-contained as possible. In particular, Chapter 2 is employed to familiarize the reader with a good deal of the mathematical material employed throughout the remainder of the text. Chapters 3 through 5 were drawn, in part, from a number of contemporary state-space and matrix algebraic references, as well as some recent research of the author, especially those portions which deal with polynomial matrices and the differential operator approach. The final three chapters which involve the synthesis of linear, multivariable systems are more recent in origin and were drawn, in large part, from the personal research of the author.

In any text of this type which is developed from class notes and lectures, feedback from students plays an important role in updating, modifying, and correcting the treatment in order to make it as palatable as possible. I would therefore like to take this opportunity to thank those numerous students who actively participated in this regard through their classroom and office discussions. I also would like to express my sincere gratitude to Mrs. Marian Regan who typed the first draft of this manuscript from my handwritten notes which were often almost illegible, and to Mrs. Katherine MacDougall for her expeditious typing of the final manuscript from which this text was prepared.

W. A. Wolovich

Providence, Rhode Island

TABLE OF CONTENTS

Page

Applied Mathematical Sciences | Volume 11

CHAPTER 1
INTRODUCTION

In the past decade control systems engineers have witnessed a rather drastic shift in the mathematical techniques employed for the analysis and synthesis of dynamical systems. The emphasis has shifted from the so-called "classical" or frequency domain methods developed prior to 1960 by investigators such as Nyquist, Bode, and Evans which have played such an important role with respect to scalar (single input/output) systems, to the more "modern" state-space approach which was popularized in this country in the early 1960's through the combined efforts of a number of individuals. This shift in emphasis is usually attributed to the increasing complexity of systems requiring control as well as the need or desire to "optimize" their performance. In the past decade, therefore, a wealth of information has been published dealing with the analysis and synthesis of "more complex" dynamical systems primarily from the more modern state-space point of view although, more recently, a small but increasing number of studies have employed frequency domain methods.

While system "complexity" can be measured in any number of ways; e.g. the number and type of nonlinearities, the order of the system, or the number of system inputs and outputs, the results given in this book will be directed primarily at this latter "measure of complexity". More specifically, the primary objective of this text will be to present a concise but comprehensive study of various techniques which can be used to analyze and control the behavior of a relatively simple class of multivariable (multi-input/output) dynamical systems, namely those which are linear, time-invariant, and finite dimensional. It might be noted that while the physical behavior of a particular multivariable system can not always be characterized in this way, in many cases it can be approximated in this manner, especially if only small perturbations about some nominal operating point are considered. This fact, of course, has long been recognized and employed with considerable success in the case of scalar systems compensated by classical methods.

In keeping with this primary objective, every attempt will be made to consolidate the most recent and significant theoretical contributions made in the field of linear multivariable systems analysis and synthesis. In order to be as general as possible, the procedures developed here will not be confined to only one domain, but will utilize both time and frequency domain methods. This permits a more efficient treatment than would be possible if the development were confined to a single domain. To illustrate this latter point, it might be noted that a particular design objective often can be most easily stated in one domain (e.g. a diagonal transfer matrix in the case of "decoupling"), although it may be attainable in terms of a compensation method which is best defined in the other domain (e.g. via linear state variable feedback). The ability to easily transfer from one domain to the other, which characterizes the approach employed here, usually can facilitate the resolution of specific synthesis questions of this type.

In general, Chapter 2 is employed primarily as a convenience to the reader in order to introduce certain well known, fundamental mathematical notions which are used throughout the remainder of the text. The material presented in Section 2.5 which deals with polynomial matrices represents the only exception, since it is not as completely or concisely accessible in any other text. The remaining first three chapters (3, 4, and 5) can perhaps best be characterized as "analysis chapters"; i.e. these chapters are devoted almost exclusively to a comprehensive study of three of the most common methods employed to describe the dynamical behavior of linear, time-invariant, multivariable systems as well as the interrelationships between any two of these three methods. In particular, a rather thorough exposition of the state-space representation is given in Chapter 3, transfer matrix methods are presented in Chapter 4, and the differential operator representation, which acts as a "bridge" between time and frequency domain methods, is discussed in Chapter 5. The various transfer relationships between any two of these three representations are developed in these three chapters and summarized at the conclusion of Chapter 5. This rather thorough approach to the analysis of multivariable systems thus permits a more comprehensive discussion of the various synthesis questions which are presented in the final three chapters. It also serves to provide a considerable amount of new insight into the structural properties of multivariable systems. More specifically, a notion of equivalence between state-space and differential operator representations is developed early (Section 5.2) and then used in a variety of ways throughout the remainder of the text; e.g. the notions of controllability and observability, previously confined to state-space systems alone, are extended in a natural way to include the more general class of differential operator systems in Section 5.3. Also, a resolution of the questions of "input function observability" and "output function controllability" via the construction of left and right inverse systems (Section 5.5) is greatly facilitated by the equivalence results obtained in Section 5.2.

As noted above, the final three chapters are devoted to a comprehensive study of the synthesis of multivariable systems, beginning in Chapter 6 with a thorough discussion of linear state variable feedback (l.s.v.f.) which is, by far, the most common contemporary method employed for the compensation of linear systems. The employment of "exponential state observers" in lieu of actual l.s.v.f. is then discussed at the conclusion of this chapter. A frequency domain characterization of the time domain notion of l.s.v.f. is then given in Chapter 7, along with a method for achieving the equivalent of a l.s.v.f compensation scheme implemented via a time domain "exponential state estimator", thus establishing a most important time/frequency domain analog. We then employ this frequency domain analog of l.s.v.f. to show that any l.s.v.f. design can be achieved by "appropriate" feedforward compensation. This latter observation then serves to motivate a rather general multivariable compensation scheme, developed in Section 7.4, which is shown to be analogous to the employment of "internal feedforward compensation" in combination with l.s.v.f. This final compensation scheme is fully exploited in Chapter 8 to resolve a number of general multivariable synthesis questions such as complete and arbitrary closed loop pole placement, decoupling, and exact model matching.

It might finally be noted that not all of the results given are formally developed in the main text; i.e. the problems presented at the conclusion of each chapter often serve as a means of formally establishing some rather important results which might otherwise cloud or unnecessarily complicate the major findings of a particular section. In this regard, informal remarks are often employed in the main text to state some rather important extensions of or clarifications to formally established results, which are then left as an exercise for the reader to formally establish in the problem section of the chapter.

CHAPTER 2

MATHEMATICAL PRELIMINARIES

2.1 INTRODUCTION

The primary purpose of this chapter is to introduce a number of preliminary mathematical concepts which are fundamental to the remainder of the text. We assume that the reader is familiar with certain basic notions in linear algebra, especially those involving matrix terminology and manipulation, but that he does not wish to consult an outside source in order to familiarize himself with certain more abstract concepts which will be required. We begin by formally defining and illustrating the concept of a "linear vector space $\mathscr{V}$ over a field $\mathscr{F}$" in Section 2.2. In this way, we are then able to introduce the notions of "linearly independent (or dependent) vectors", a "basis" for an n-dimensional vector space, and the use of scalar matrices to represent a change of basis. In Section 2.3, we introduce "linear operators" and illustrate the use of scalar matrices to represent finite dimensional linear operators in a given basis as well as an altered basis. A number of important and useful definitions and results involving scalar matrices are then presented in Section 2.4. In particular, "matrix rank" is formally defined, and various tests for rank determination are outlined. The notions of "eigenvalues" and the "characteristic polynomial" of scalar matrices with elements in the "complex field" $\mathscr{C}$ are also presented along with the very important Cayley-Hamilton Theorem. Section 2.5 is employed to introduce a number of less familiar results (some of which are new) pertaining to polynomial matrices, which play a paramount role in the development of a number of various analysis and synthesis techniques developed in the subsequent chapters. Certain of the definitions and results given are illustrated by example, and a number of selected problems are used to enable the reader to develop proficiency with the various techniques and procedures presented.

2.2 LINEAR VECTOR SPACES

Since the theory of linear dynamical systems is so highly dependent on linear algebra, we begin this chapter by reviewing several concepts which will prove to be fundamental to the development which follows. It is assumed that the reader is already familiar with certain basic algebraic concepts and, in particular, the notions of a set, a matrix, the transpose of a matrix A, denoted by A^T, the determinant of a square matrix A, denoted by $|A|$, and the inverse of a nonsingular square matrix A, denoted by A^{-1}, which is expressible as the quotient of the adjoint of A, denoted by A^+, and the determinant of A; i.e. $A^{-1} = A^+ \div |A|$. We further assume that the reader can compute the product of two matrices as well as any of the matrix functions indicated above. Since these concepts are often acquired without the formalism of "vector spaces over fields" and "linear operators", we will begin by formally defining and illustrating these notions.

DEFINITION 2.2.1: A field $\mathcal{F}$ consists of a set of objects or elements-- $a_1, a_2, a_3, \ldots$ called scalars which satisfy the following axioms:

(a) To every pair of scalars $\{a_i, a_j\}$ in $\mathcal{F}$, there corresponds a unique element, $a_i + a_j$, in $\mathcal{F}$ called the sum of a_i and a_j. Furthermore, the sum satisfies the commutative property, namely $a_i + a_j = a_j + a_i$, as well as the associative property, namely $a_i + (a_j + a_k) = (a_i + a_j) + a_k$.

(b) To every pair of scalars $\{a_i, a_j\}$ in $\mathcal{F}$, there corresponds a unique scalar, $a_i \cdot a_j$ or simply $a_i a_j$, in $\mathcal{F}$ called the product of a_i and a_j. Furthermore, the product satisfies the commutative property as well as the associative property.

(c) There is a unique scalar in $\mathscr{F}$, denoted by 0, such that $a + 0 = a$ for every scalar a in $\mathscr{F}$.

(d) There is a unique scalar in $\mathscr{F}$, denoted by 1, such that $a \cdot 1 = 1 \cdot a = a$ for every scalar a in $\mathscr{F}$.

(e) For every scalar a in $\mathscr{F}$, there corresponds a unique scalar $(-a)$ in $\mathscr{F}$ such that $a + (-a) = 0$.

(f) Multiplication is <u>distributive</u> with respect to addition; i.e. $a_i(a_j + a_k) = (a_i a_j) + (a_i a_k)$ for all elements a_i, a_j, and a_k in $\mathscr{F}$.

(g) To every nonzero element a in $\mathscr{F}$, there corresponds a unique element, a^{-1} or $1/a$, in $\mathscr{F}$, called the <u>inverse</u> of a, such that $aa^{-1} = 1$.

The following examples serve to illustrate the notion of a field:

EXAMPLE 2.2.2: The set $\mathscr{R}$ of all real numbers and the set $\mathscr{C}$ of all complex numbers, with the usual (base 10) rules of addition and multiplication constitute two of the most common fields. It might further be noted that $\mathscr{R}$ is a <u>subfield</u> of $\mathscr{C}$; i.e. a field whose elements also belong to $\mathscr{C}$.

EXAMPLE 2.2.3: The set of integers: $\ldots,-2,-1,0,1,2,\ldots$ with the usual (base 10) rules of addition and multiplication does not constitute a field since axiom (g) is violated.

EXAMPLE 2.2.4: The set $\{0,1\}$ of binary numbers is a field if the rules of addition and multiplication are appropriately defined, i.e. if $0 + 0 = 1 + 1 = 0 \cdot 1 = 0 \cdot 0 = 0$ and $1 + 0 = 1 \cdot 1 = 1$. Note that if the (base 10) sum, $1 + 1 = 2$ is employed in defining addition, axiom (a) is violated and the set (thus defined) does not constitute a field.

EXAMPLE 2.2.5: The set of all rational polynomial functions or simply the set of rational functions: $f(s) = \frac{a(s)}{b(s)}$, where $b(s) \neq 0$ and both $a(s)$ and $b(s)$ are polynomials of arbitrary but finite degree (m and n respectively) with coefficients in the field $\mathscr{R}$; i.e. $a(s) = a_o + a_1 s + \ldots + a_m s^m$ and $b(s) = b_o + b_1 s + \ldots + b_n s^n$ with all the a_i and b_j in $\mathscr{R}$, along with the usual rules of polynomial addition and multiplication, constitute the important field of rational functions which we will denote as $\mathscr{P}$. It might be noted that the set of all polynomials does not constitute a field since axiom (g) is violated as in Example 2.2.3.

In view of the above, it should be clear that a field may consist of either a finite or infinite number of scalars and, in general, is dependent upon a defined set of arithmetic rules for addition and multiplication. In view of this definition of a field, we now have:

DEFINITION 2.2.6: A <u>vector space (linear space) $\mathscr{V}$ over a field $\mathscr{F}$</u> consists of the following:

(a) A field $\mathscr{F}$ of scalars;

(b) A set $\mathscr{V}$ of objects, called <u>vectors</u>;

(c) An operation called <u>vector addition</u>, which associates with each pair of vectors $\{v_1, v_2\}$ in $\mathscr{V}$, a vector $v_1 + v_2$ in $\mathscr{V}$, called the <u>sum</u> of v_1 and v_2. Furthermore,

- (i) addition is commutative; i.e. $v_1 + v_2 = v_2 + v_1$,
- (ii) addition is associative; i.e. $v_1 + (v_2+v_3) = (v_1+v_2)+v_3$,
- (iii) there is a unique vector 0 in $\mathscr{V}$, called the <u>zero vector</u>, such that $v + 0 = v$ for all v in $\mathscr{V}$, and
- (iv) for each vector v in $\mathscr{V}$ there is a unique vector $-v$ in $\mathscr{V}$ such that $v + (-v) = 0$.

(d) An operation called <u>scalar multiplication</u>, which associates with each scalar a in $\mathscr{F}$ and vector v in $\mathscr{V}$, a vector av in $\mathscr{V}$, called the <u>product</u> of a and v. Furthermore,

- (i) $1 \cdot v = v$,

(ii) $(a_i a_j)v = a_i(a_j v)$

(iii) $a(v_i + v_j) = av_i + av_j$, and

(iv) $(a_i + a_j)v = a_i v + a_j v$.

It is important to note that a vector space should always be defined over some particular field $\mathscr{F}$ and is therefore a composite object, sometimes denoted by the pair $\{\mathscr{V}, \mathscr{F}\}$. In certain cases, however, when there is no possibility of confusion, a vector space may be defined without explicit reference to the field over which it is defined. In these cases, elements which constitute the vectors are generally assumed to define the unspecified field. The following examples help to illustrate these points.

EXAMPLE 2.2.7: The set of all n-tuples in some field $\mathscr{F}$, defined as the row vectors $v_i = (a_{i1}, a_{i2}, \ldots, a_{in})$ with scalars a_{ij} in $\mathscr{F}$, represents a vector space $\mathscr{V}$ over the field $\mathscr{F}$ provided vector addition and scalar multiplication are defined in the usual way; i.e. provided $v_i + v_j = (a_{i1}+a_{j1}, a_{i2}+a_{j2}, \ldots, a_{in}+a_{jn})$ and $av_i = (aa_{i1}, aa_{i2}, \ldots, aa_{in})$. It might be noted that the vectors which represent the n-tuples over $\mathscr{F}$ could have been chosen, just as easily, to be column vectors. Furthermore, this example serves to illustrate the final point made prior to the example--namely that the elements a_{ij} of each vector v_i define the field $\mathscr{F}$.

EXAMPLE 2.2.8: The set of all polynomials with coefficients in the field $\mathscr{R}$ is defined as the set of all vectors $a(s) = a_o + a_1 s + \ldots + a_n s^n$ with the a_i in $\mathscr{R}$ and n any positive finite integer. This set constitutes a vector space over $\mathscr{R}$ provided vector addition and scalar multiplication of polynomials are defined in the usual way. It might be noted that the set of all polynomials with coefficients in $\mathscr{R}$ and degree less than or equal to some specified integer n represents a "subspace" of this vector space; i.e. subspace of a vector space is defined as any portion of the vector space which itself retains all of the properties of the vector space. In view of the definition of a vector space over $\mathscr{F}$, we have generalized the notion of a vector to include more than just an n-tuple or a magnitude and direction in the "visual" two and three dimensional spaces. This is sometimes a difficult adjustment for students to make, especially if the word vector has become synonymous with a visual object or direction, and is thus worthy of mention.

DEFINITION 2.2.9: A set of vectors $v_1, v_2, \ldots, v_n$ in $\{\mathscr{V}, \mathscr{F}\}$ is said to be <u>linearly dependent over $\hat{\mathscr{F}}$</u> or simply <u>dependent over $\hat{\mathscr{F}}$</u> if there exist scalars $\hat{a}_1, \hat{a}_2, \ldots, \hat{a}_n$ in $\hat{\mathscr{F}}$, <u>not all zero</u>, such that

$$\hat{a}_1 v_1 + \hat{a}_2 v_2 + \ldots + \hat{a}_n v_n = \sum_1^n \hat{a}_i v_i = 0. \qquad 2.2.10$$

Otherwise the set $v_1, v_2, \ldots, v_n$ is said to be <u>linearly independent over $\hat{\mathscr{F}}$</u> or <u>simply independent over $\hat{\mathscr{F}}$</u>. Note that $\hat{\mathscr{F}}$ need not correspond to $\mathscr{F}$. The following examples illustrate the notions of linear dependence and independence:

EXAMPLE 2.2.11: The (column) vectors $v_1 = \begin{bmatrix} 1-i \\ 3+4i \end{bmatrix}$, $v_2 = \begin{bmatrix} i \\ 1 \end{bmatrix}$, and $v_3 = \begin{bmatrix} 0 \\ 1 \end{bmatrix}$ are linearly dependent over the field $\mathscr{C}$ of complex numbers (the field in which the elements of the vectors lie) since if $a_1 = -1$, $a_2 = -1-i$, and $a_3 = 4 + 5i$, then $\sum_1^3 a_i v_i = 0$. However, these same vectors are linearly independent over the field $\mathscr{R}$ (see Problem 2-4).

EXAMPLE 2.2.12: The vectors $1, s, s^2, \ldots, s^n$ are clearly independent over $\mathscr{R}$ but are not linearly independent over the field $\mathscr{P}$ of rational functions. It is thus seen that <u>the notions of linear dependence and independence are highly dependent on the field over which dependence (or independence) is defined</u>, an important point to note. If no specific mention of a field $\hat{\mathscr{F}}$ is made when discussing dependence and independence of vectors in $\{\mathscr{V}, \mathscr{F}\}$, it will be assumed that the field $\mathscr{F}$ is implicit in the definition. This being the case, the vectors v_1, v_2, and v_3 of Example 2.2.11 would be linearly dependent while the vectors of Example 2.2.12 would be linearly independent.

Associated with the notion of independent vectors is the maximum number of independent vectors which can belong to the same space $\{\mathscr{V}, \mathscr{F}\}$, and we now formalize this notion.

DEFINITION 2.2.13: The maximum number of linearly independent vectors (over $\mathscr{F}$) in a vector space $\{\mathscr{V}, \mathscr{F}\}$ will be called the dimension of the vector space. To illustrate, consider the following examples:

EXAMPLE 2.2.14: The dimension of the space of n-tuples over $\mathscr{F}$ is clearly n, since it is possible to represent any member of this space as a linear combination of the n linearly independent vectors e_1, $e_2,\ldots,e_n$, where e_i is an n-tuple with zero entries everywhere except for a 1 in the i-th entry.

EXAMPLE 2.2.15: The dimension of the space of polynomials with coefficients in $\mathscr{R}$ is equal to one more than the degree of the polynomial of highest degree in this space, since it is possible to express any member of this space as a linear combination of the linearly independent vectors $1,s,s^2,\ldots,s^n,s^{n+1},\ldots$. Since the degree of the polynomial of highest degree in this space is an unbounded integer, this space has unbounded (infinite) dimension. It might be noted that, for the most part, the vector spaces which we shall encounter in this text will be finite dimensional.

DEFINITION 2.2.16: A set of linearly independent vectors (over $\mathscr{F}$) in $\{\mathscr{V}, \mathscr{F}\}$ is said to be a basis of $\mathscr{V}$ if every vector in $\mathscr{V}$ can be expressed as a linear combination of the set. It might be noted that the vectors $e_1,e_2,\ldots,e_n$ defined in Example 2.2.14 clearly represent a basis of the space of n-tuples over $\mathscr{F}$. Furthermore, this particular basis is known as the standard basis for the space of n-tuples. Also the vectors $1,s,s^2,\ldots,s^n,s^{n+1},\ldots$ represent a basis of the space of polynomials with coefficients in $\mathscr{R}$ --see Example

2.2.15.

It is often important to determine a basis of a vector space or to ascertain whether or not a particular set of linearly independent vectors qualifies as a basis for the space. The following theorem provides a means of resolving these questions.

THEOREM 2.2.17: Any set of n linearly independent vectors (over $\mathscr{F}$) in $\{\mathscr{V}, \mathscr{F}\}$ qualifies as a basis of the n-dimensional vector space $\{\mathscr{V}, \mathscr{F}\}$.

Proof: Let $v_1, v_2, \ldots, v_n$ represent any set of linearly independent vectors in $\{\mathscr{V}, \mathscr{F}\}$, an n-dimensional vector space. If v_o is any vector in $\mathscr{V}$, then $\sum_o^n a_i v_i = 0$ must imply that not all of the $a_i = 0$ since if the converse were true, all $n + 1$ vectors $v_o, v_1, v_2, \ldots, v_n$ would be linearly independent and $\{\mathscr{V}, \mathscr{F}\}$ would have dimension greater than n. Furthermore, $a_o \neq 0$, since if $a_o = 0$, $\sum_o^n a_i v_i = 0$ would imply that $\sum_1^n a_i v_i = 0$ for not all the $a_i = 0$, contrary to the assumption that $v_1, v_2, \ldots, v_n$ are linearly independent. Therefore, any v_o in $\{\mathscr{V}, \mathscr{F}\}$ can be expressed as a linear combination of the v_i for $i = 1,2,\ldots,n$, which qualifies this set of vectors as a basis of $\mathscr{V}$ and thus establishes the theorem.

We now show that the representation of any member v_o of $\{\mathscr{V}, \mathscr{F}\}$ is uniquely specified in terms of a particular basis v_1, $v_2, \ldots, v_n$ of $\mathscr{V}$. In particular, assume that both

$$a_1 v_1 + a_2 v_2 + \ldots + a_n v_n = v_o \qquad 2.2.18$$

and

$$b_1 v_1 + b_2 v_2 + \ldots + b_n v_n = v_o \qquad 2.2.19$$

are representations of the same vector v_o and that the a_i and b_i for $i = 1,2,\ldots,n$ belongs to $\mathscr{F}$. If we subtract 2.2.19 from 2.2.18 we obtain

$$(a_1-b_1)v_1 + (a_2-b_2)v_2 + \ldots + (a_n-b_n)v_n = 0 \qquad 2.2.20$$

and since the v_i are linearly independent, $a_i = b_i$ for all $i = 1,2,\ldots,n$, thus establishing the desired result.

It should be noted that the ordered n-tuple $(a_1,a_2,\ldots,a_n)$ which defines the vector v_o is clearly dependent on the particular basis chosen as well as the ordering of the basis vectors, and it is now enlightening to investigate the relationship between the representations of the same vector v_o in two different bases. In particular, let $(a_1,a_2,\ldots,a_n)$ be the representation of v_o with respect to the basis $v_1v_2,\ldots,v_n$; i.e.

$$v_o = a_1v_1 + a_2v_2 + \ldots + a_nv_n = [v_1,v_2,\ldots,v_n]\begin{bmatrix} a_1 \\ a_2 \\ \vdots \\ a_n \end{bmatrix} \qquad 2.2.21$$

if we employ the standard <u>inner product</u> of the row vector $[v_1,v_2,\ldots,v_n]$ and the column vector $[a_1,a_2,\ldots,a_n]^T$ which will be denoted as A for convenience. If $\hat{v}_1,\hat{v}_2,\ldots,\hat{v}_n$ represents a different basis of $\mathscr{V}$, then in view of the above, every element of the basis $v_1,v_2,\ldots,v_n$ can be expressed <u>uniquely</u> in terms of the basis $\hat{v}_1,\hat{v}_2,\ldots,\hat{v}_n$ by an appropriate n-tuple in $\mathscr{F}$; i.e.

$$v_i = b_{1i}\hat{v}_1 + b_{2i}\hat{v}_2 + \ldots + b_{ni}\hat{v}_n = [\hat{v}_1,\hat{v}_2,\ldots,\hat{v}_n]\begin{bmatrix} b_{1i} \\ b_{2i} \\ \vdots \\ b_{ni} \end{bmatrix} \qquad 2.2.22$$

for all $i = 1,2,\ldots,n$. Therefore

$$[v_1,v_2,\ldots,v_n] = [\hat{v}_1,\hat{v}_2,\ldots,\hat{v}_n] \begin{bmatrix} b_{11} & b_{12} & \cdots & b_{1n} \\ b_{21} & & & \vdots \\ \vdots & & & \\ b_{n1} & \cdots & & b_{nn} \end{bmatrix}, \qquad 2.2.23$$

or

$$[v_1,v_2,\ldots,v_n] = [\hat{v}_1,\hat{v}_2,\ldots,\hat{v}_n]B \qquad 2.2.24$$

where B denotes the square, n-dimensional matrix $[b_{ij}]$. If we now recall 2.2.21, it is clear (in view of 2.2.24) that v_o can be expressed in terms of the basis $\hat{v}_1,\hat{v}_2,\ldots,\hat{v}_n$; i.e.

$$v_o = [\hat{v}_1,\hat{v}_2,\ldots,\hat{v}_n]\hat{A} \qquad 2.2.25$$

where

$$\hat{A} = BA \qquad 2.2.26$$

This latter equation represents the desired result; i.e. given the representation A of a vector v_o in a basis $v_1,v_2,\ldots,v_n$ as well as the relationship B between two different bases, the representation $\hat{A} = [\hat{a}_1,\hat{a}_2,\ldots,\hat{a}_n]^T$ of v_o in the basis $\hat{v}_1,\hat{v}_2,\ldots,\hat{v}_n$ can readily be ascertained using 2.2.26. It should be noted that if we initially knew $\hat{A}$, the representation of v_o with respect to the basis $\hat{v}_1,\hat{v}_2,\ldots,\hat{v}_n$, and wished to determine its n-tuple representation A with respect to the ordered basis $v_1,v_2,\ldots,v_n$, then in view of 2.2.16,

$$[\hat{v}_1,\hat{v}_2,\ldots,\hat{v}_n] = [v_1,v_2,\ldots,v_n]\hat{B} \qquad 2.2.27$$

for some <u>unique</u> $n \times n$ matrix $\hat{B} = [\hat{b}_{ij}]$ with elements in $\mathscr{F}$. In view of 2.2.25 and 2.2.27 it then follows that

$$v_o = [v_1,v_2,\ldots,v_n]\hat{B}\hat{A}, \qquad 2.2.28$$

and equating 2.2.28 to 2.2.21 we see that $\hat{B}\hat{A} = A$ or, in view of 2.2.26, that

$$\hat{B}BA = A \qquad 2.2.29$$

which implies that $\hat{B} = B^{-1}$, since A can represent any arbitrary vector in $\mathscr{V}$.

2.3 LINEAR OPERATORS

The results of the previous section will now be employed to study linear mappings (transformations) from one finite dimensional vector space into another. In particular, consider the n and m dimensional vector spaces $\mathscr{V}$ and $\mathscr{W}$ respectively, defined over the same field $\mathscr{F}$. If $v_1, v_2, \ldots, v_n$ is a basis of $\mathscr{V}$ and $w_1, w_2, \ldots, w_m$ a basis for $\mathscr{W}$, then any vector v in $\mathscr{V}$ can be represented by an n-tuple $(a_1, a_2, \ldots, a_n)$ in $\mathscr{F}$, and any vector w in $\mathscr{W}$ can be represented by an m-tuple $(b_1, b_2, \ldots, b_m)$ in $\mathscr{F}$; i.e.

$$v = a_1 v_1 + a_2 v_2 + \cdots + a_n v_n = [v_1, v_2, \ldots, v_n]A, \qquad 2.3.1$$

where $A^T = [a_1, a_2, \ldots, a_n]$, and

$$w = b_1 w_1 + b_2 w_2 + \cdots + b_m w_m = [w_1, w_2, \ldots, w_m]B, \qquad 2.3.2$$

where $B^T = [b_1, b_2, \ldots, b_m]$.

DEFINITION 2.3.3: A linear transformation from $\mathscr{V}$ into $\mathscr{W}$ is a function L from $\mathscr{V}$ into $\mathscr{W}$ such that $L(av + \hat{v}) = a(Lv) + L\hat{v}$ for all v and $\hat{v}$ in $\mathscr{V}$ and scalars a in $\mathscr{F}$. It should be noted that $a(Lv)$ and $L\hat{v}$ are both members of $\mathscr{W}$. We now establish an important and useful fact regarding the representation of linear operators, namely:

THEOREM 2.3.4: Consider the n and m dimensional vector spaces $\mathscr{V}$ and $\mathscr{W}$ respectively over the same field $\mathscr{F}$. Any linear transformation L from $\mathscr{V}$ into $\mathscr{W}$ is specified by the (n) individual mappings $Lv_1, Lv_2, \ldots, Lv_n$ of the basis vectors of $\mathscr{V}$ into $\mathscr{W}$. Furthermore, with respect to the ordered basis $v_1, v_2, \ldots, v_n$ of $\mathscr{V}$ and

<u>$w_1, w_2, \ldots, w_m$ of $\mathscr{W}$, L has a unique $m \times n$ matrix representation $C = [c_{ij}]$ with elements in $\mathscr{F}$.</u>

<u>Proof</u>: Since any vector v in $\mathscr{V}$ can be represented uniquely by an ordered n-tuple $(a_1, a_2, \ldots, a_n)$ in $\mathscr{F}$ as indicated by 2.3.1, it is clear in view of the linearity of L, that

$$Lv = a_1 Lv_1 + a_2 Lv_2 + \ldots + a_n Lv_n = [Lv_1, Lv_2, \ldots, Lv_n] \begin{bmatrix} a_1 \\ a_2 \\ \vdots \\ a_n \end{bmatrix} \quad 2.3.5$$

where the (n) individual mappings, Lv_i for $i = 1,2,\ldots,n$, are each vectors in $\mathscr{W}$. Since $w_1, w_2, \ldots, w_m$ is a basis of $\mathscr{W}$, it is clear that each Lv_i can be uniquely represented by some linear combination (an m-tuple) of these (m) basis vectors; i.e. for $i = 1,2,\ldots,n$,

$$Lv_i = c_{1i}w_1 + c_{2i}w_2 + \ldots + c_{mi}w_m = [w_1, w_2, \ldots, w_m] \begin{bmatrix} c_{1i} \\ c_{2i} \\ \vdots \\ c_{mi} \end{bmatrix} \quad 2.3.6$$

or, in general,

$$[Lv_1, Lv_2, \ldots, Lv_n] = [w_1, w_2, \ldots, w_m]C, \quad 2.3.7$$

where $C = [c_{ij}]$ is an $m \times n$ matrix with elements in $\mathscr{F}$. Theorem 2.3.4 is thus constructively established.

In view of 2.3.5 it is now clear that any vector v transformed from $\mathscr{V}$ into $\mathscr{W}$ under L can be represented in $\mathscr{W}$ by the m-tuple CA; i.e.

$$Lv = [w_1, w_2, \ldots, w_m]CA, \quad 2.3.8$$

where $A^T = [a_1, a_2, \ldots, a_n]$ and is defined by 2.3.5. Equating 2.3.8 to 2.3.2 it is also clear that B, the m-tuple representation of Lv in $\mathscr{W}$ is expressible as the product CA, where C is the unique matrix representation of L with respect to the defined ordered bases of $\mathscr{V}$ and $\mathscr{W}$, and A is the n-tuple representation of v in $\mathscr{V}$; i.e.

$$B = CA \qquad 2.3.9$$

To illustrate Theorem 2.3.4, consider the following:

EXAMPLE 2.3.10: Consider the n + 1 dimensional vector space consisting of all polynomials $a(s) = a_o + a_1 s + \ldots + a_n s^n$ of degree less than or equal to n with coefficients in $\mathscr{R}$. An obvious basis for this space is the ordered set of vectors $1, s, s^2, \ldots, s^n$. The differential operator $\frac{d}{ds}$ associates with each polynomial in this space, a polynomial of lower (by one) degree. Furthermore, the differential operator is linear since $\frac{d}{ds}[ca(s) + b(s)] = c\frac{da(s)}{ds} + \frac{db(s)}{ds}$, and therefore represents a linear transformation from one vector space (of dimension n + 1) into another (of dimension n). If $1, s, s^2, \ldots, s^{n-1}$ is chosen as the ordered basis of the latter space, the matrix representation C of the differential operator is readily determined; i.e. using 2.3.7, $[Lv_1, Lv_2, \ldots, Lv_{n+1}] = [0, 1, 2s, \ldots, ns^{n-1}] =$

$$[1, s, s^2, \ldots, s^{n-1}]C, \text{ where } C = \begin{bmatrix} 0 & 1 & 0 & \ldots & 0 \\ 0 & 0 & 2 & \ldots & 0 \\ 0 & 0 & 0 & & \cdot \\ \vdots & \vdots & \vdots & \ddots & \vdots \\ 0 & 0 & 0 & \ldots & n \end{bmatrix}.$$

It should be clear at this point that L need not map $\mathscr{V}$ into all of $\mathscr{W}$; e.g. $\mathscr{W}$ can be of larger dimension than $\mathscr{V}$. We now formalize this observation via the following:

DEFINITION 2.3.11: Let L be a linear transformation from $\mathscr{V}$ into $\mathscr{W}$, vector spaces of dimensions n and m respectively over $\mathscr{F}$.

The range space of L is the set of all vectors in $\mathscr{W}$ which can be expressed as some linear combination (sum) of the vectors $Lv_1, Lv_2, \ldots, Lv_n$. Also the set of all vectors v in $\mathscr{V}$ such that $Lv = 0$ is called the null space of L. It can readily be verified that the range space of L is a subspace of $\mathscr{W}$ and the null space of L is a subspace of $\mathscr{V}$, and the proof of these facts is left as an exercise for the reader (see Problem 2-9).

If $\mathscr{V}$ and $\mathscr{W}$ represent the same n-dimensional space, we have an important subclass of linear operators, which we will now discuss in further detail. In particular, consider any linear mapping L of $\mathscr{V}$ into itself. If $v_1, v_2, \ldots, v_n$ is an ordered basis of $\mathscr{V}$, then L has a matrix representation C which can be found by simply determining the n-tuple representation in $\mathscr{F}$ of each of the (n) basis vectors under L; i.e.

$$[Lv_1, Lv_2, \ldots, Lv_n] = [v_1, v_2, \ldots, v_n]C \tag{2.3.12}$$

Once C has been obtained it is often important to determine the effect on this matrix representation for L if a new basis $\hat{v}_1, \hat{v}_2, \ldots, \hat{v}_n$ of $\mathscr{V}$ is employed. It is clear that the matrix representation $\hat{C}$ for L in this new basis can be found if the effect of L on each new basis element is known: i.e. as in 2.3.12,

$$[L\hat{v}_1, L\hat{v}_2, \ldots, L\hat{v}_n] = [\hat{v}_1, \hat{v}_2, \ldots, \hat{v}_n]\hat{C}. \tag{2.3.13}$$

Often, however, the relationships implied by 2.3.13 are not explicitly given, but the matrix C (in 2.3.12) is known. To determine $\hat{C}$ we now recall 2.2.24, an expression for the matrix relationship B between the two bases of $\mathscr{V}$. In particular, if we substitute the right side of 2.2.24 for $[v_1, v_2, \ldots, v_n]$ in 2.3.12 and postmultiply both sides of the resulting expression by the inverse of B, we obtain the desired relationship; i.e.

$$[L\hat{v}_1 L\hat{v}_2,\ldots,L\hat{v}_n] = [\hat{v}_1,\hat{v}_2,\ldots,\hat{v}_n]BCB^{-1} \qquad 2.3.14$$

Comparing 2.3.14 with 2.3.13, it is clear, in view of Theorem 2.3.4, that

$$\hat{C} = BCB^{-1} \qquad 2.3.15$$

Any pair of matrices $\{C,\hat{C}\}$ which satisfies an expression such as 2.3.15 is said to be <u>similar</u>. Consequently, it is now clear that <u>any two matrix representations of the same linear transformation are similar</u>. To illustrate linear mappings of $\mathscr{V}$ into itself and a change of basis in $\mathscr{V}$, consider the following:

EXAMPLE 2.3.16: Consider the two dimensional vector space $\mathscr{V}$ of all 2-tuples over $\mathscr{R}$ with basis vectors $v_1 = \begin{bmatrix}1\\1\end{bmatrix}$ and $v_2 = \begin{bmatrix}-1\\0\end{bmatrix}$. If L represents a linear transformation of $\mathscr{V}$ into itself, defined by $Lv_1 = \begin{bmatrix}-1\\1\end{bmatrix}$ and $Lv_2 = \begin{bmatrix}1\\-1\end{bmatrix}$, the matrix representation C of L with respect to the ordered basis v_1,v_2 can readily be determined via 2.3.12; i.e. $[Lv_1,Lv_2] = \begin{bmatrix}-1 & 1\\1 & -1\end{bmatrix} = [v_1,v_2]C = \begin{bmatrix}1 & -1\\1 & 0\end{bmatrix}C$ which, by inverting $[v_1,v_2]$, implies that $C = \begin{bmatrix}1 & -1\\2 & -2\end{bmatrix}$. If we now require the matrix representation $\hat{C}$ of L with respect to the ordered basis $\hat{v}_1,\hat{v}_2 = \begin{bmatrix}0\\1\end{bmatrix}, \begin{bmatrix}1\\1\end{bmatrix}$, we simply employ 2.2.24; i.e. $[v_1,v_2] = [\hat{v}_1,\hat{v}_2]B$, or $\begin{bmatrix}1 & -1\\1 & 0\end{bmatrix} = \begin{bmatrix}0 & 1\\1 & 1\end{bmatrix}B$, which implies that $B = \begin{bmatrix}0 & 1\\1 & -1\end{bmatrix}$. $\hat{C}$ can now be found by using 2.3.15; i.e. $\hat{C} = BCB^{-1} = \begin{bmatrix}0 & 2\\0 & -1\end{bmatrix}$. With respect to the standard basis $e_1 = \begin{bmatrix}1\\0\end{bmatrix}$, $e_2 = \begin{bmatrix}0\\1\end{bmatrix}$ of $\mathscr{V}$, the matrix representation C_e of L would be $\begin{bmatrix}-1 & 0\\1 & 0\end{bmatrix}$ as the reader can readily verify. Thus, L maps e_1 into the vector $\begin{bmatrix}-1\\1\end{bmatrix}$ and e_2 into the null vector. The latter matrix, $C_e = \begin{bmatrix}-1 & 0\\1 & 0\end{bmatrix}$ is the one usually associated with the linear operator L, since the mapping Lv of a vector $v = \begin{bmatrix}a\\b\end{bmatrix}$ in $\mathscr{V}$

under L is most frequently represented by $C_e v$ or $\begin{bmatrix} -1 & 0 \\ 1 & 0 \end{bmatrix}\begin{bmatrix} a \\ b \end{bmatrix}$; i.e. $[Le_1, Le_2] = L[e_1, e_2] = [e_1, e_2]C_e$, and since any vector $\begin{bmatrix} a \\ b \end{bmatrix}$ in $\mathscr{V}$ is represented as $ae_1 + be_2 = [e_1, e_2]\begin{bmatrix} a \\ b \end{bmatrix}$, where $[e_1, e_2] = \begin{bmatrix} 1 & 0 \\ 0 & 1 \end{bmatrix}$, it is clear that $Lv = L[e_1, e_2]\begin{bmatrix} a \\ b \end{bmatrix} = [e_1, e_2]C_e v = C_e v$. C_e will be called the <u>standard matrix representation</u> for L in $\mathscr{V}$. It should be noted that in this example, the range space of L is the set of all vectors expressible as the product $a\begin{bmatrix} 1 \\ -1 \end{bmatrix}$ for any scalar a in $\mathscr{R}$. Furthermore, the null space of L is the set of all vectors expressible as the product $a\begin{bmatrix} 0 \\ 1 \end{bmatrix}$. It is well known fact in linear algebra that <u>the dimension of the range space</u> (the <u>rank</u>) <u>when added to the dimension of the null space</u> (the <u>nullity</u>) <u>of a linear operator L, which maps the n dimensional space $\mathscr{V}$ into the m dimensional space $\mathscr{W}$, is equal to the dimension n of the space $\mathscr{V}$</u>. This fact will not be formally established here but is left as an exercise for the reader (see Problem 2-11). Example 2.3.16 also serves to illustrate this fact. In particular, both the rank and nullity of L equal one, and when added together, produce two, the dimension of $\mathscr{W} = \mathscr{V}$.

2.4 SCALAR MATRICES

It should now be clear that matrices with elements in a field $\mathscr{F}$, which are called <u>scalar matrices</u> play a fundamental role in the study of linear algebra. As we have shown, they can represent a change of basis in $\{\mathscr{V}, \mathscr{F}\}$ or they can be used to denote an ordered set of basis vectors. Matrices are also commonly used to represent linear operators which map one vector space into another and are generally employed to simplify the notation required to illustrate various mathematical relationships. There are certain important properties of scalar matrices which can be defined irrespective of their specific employment, and the purpose of this section will be to discuss certain of these general properties.

DEFINITION 2.4.1: The rank ρ of a matrix A, denoted as $\rho[A]$, is equal to the maximum number of linearly independent columns (or rows) of A over the smallest field $\mathcal{F}$ which contains the elements of A.[†] It might be noted that whenever a matrix A is used to represent a linear operator L, the rank of A is identical to the dimension of the range space (the rank) of L. As defined here, however, the notion of matrix rank is introduced without mention of linear operators. It might further be noted that the number of linearly independent columns of a matrix A is always equal to the number of independent rows of A, provided the notion of independence is defined with respect to the smallest field $\mathcal{F}$ which contains the elements of A. To illustrate these points, consider the following examples:

EXAMPLE 2.4.2: The matrix $A = \begin{bmatrix} 1 & 0 & -1 \\ 2 & 1 & -1 \\ 0 & 1 & 1 \end{bmatrix}$ with elements in the field $\mathcal{R}$ has rank 2, since only two columns (or two rows) of A are linearly independent; i.e. the third column of A can be expressed as the difference between the first and second columns.

EXAMPLE 2.4.3: The polynomial matrix $P(s) = \begin{bmatrix} s+1 \\ s+2 \end{bmatrix}$ with elements in the vector space of polynomials with coefficients in $\mathcal{R}$ has rank 1, since the smallest field which includes the elements of P(s) is $\mathcal{P}$, the field of rational functions with coefficients in $\mathcal{R}$, and the scalar $\frac{s+2}{s+1}$ in $\mathcal{P}$ times the first row of P(s) equals s+2, the second row. Also, since P(s) has only one column, it cannot have rank greater than one. It is interesting to note that in this example, the two rows of P(s) are linearly independent over $\mathcal{R}$ although P(s) has rank 1. Perhaps the easiest way to determine the rank of a matrix is to employ the following well-known fact: The rank of a matrix A is equal to the order of the largest order nonzero minor of A (The

---

[†]Thus defined, the notion of matrix rank does not apply strictly to scalar matrices.

dimension of the largest "submatrix" of A obtained by column/row deletion with nonzero determinant), and as a consequence of this fact: any n-dimensional square matrix A has full rank n if and only if its determinant is nonzero. Both of these statements apply to polynomial matrices as well as matrices with elements in some field $\mathscr{F}$.[†] Note that the determinant of the rank 2 matrix A of Example 2.4.2 is zero, which is consistent with these two statements.

Associated with every $(n \times n)$ matrix A with elements in $\mathscr{C}$ (or $\mathscr{R}$) is a set of n scalars in $\mathscr{C}$ known as the "eigenvalues" of A. These scalars play an important role in linear system theory, as we will later show, and they are now formally defined.

DEFINITION 2.4.4: An eigenvalue of a square matrix A with elements in $\mathscr{C}$ is any scalar λ (in $\mathscr{C}$) such that the determinant of the $(n \times n)$ matrix $(\lambda I-A)$ is zero; i.e. the solutions of $|\lambda I-A| = 0$ define the eigenvalues of A. The determinant of the polynomial matrix $(\lambda I-A)$ is called the characteristic polynomial of A and will be denoted by $\Delta(\lambda)$. It should be noted that $\Delta(\lambda)$, thus defined, is a monic polynomial in λ of degree n; i.e.

$$|\lambda I-A| = \Delta(\lambda) = a_o + a_1\lambda + \ldots + a_{n-1}\lambda^{n-1} + \lambda^n \qquad 2.4.5$$

whose (n) zeros are the eigenvalues of A. To illustrate these notions consider the following:

EXAMPLE 2.4.6: The three eigenvalues of the (3×3) matrix A of Example 2.4.2 can be determined by equating the characteristic polynomial of A to zero; i.e. $\Delta(\lambda) = |\lambda I-A| = \left|\begin{bmatrix} \lambda-1, & 0, & 1 \\ -2, & \lambda-1, & 1 \\ 0, & -1, & \lambda-1 \end{bmatrix}\right| =$

[†] Actually, polynomial matrices may be considered to have elements in the field $\mathscr{P}$ for the purposes of this section. Strictly speaking, however, the elements of a polynomial matrix belong to something less than a field; in particular, a "commutative ring with identity"-- see Section 2.5.

$\lambda^3 - 3\lambda^2 + 4\lambda = \lambda(\lambda - \frac{3}{2} - j\frac{\sqrt{7}}{2})(\lambda - \frac{3}{2} + j\frac{\sqrt{7}}{2}) = 0$. The scalars: 0, $\frac{3}{2} + j\frac{\sqrt{7}}{2}$, and $\frac{3}{2} - j\frac{\sqrt{7}}{2}$ clearly represent the three eigenvalues of A, and $\Delta(\lambda) = \lambda^3 - 3\lambda^2 + 4\lambda$ is the characteristic polynomial of A. It might be noted that the relation $\Delta(\lambda) = 0$ is called the characteristic equation of the matrix A. In view of the proceding, we are now in a position to state and establish a fundamental result in linear algebra, namely:

THEOREM 2.4.7 (Cayley-Hamilton): Any $n \times n$ matrix A with elements in $\mathscr{C}$ satisfies its own characteristic equation; i.e. $\Delta(A) = 0$.

Proof: We first express the inverse of $(\lambda I-A)$ as the quotient of the adjoint of $(\lambda I-A)$, a polynomial matrix with elements of degree $\leq n-1$, and the determinant $\Delta(\lambda)$ of $(\lambda I-A)$; i.e.

$$(\lambda I-A)^{-1} = \frac{(\lambda I-A)^+}{\Delta(\lambda)}, \qquad 2.4.8$$

where $(\lambda I-A)^+$ can be expressed as

$$(\lambda I-A)^+ = B_o + B_1\lambda + \ldots + B_{n-1}\lambda^{n-1} \qquad 2.4.9$$

If we now multiply both sides of 2.4.8 by $\Delta(\lambda)(\lambda I-A)$, while simultaneously employing the expression 2.4.9 for $(\lambda I-A)^+$, we obtain

$$a_o + a_1\lambda + \ldots + a_{n-1}\lambda^{n-1} + \lambda^n = -AB_o + (B_o - AB_1)\lambda + \ldots + (B_{n-2} - AB_{n-1})\lambda^{n-1} + B_{n-1}\lambda^n. \qquad 2.4.10$$

If we then equate the coefficients of identical powers of λ in 2.4.10, the following recursive relationships result:

$$\begin{aligned} B_{n-1} &= I, \\ B_{n-2} &= AB_{n-1} + a_{n-1}I, \\ B_{n-3} &= AB_{n-2} + a_{n-2}I, \\ &\vdots \\ B_o &= AB_1 + a_1 I, \end{aligned} \qquad 2.4.11$$

and

$$AB_o = -a_oI.$$

Substituting recursively for B_o, then B_1, then B_2, etc. up to and including B_{n-1}, their respective values as given by the relations 2.4.11, we establish the fact that $A^n + a_{n-1}A^{n-1} + \ldots + a_1A + a_oI = \Delta(A) = 0$, and thus establish the Cayley-Hamilton Theorem.

EXAMPLE 2.4.12: We can readily verify that the matrix A given in Examples 2.4.2 and 2.4.6 satisfies its own characteristic equation $\Delta(\lambda) = \lambda^3-3\lambda^2+4\lambda = 0$; i.e. $A^3 = \begin{bmatrix} -1 & -3 & -2 \\ 4 & -4 & -8 \\ 6 & 2 & -4 \end{bmatrix}$, $-3A^2 = \begin{bmatrix} -3 & 3 & 6 \\ -12 & 0 & 12 \\ -6 & -6 & 0 \end{bmatrix}$, $4A = \begin{bmatrix} 4 & 0 & -4 \\ 8 & 4 & -4 \\ 0 & 4 & 4 \end{bmatrix}$ and the sum $A^3-3A^2+4A = \Delta(A) = 0$, as the reader can readily verify.

It should be noted that although A does satisfy its own characteristic equation, $\Delta(A)$ is not always the monic polynomial Δ_m, of least degree, with the property that $\Delta_m(A) = 0$. We therefore have:

DEFINITION 2.4.13: The monic polynomial Δ_m of least degree with the property that $\Delta_m(A) = 0$ is called the minimal polynomial of A. It might be noted that the minimal polynomial of A always divides the characteristic polynomial Δ of A, and that all roots of Δ_m are also roots of Δ.

2.5 POLYNOMIAL MATRICES

We now consider a most useful class of matrices in linear systems theory, namely those whose elements are polynomials of finite, but unbounded, degree with coefficients in the field $\mathscr{R}$. These polynomial matrices were introduced briefly in the previous section and differ from scalar matrices in one fundamental way. In particular, the polynomial elements of this class of matrices do not belong to a field $\mathscr{F}$, but rather belong to a "commutative ring with identity", which has all the properties of a field except for the inverse, i.e. the inverse of a polynomial of degree one or more is not a polynomial. Since the only class of matrices with elements in a commutative ring which we will employ in our subsequent discussions are polynomial matrices, the notion of rings will not be discussed any further, and we will direct our attention exclusively to some important properties of polynomial matrices without attempting to generalize these properties to include all matrices with elements in a commutative ring. In particular, we now have:

DEFINITION 2.5.1: The following three elementary row (column) operations on the polynomial matrix $P(s)$ with coefficients in $\mathscr{R}$ are defined:

(i) Interchange of rows (columns) i and j.

(ii) Multiplication of row (column) i by a nonzero scalar in $\mathscr{R}$.

(iii) Replacement of row (column) i by itself plus any polynomial multiplied by any other row (column) j.

DEFINITION 2.5.2: A unimodular matrix $U(s)$ is defined as any square matrix which can be obtained from the identity matrix I by a finite number of elementary row and column operations on I. The determinant of a unimodular matrix is therefore a nonzero scalar in $\mathscr{R}$, and conversely, any polynomial matrix whose determinant is a nonzero scalar in $\mathscr{R}$ is a unimodular matrix. In view of the above, it follows that any sequence of elementary row operations on $P(s)$ is equivalent to premultiplication (left multiplication) of $P(s)$ by an appropriate unimodular matrix $U_L(s)$. Similarly, any sequence of elementary column operations on $P(s)$ is equivalent to postmultiplication (right multiplication) of $P(s)$ by an appropriate unimodular matrix $U_R(s)$.

DEFINITION 2.5.3: Two polynomial matrices $P(s)$ and $Q(s)$ will be called (a) row equivalent, (b) column equivalent, or (c) equivalent if and only if one of them can be obtained from the other by a sequence of elementary (a) row, (b) column, or (c) row and column operations respectively. $P(s)$ is thus (a) row equivalent, (b) column equivalent, or (c) equivalent to $Q(s)$ if and only if (a) $P(s) = U_L(s)Q(s)$, (b) $P(s) = Q(s)U_R(s)$, or (c) $P(s) = U_L(s)Q(s)U_R(s)$ respectively, where $U_L(s)$ and $U_R(s)$ are unimodular matrices.

The degree of a polynomial matrix $P(s)$ is equal to the degree of the polynomial element of highest degree in $P(s)$. The degree of the i-th column of $P(s)$, denoted as $\partial_{ci}[P(s)]$, or simply ∂_{ci}, is defined by applying the definition of degree to the i-th column of $P(s)$ (a matrix with only one column). We will denote by $\Gamma_c[P(s)]$, or simply Γ_c, the scalar matrix with elements in $\mathscr{R}$ consisting of the coefficients of the highest degree s terms in each column of $P(s)$. To illustrate, consider the following:

EXAMPLE 2.5.4: If $P(s) = \begin{bmatrix} s^2-3, & 1, & 2s \\ 4s+2, & 2, & 0 \\ -s^2, & s+3, & -3s+2 \end{bmatrix}$, then $\partial_{c1} = 2$, $\partial_{c2} = \partial_{c3} = 1$, and for this $P(s)$, $\Gamma_c = \begin{bmatrix} 1 & 0 & 2 \\ 0 & 0 & 0 \\ -1 & 1 & -3 \end{bmatrix}$. The notion of the degree of the i-th row of $P(s)$ is similarly defined; i.e. using the same example to illustrate "row degree", $\partial_{r1} = \partial_{r3} = 2$, $\partial_{r2} = 1$, and for this $P(s)$, $\Gamma_r = \begin{bmatrix} 1 & 0 & 0 \\ 4 & 0 & 0 \\ -1 & 0 & 0 \end{bmatrix}$. If $P(s)$ is a $q \times q$ nonsingular matrix, it can therefore be represented as

$$P(s) = \left[\begin{array}{c|c|c|c} p_{11}s^{\partial_{c1}}+\dots & p_{12}s^{\partial_{c2}}+\dots & & p_{1q}s^{\partial_{cq}}+\dots \\ p_{21}s^{\partial_{c1}}+\dots & \cdot & & \cdot \\ \cdot & \cdot & \cdots\cdots & \cdot \\ \vdots & \cdot & & \cdot \\ p_{q1}s^{\partial_{c1}}+\dots & p_{q2}s^{\partial_{c2}}+\dots & & p_{qq}s^{\partial_{cq}}+\dots \end{array}\right], \quad 2.5.5$$

where the +... denotes lower degree s-terms in each (i-th) column of $P(s)$. It is clear that $\Gamma_c[P(s)] = [p_{ij}]$. Let $\text{diag}[s^{\partial_{ci}}]$ be defined as the $q \times q$ diagonal matrix with diagonal entries $s^{\partial_{ci}}$, and define $P^c(s)$ as $\Gamma_c \times \text{diag}[s^{\partial_{ci}}]$. It can then easily be verified by induction that $|P(s)| = |\Gamma_c|s^p$ + lower degree terms in s, where $p = \sum_1^q \partial_{ci}$, and that $|P^c(s)| = \gamma_c s^p$, where $\gamma_c = |\Gamma_c|$.

DEFINITION 2.5.6: A $(q_1 \times q_2)$ polynomial matrix, $P(s)$, will be called <u>column (row) proper</u> if and only if $\Gamma_c[P(s)]$ $(\Gamma_r[P(s)])$ has full rank $(= \min\{q_1,q_2\})$. In view of the above, it is thus clear that <u>a square polynomial matrix is column (row) proper if and only if</u> $\gamma_c = |\Gamma_c|$ $(\gamma_r = |\Gamma_r|) \neq 0$. In light of this definition, we can now constructively establish:

THEOREM 2.5.7: <u>Any nonsingular polynomial matrix $P(s)$ is column (row) equivalent to a column (row) proper matrix; i.e. one can always find a unimodular matrix</u> $U_R(s)$ $(U_L(s))$ <u>which reduces $P(s)$ to column (row) proper form</u>.

<u>Proof</u>: If $|P^c(s)| \neq 0$, $P(s)$ would, by definition, be column proper and we would be done. Therefore, suppose that $|P^c(s)| = \gamma_c s^p = 0$; i.e. suppose that $\gamma_c = 0$. This would imply that the (q) column vectors $p_i^c(s)$ which comprise $P^c(s)$ were linearly dependent over $\mathscr{P}$,

the field of rational functions and, in particular, that

$$\sum_1^q p_i(s)P_i^c(s) = 0 \qquad 2.5.8$$

for two or more nonzero monomials $p_i(s)$, for $i = 1,2,\ldots,q$. At least one of these monomials can be made unity by dividing 2.5.8 by a nonzero monomial $p_k(s)$ of lowest degree; i.e.

$$\sum_1^q (p_i(s)/p_k(s))\; P_i^c(s) = \sum_1^q \tilde{p}_i(s)P_i^c(s) = 0, \qquad 2.5.9$$

where $\tilde{p}_k(s) = 1$. The replacement of column k of $P^c(s)$ by $\sum_1^q \tilde{p}_i(s)P_i^c(s)$ is analogous to postmultiplying $P^c(s)$ by the unimodular matrix $U_1(s)$, where $U_1(s)$ is the identity matrix with an "appropriately altered" k-th column; i.e.

$$U_1(s) = \begin{bmatrix} 1 & 0 & \cdots & \tilde{p}_1(s) & 0 & \cdots & 0 \\ 0 & 1 & \cdots & \tilde{p}_2(s) & 0 & \cdots & 0 \\ \vdots & & \ddots & & & & \\ 0 & & & 1 & 0 & \cdots & 0 \\ 0 & & & \tilde{p}_{k+1}(s) & 0 & \cdots & 0 \\ \vdots & & & \vdots & & \ddots & \vdots \\ 0 & & & \tilde{p}_q(s) & 0 & & 1 \end{bmatrix}, \qquad 2.5.10$$

where at least one of the $\tilde{p}_i(s)$, $i \neq k$, is nonzero. Clearly, if $P^c(s)$ is postmultiplied by $U_1(s)$, the k-th column of $P^c(s)U_1(s)$ will be identically zero. Now if $\overline{P}(s) = P(s)U_1(s)$, $\overline{P}(s)$ is identical to $P(s)$ with the exception of an altered k-th column. More specifically, the degree of the k-th column of $\overline{P}(s)$ is strictly less than the degree ∂_{ck} of the k-th column of $P(s)$. $\overline{P}(s)$ is therefore a new candidate (of lower column degree) for a column proper matrix; i.e. $|P^c(s)| = \gamma_{1c}s^{p1}$ + lower degree terms in s, where $p1$ is strictly less than p. If $\gamma_{1c} \neq 0$, the reduced matrix $\overline{P}(s)$ is column proper and we're done. If $\gamma_{1c} = 0$, the above procedure can be repeated again and as many times as necessary to produce (for some j) a nonzero γ_j. Then $\hat{P}(s) = P(s) \prod_1^j U_i(s)$ would be column proper and $P(s)$ would be column equivalent to $\hat{P}(s)$, since $U_R(s) = \prod_1^j U_i(s)$ is a unimodular matrix. The first half of Theorem 2.5.7 is thus constructively established and the second half follows directly by "duality" (i.e. by transposing matrices and repeating the same arguments). To illustrate the algorithm, we recall the polynomial matrix $P(s)$ of Example 2.5.4; i.e. if $P(s) = \begin{bmatrix} s^2-3 & 1, & 2s \\ 4s+2, & 2, & 0 \\ -s^2, & s+3, & -3s+2 \end{bmatrix}$, then $\Gamma_c = \begin{bmatrix} 1 & 0 & 2 \\ 0 & 0 & 0 \\ -1 & 1 & -3 \end{bmatrix}$, and $|P(s)| = 6s^3 + 44s^2 + 28s - 16 \neq 0$. Clearly, this nonsingular $P(s)$ is not column proper since $|\Gamma_c| = \gamma_c = 0$. To reduce $P(s)$ to column proper form, we first determine $P^c(s)$; i.e.

$$P^c(s) = \begin{bmatrix} s^2 & 0 & 2s \\ 0 & 0 & 0 \\ -s^2 & s & -3s \end{bmatrix} = \Gamma_c \times \text{diag}[s^{\partial ci}] \quad \text{with} \quad s^{\partial ci} = s^2 \text{ and } s^{\partial c2} =$$

$s^{\partial c3} = s$. We next determine any three monomials $p_i(s)$ which satisfy 2.5.8. For this example, $p_1(s) = -2$, $p_2(s) = s$, and $p_3(s) = s$ represent an appropriate choice. Since $p_1(s)$ represents the monomial of least degree, we divide all three monomials by -2 to obtain $U_1(s)$; i.e.

$$U_1(s) = \begin{bmatrix} 1 & 0 & 0 \\ -\frac{s}{2} & 1 & 0 \\ -\frac{s}{2} & 0 & 1 \end{bmatrix}. \quad \overline{P}(s) = P(s)U_1(s) = \begin{bmatrix} -\frac{s}{2}-3, & 1, & 2s \\ 3s+2, & 2, & 0 \\ -\frac{5}{2}s, & s+3, & -3s+2 \end{bmatrix}$$

is then determined and found to be column proper. Therefore $U_1(s)$ represents an appropriate unimodular matrix; which "reduces" $P(s)$ to column proper form, and no further reduction is necessary.

We now establish another theorem of considerable importance, namely:

THEOREM 2.5.11: Any $(p \times q)$ polynomial matrix $P(s)$ is row (column) equivalent to one of the two upper right triangular (lower left triangular) matrices shown below; i.e. one can always find a unimodular matrix $U_L(s)$ $(U_R(s))$ which reduces $P(s)$ to upper right triangular (lower left triangular) form $\hat{P}(s) = U_L(s)P(s)$ $(P(s)U_R(s))$ where[†]

(i) when $p \le q$ (ii) when $p \ge q$

$$\hat{P}(s) = \begin{bmatrix} \hat{p}_{11}(s) & \hat{p}_{12}(s) & \cdots & & \hat{p}_{1q}(s) \\ 0 & \hat{p}_{22}(s) & \cdots & & \hat{p}_{2q}(s) \\ \vdots & \vdots & \ddots & & \vdots \\ 0 & 0 & \cdots & \hat{p}_{pp}(s) \cdots & \hat{p}_{pq}(s) \end{bmatrix},$$

or

$$\begin{bmatrix} \hat{p}_{11}(s) & \hat{p}_{12}(s) & \cdots & \hat{p}_{1q}(s) \\ 0 & \hat{p}_{22}(s) & \cdots & \vdots \\ \vdots & & \ddots & \vdots \\ 0 & \cdots & & \hat{p}_{qq}(s) \\ 0 & \cdots & & 0 \\ \vdots & & & \vdots \\ 0 & \cdots & & 0 \end{bmatrix}$$

Furthermore, in both of the above forms, the polynomials $\hat{p}_{1k}(s)$, $\hat{p}_{2k}(s), \ldots, \hat{p}_{k-1,k}(s)$, are of lower degree than $\hat{p}_{kk}(s)$ for all $k = 1,2,\ldots,p$ if $\partial[\hat{p}_{kk}(s)] > 0$, and are all zero if $\hat{p}_{kk}(s)$ is a non-zero scalar in $\mathscr{R}$.

Proof: If the first column of $P(s)$ is not identically zero we can choose a polynomial of least degree from its elements and, by a permutation of the rows, make it the new 1,1 entry, $\tilde{p}_{11}(s)$. We then

[†]It should be noted that only the upper right triangular form is displayed.

apply the division algorithm to every other nonzero entry in the first column; i.e. we divide every other nonzero element $\tilde{p}_{i1}(s)$ in the first column by $\tilde{p}_{11}(s)$, obtaining the quotients $\tilde{q}_{i1}(s)$ and remainders $\tilde{r}_{i1}(s)$ according to the relationship:

$$\tilde{p}_{i1}(s) = \tilde{p}_{11}(s)\tilde{q}_{i1}(s) + \tilde{r}_{i1}(s), \qquad 2.5.12$$

where either $\tilde{r}_{i1}(s) = 0$ or $\partial(\tilde{r}_{i1}(s)) < \partial(\tilde{p}_{11}(s))$. We then subtract from each nonzero i-th row, the first row multiplied by $\tilde{q}_{i1}(s)$. If not all of the remainders $\tilde{r}_{i1}(s)$ are zero, we choose one of least degree and make it the new 1,1 entry by another permutation of the rows. The net result of repeating this process "as many times as necessary" is to continually reduce the degree of the polynomial element in the 1,1 entry. Since the degree of the 1,1 entry is finite, this repeated process must end at some stage; in particular, when all of the remaining elements of the first column are identically zero.

We next consider the second column of this altered matrix and, ignoring the first row for the moment, apply the above procedure to the elements beginning with the second row and second column. In this way we zero all of the elements below the 2,2 entry. If the 1,2 element is of equal or higher degree than that of the 2,2 element, the division algorithm can be employed to reduce the 1,2 element to the remainder term associated with the division of the 1,2 entry by the 2,2 entry or to zero if both elements are scalars. Continuing in this manner, with the elements beginning with the third row and third column next, we eventually reduce $P(s)$ to the appropriate form. To illustrate this constructive proof of the theorem, consider the following:

EXAMPLE 2.5.13: Suppose we are required to reduce the polynomial matrix $P(s) = \begin{bmatrix} s+1, & 1 \\ s^2, & 2 \end{bmatrix}$ to upper right triangular form $\hat{P}(s)$. We first note that the 1,1 entry $s+1$ is of lower degree than the 2,1 entry s^2. Therefore, an initial permutation of rows is unnecessary. The division algorithm 2.5.12 is then employed; i.e. $\frac{s^2}{s+1} = (s-1) + \frac{1}{s+1}$, where $\tilde{q}_{21}(s) = (s-1)$ and $\tilde{r}_{21}(s) = 1$. Subtracting $\tilde{q}_{21}(s) = (s-1)$ times the first row of $P(s)$ from the second, we obtain: $\begin{bmatrix} s+1, & 1 \\ 1, & -s+3 \end{bmatrix}$. We now interchange the rows since $\tilde{r}_{21}(s) = 1$ is of lower degree than $\tilde{p}_{11}(s) = s+1$. If we then multiply the resulting first row by s+1 and subtract the resulting expression from the second row, we obtain the desired result; i.e. summarizing these various steps as a sequence of consecutive premultiplications of $P(s)$ by the unimodular matrices $U_3(s)U_2(s)U_1(s) = U_L(s)$, we have:

$$\underbrace{\begin{bmatrix} 1, & 0 \\ -s-1, & 1 \end{bmatrix} \begin{bmatrix} 0 & 1 \\ 1 & 0 \end{bmatrix} \begin{bmatrix} 1, & 0 \\ -s+1, & 1 \end{bmatrix}}_{\begin{bmatrix} -s+1, & 1 \\ s^2, & -s-1 \end{bmatrix}} \begin{bmatrix} s+1, & 1 \\ s^2, & 2 \end{bmatrix} = \begin{bmatrix} 1, & -s+3 \\ 0, & s^2-2s-2 \end{bmatrix}$$

$$\begin{bmatrix} -s+1, & 1 \\ s^2, & -s-1 \end{bmatrix} = U_L(s) \times P(s) = \hat{P}(s)$$

where $\hat{P}(s)$ is upper right triangular and column proper; i.e. $\Gamma_c[P(s)] = \begin{bmatrix} 1 & 0 \\ 0 & 1 \end{bmatrix}$, a nonsingular matrix.

We now note that whenever $P(s)$ in Theorem 2.5.11 is nonsingular, $\hat{P}(s)$ will be column proper. Furthermore, any nonsingular polynomial matrix $P(s)$ can also be reduced to lower left row proper form by elementary column operations; i.e. one can always find a unimodular matrix $U_R(s)$ such that $P(s)U_R(s) = \hat{P}(s)$, where $\hat{P}(s)$ is in lower left row proper form. In particular, we can either find an appropriate $U_R(s)$ directly or by duality; i.e. we can first transpose $P(s)$ and then reduce $P^T(s)$ to upper right column proper form via $U_L(s)$ in accordance with the various steps which were just outlined. $U_R(s) = U_L^T(s)$ then represents an appropriate unimodular matrix which reduces $P(s)$ to lower left row proper form.

In view of the results presented thus far in this section and, in particular, by combining Theorems 2.5.7 and 2.5.11, it is now clear that:

THEOREM 2.5.14: Any nonsingular polynomial matrix $P(s)$ can be reduced to column (row) proper form by either elementary row or elementary column operations. This result will prove to be most useful in certain of our subsequent discussions.

The notion of a pair of "relatively prime polynomials" is a fundamental one in linear algebra, and one which can now be extended to include the matrix case. We begin by employing:

DEFINITION 2.5.15: If three polynomial matrices satisfy the relation: $P(s) = H(s)G(s)$, then $G(s)$ $(H(s))$ is called a right (left) divisor of $P(s)$, and $P(s)$ is called a left (right) multiple of $G(s)(H(s))$. A greatest common right divisor (g.c.r.d.) of two polynomial matrices $P(s)$ and $R(s)$ is a common right divisor which is a left multiple of every common right divisor of $P(s)$ and $R(s)$. Similarly, a greatest common left divisor (g.c.l.d.) of two polynomial matrices $P(s)$ and $Q(s)$ is a common left divisor which is a right multiple of every common left divisor of $Q(s)$ and $P(s)$. In view of these definitions, we can now state and establish:

THEOREM 2.5.16: Consider the pair $\{P(s),R(s)\}$ $(\{P(s),Q(s)\})$ of polynomial matrices which have the same number of columns (rows). If the composite matrix $\begin{bmatrix} P(s) \\ \hdashline R(s) \end{bmatrix}$ $([P(s) \vdots Q(s)])$ is reduced to upper right (lower left) triangular form $\begin{bmatrix} T_R(s) \\ \hdashline 0 \end{bmatrix}$ $([T_L(s) \vdots 0])$[†] as in Theorem 2.5.11, then $T_R(s)(T_L(s))$ is a g.c.r.d. $G_R(s)$ (a g.c.l.d. $G_L(s)$) of $\{P(s),R(s)\}$ $(\{P(s),Q(s)\})$.

[†]The two triangular forms displayed assume more rows (columns) of the composite matrix than columns (rows), an assumption which need not hold (see Problems 2-27 and 2-28).

Proof: We need only establish the first half of the theorem; i.e. the part which is not in parenthesis, since the other half follows directly by duality. In particular, consider the composite matrix $\begin{bmatrix} P(s) \\ \hline R(s) \end{bmatrix}$. In light of Theorem 2.5.11 we can clearly reduce this matrix, via some unimodular matrix $U_L(s)$, to upper right triangular form; i.e.

$$U_L(s) \begin{bmatrix} P(s) \\ \hline R(s) \end{bmatrix} = \begin{bmatrix} T(s) \\ \hline 0 \end{bmatrix} \qquad 2.5.17$$

where $\begin{bmatrix} T(s) \\ \hline 0 \end{bmatrix}$ is upper right triangular. Equation 2.5.17 implies, in turn, that:

$$\begin{bmatrix} P(s) \\ \hline R(s) \end{bmatrix} = U_L^{-1}(s) \begin{bmatrix} T(s) \\ \hline 0 \end{bmatrix} = \left[\begin{array}{c|c} L_1(s) & L_2(s) \\ \hline L_3(s) & L_4(s) \end{array}\right] \begin{bmatrix} T(s) \\ \hline 0 \end{bmatrix} = \begin{bmatrix} L_1(s)T(s) \\ L_3(s)T(s) \end{bmatrix} \qquad 2.5.18$$

for some pair $\{L_1(s), L_3(s)\}$ of polynomial matrices, since the inverse of a unimodular matrix is itself a unimodular matrix. Thus, it is clear that:

$$P(s) = L_1(s)T(s) \quad \text{and} \quad R(s) = L_3(s)T(s), \qquad 2.5.19$$

or, equivalently, that $T(s)$ is a common right divisor of $P(s)$ and $R(s)$. To show that it is a greatest common right divisor, note that if $U_L(s)$ is partitioned as $\left[\begin{array}{c|c} L_5(s) & L_6(s) \\ \hline L_7(s) & L_8(s) \end{array}\right]$, then 2.5.17 directly implies that

$$L_5(s)P(s) + L_6(s)R(s) = T(s) \qquad 2.5.20$$

Therefore, it is clear that every common right divisor of $P(s)$ and $R(s)$ is also a right divisor of $T(s)$; i.e. that $T(s)$ is also a left multiple of every common right divisor of $P(s)$ and $R(s)$. $T(s)$ is thus a g.c.r.d. $G_R(s)$ of $P(s)$ and $R(s)$, which establishes the theorem.

COROLLARY 2.5.21: <u>If $P(s)$ is nonsingular, then $G_R(s)$ is nonsingular</u>. This follows directly from 2.5.19 since $G_R(s) = T(s)$. It can also be shown that <u>if</u> $G_R(s)$ <u>is a g.c.r.d. of $P(s)$ and $R(s)$, with $P(s)$ nonsingular, then any other g.c.r.d.</u> $\hat{G}_R(s)$ <u>of $P(s)$ and $R(s)$ is row equivalent to</u> $G_R(s)$; i.e. $\hat{G}_R(s) = U_L(s)G_R(s)$. This statement will not be formally established here but is left as an exercise for the reader (see Problem 2-22). To illustrate the notions of a g.c.r.d. and a g.c.l.d. of two polynomial matrices, we employ the following:

EXAMPLE 2.5.22: Consider the polynomial matrices $P(s) = \begin{bmatrix} s^2, & -1 \\ -s\ , & s^2 \end{bmatrix}$ and $R(s) = \begin{bmatrix} s & -s \\ 0 & 1 \end{bmatrix}$. To find a g.c.r.d. $G_R(s)$ of $P(s)$ and $R(s)$ we reduce the composite matrix $\begin{bmatrix} P(s) \\ \hline R(s) \end{bmatrix} = \begin{bmatrix} s^2 & -1 \\ -s & s^2 \\ s & -s \\ 0 & 1 \end{bmatrix}$ to upper right triangular form. For this example, it is clear that by multiplying the last row $[0 \quad 1]$ of the composite matrix by the appropriate monomial and adding the resultant expressions to the remaining rows, all other elements in the second column can be zeroed. The first column terms can also be set equal to zero, with the exception of an s, by employing an analogous procedure. Therefore, it is clear that $G_R(s) = \begin{bmatrix} s & 0 \\ 0 & 1 \end{bmatrix}$ is a g.c.r.d. of $R(s)$ and $P(s)$.

To find a g.c.l.d. $G_L(s)$ and $P(s)$ and $R(s)$ we reduce the composite matrix $[P(s)\ R(s)] = \begin{bmatrix} s^2, & -1, & s, & -s \\ -s, & s^2, & 0, & 1 \end{bmatrix}$ to lower left triangular form. By adding the third and fourth columns of the composite matrix, we obtain the column vector $\begin{bmatrix} 0 \\ 1 \end{bmatrix}$ which can be used to zero all other second row entries. The column vector $\begin{bmatrix} -1 \\ 0 \end{bmatrix}$ is left as the second column of the remaining matrix and can be used to zero the remaining first row entries. It is therefore clear that the two-dimensional identity matrix is a g.c.l.d. of $P(s)$ and $R(s)$. This example also serves to illustrate an important point, namely _if a g.c.r.d. of two polynomial matrices is not a unimodular matrix, then it does not necessarily follow that a g.c.l.d. of the two matrices is also a nonunimodular matrix, and vice versa_.

We have now defined and demonstrated the notions of a g.c.r.d. and a g.c.l.d. of two polynomial matrices. These two concepts now enable us to extend to the matrix case the well known notion of relatively prime polynomials. In particular, we now have:

DEFINITION 2.5.23: A pair $\{P(s),R(s)\}$ ($\{P(s),Q(s)\}$) of polynomial matrices which has the same number of columns (rows) is said to be _relatively right prime (relatively left prime)_ if and only if their g.c.r.d. (g.c.l.d.) are unimodular matrices. It is clear in view of Example 2.5.22 that _two polynomial matrices may be relatively right prime but not relatively left prime and vice versa_. As in the case of scalar polynomials, the notion of a pair of relatively right (or left) prime polynomial matrices thus implies the inability to factor some nonunimodular matrix from the right (or left) side of both members of the pair.

2.6 CONCLUDING REMARKS AND REFERENCES

The material presented in this chapter was drawn, for the most part, from three primary sources, namely references [H1], [G1], and [M1]. Sections 2.2 through 2.4 contain standard definitions and theorems which can be found in far more detail in a number of linear algebra texts such as Hoffman and Kunze [H1] and Gantmacher [G1]. It might have been noted by the more astute reader that a number of seemingly important results in linear algebra were not presented, most notably perhaps, the notions of "eigenvectors" of scalar matrices and the "Jordan canonical form" representation for matrices with elements in $\mathscr{C}$. The reason why these notions were omitted is that they simply will not be utilized in any subsequent portions of the text.

The material on polynomial matrices presented in Section 2.5 is undoubtedly less familiar to the average reader than the preceding results. In particular, the notions of row and column proper polynomial matrices as well as the algorithm for reducing a nonsingular polynomial matrix to row and column proper form (Theorem 2.5.7) have not appeared as yet in any other text and were first introduced by the author in some earlier reports, i.e. [W1], [W2], and [W3]. The other results pertaining to polynomial matrices and, in particular, the notions of column and row equivalence and a g.c.l.d. and g.c.r.d. of two polynomial matrices were obtained primarily from material contained in a text by MacDuffee [M1], although the algorithm for determining a g.c.l.d. or g.c.r.d. of two polynomial matrices (Theorem 2.5.11) was found in Gantmacher [G1], and the definition of relatively right and left prime polynomial matrices was motivated by an equivalent definition in Rosenbrock [R1].

The intent of this chapter, hopefully fulfilled, has been to present only those results which are both required for subsequent developments and are not generally familiar to the average reader (seniors and first year graduate students). Additional mathematical notions will be developed subsequently as required by building on the rather general results presented in this chapter.

PROBLEMS - CHAPTER 2

2-1 Determine the transpose and rank (specify the field) of each of the following matrices:

(a) $\begin{bmatrix} j \\ 3j \\ -1 \end{bmatrix}$ (b) $\begin{bmatrix} 1 & 4 & -5 \\ 7 & 0 & 2 \end{bmatrix}$ (c) $\begin{bmatrix} s+4 & -2 \\ s^2-1 & 6 \\ 0 & 2s+3 \\ s & -s+4 \end{bmatrix}$ (d) $\begin{bmatrix} \frac{s+1}{s^2} \end{bmatrix}$

2-2 Determine the determinant, the adjoint, and the inverse (if it exists) of the following matrices:

(a) $\begin{bmatrix} 1 & 2 & 0 \\ 0 & 2 & 4 \\ 1 & 0 & -1 \end{bmatrix}$ (b) $\begin{bmatrix} \frac{s^2-3}{s} & 4s+3 \\ \frac{1}{s^2-2} & 3 \end{bmatrix}$ (c) $\begin{bmatrix} 0 & j & -2-4j \\ 1 & 0 & 1+j \\ 2 & 2j & -2-6j \end{bmatrix}$

2-3 Under what conditions, if any, will the set $\{0, \frac{1}{2}, 1, 2\}$ represent a field? Explain.

2-4 Show that the vectors employed in Example 2.2.11 are linearly independent over $\mathscr{R}$.

2-5 Find the (unique) representation of the vector $v_o = [1, 4, 0]$ in the ordered basis: $v_1 = [1, -1, 0]$, $v_2 = [1, 0, -1]$, and $v_3 = [0, 1, 0]$. Find the representation of $\hat{v}_o = [s+2, \frac{1}{s}, 2]$ in this same ordered basis.

2-6 If $v_1 = \begin{bmatrix} 2 \\ 1 \\ 0 \end{bmatrix}$, $v_2 = \begin{bmatrix} 1 \\ 0 \\ -1 \end{bmatrix}$, $v_3 = \begin{bmatrix} 1 \\ 0 \\ 0 \end{bmatrix}$, and $\hat{v}_1 = \begin{bmatrix} 1 \\ 0 \\ 0 \end{bmatrix}$, $\hat{v}_2 = \begin{bmatrix} 0 \\ 1 \\ -1 \end{bmatrix}$, $\hat{v}_3 = \begin{bmatrix} 0 \\ 1 \\ 1 \end{bmatrix}$ represent two different ordered bases of $\mathscr{V}$, the set of 3-tuples over $\mathscr{R}$, find the matrix relationship, B, between these bases. Determine the representation of the vector

$e_2 = \begin{bmatrix} 0 \\ 1 \\ 0 \end{bmatrix}$ in both of these bases and verify that 2.2.26 holds.

2-7 Show that an infinite set of linearly independent vectors in $\{\mathscr{V}, \mathscr{F}\}$ need not qualify as a basis of the infinite dimensional vector space $\{\mathscr{V}, \mathscr{F}\}$.

2-8 Let $\mathscr{V}$ be the vector space of polynomials of degree 2 or less with coefficients in $\mathscr{R}$ and $\mathscr{W}$ the space of all real 2-tuples. If the linear transformation L maps the vectors s+1, s-1, and s^2 in $\mathscr{V}$ into the respective vectors [2, -1], [0 1], and [0, 0] in $\mathscr{W}$, find its matrix representation C with respect to the basis 1, s, s^2 in $\mathscr{V}$ and the standard basis in $\mathscr{W}$. Verify that 2.3.9 is satisfied for each of the three individual mappings defined above.

2-9 Show that both the range space and the null space of a linear transformation are vector spaces.

2-10 Find the range and null spaces of the linear transformation L defined in Problem 2-8 and the range and null spaces of the linear operator $\frac{d}{ds}$ defined in Example 2.3.10.

2-11 Show that the sum of the rank and nullity of any linear operator which maps a finite dimensional vector space $\mathscr{V}$ into a finite dimensional vector space $\mathscr{W}$ is equal to n, the dimension of $\mathscr{V}$. (Hint: Let $\bar{v}_1, \bar{v}_2, \ldots, \bar{v}_k$ be any basis of the null space of L, and $\bar{v}_{k+1}, \ldots, \bar{v}_n$ an extension of the null space basis to include all of $\mathscr{V}$. Then show that $L\bar{v}_{k+1}, \ldots, L\bar{v}_n$ is a basis of the range of L.)

2-12 Can the set: $1, s, s^2, \ldots$ defined over $\mathscr{P}$ (rather than $\mathscr{R}$) belong to a vector space? If so, find its dimension and a basis. If not, explain why it cannot.

2-13 Find the eigenvalues, Δ, and Δ_m of the following matrices:

(a) $\begin{bmatrix} 1 & 1 \\ -1 & 3 \end{bmatrix}$ (b) $\begin{bmatrix} 2 & 0 \\ 0 & 2 \end{bmatrix}$ (c) $\begin{bmatrix} 1 & 0 & 0 \\ 1 & -1 & -1 \\ 0 & 0 & 1 \end{bmatrix}$

2-14 Is the integration operator, $\int_0^s ds$, a linear operator on the vector space of polynomials of degree n or less with coefficients in $\mathscr{R}$? If it is, what are $\mathscr{W}$ and the range and null spaces of L? If not, explain why it isn't.

2-15 Consider the two bases of the set of all 3-tuples over $\mathscr{R}$ which were defined in Problem 2-6. If the linear operator L maps v_1, v_2, and v_3 into e_1, e_2, and e_3 respectively, find its matrix representations (C and $\hat{C}$) in the two bases.

2-16 Prove that if A and B are $(q \times n)$ and $(n \times r)$ matrices respectively, with elements in the same field , then the relationship:

$$\rho[A] + \rho[B] - n \leq \rho(AB) \leq \min(\rho[A], \rho[B]),$$

which is known as Sylvester's Inequality, holds. (Hint: Assume that the matrix product, AB, represents two successive linear transformations, namely L_B, which maps an r dimensional space into an n dimensional space and L_A, which then maps the n dimensional space into a q dimensional space and then use the fact, established in Problem 2-11, that rank (L) + nullity (L) = dimension $(\mathscr{V})$).

2-17 Show that the determinant of a unimodular matrix, $U(s)$, is always nonzero and independent of s.

2-18 Can two equivalent polynomial matrices have different degrees? Explain.

2-19 Can a singular polynomial matrix be either column or row proper? Explain.

2-20 Determine whether or not the following polynomial matrices are column (row) proper. If not, find a unimodular matrix, if one exists, which reduces the given matrix to (i) a column equivalent column (row) proper form and (ii) a row equivalent column (row) proper form.

(a) $\begin{bmatrix} 2s^2+3s-1, & 4s-1, & 0 \\ 0, & -3, & s^3-4s \\ 2s+3, & -s^2-s+3, & -s^2-2 \end{bmatrix}$ (b) $\begin{bmatrix} -2s+4, & 2s, & 3 \\ 0, & -s^2+1, & -s-4 \\ 3, & 1, & 1 \end{bmatrix}$

(c) $\begin{bmatrix} s+1, & -1, & s+2 \\ 2s^2+s-3, & 2s^2-1, & s-2 \\ -4s+4, & -4s+3, & 1 \end{bmatrix}$

2-21 A square n-<u>th</u> degree polynomial matrix, $P(s)$, represented as $P_o+P_1s+\ldots+P_ns^n$, where the P_i are square scalar matrices with elements in $\mathscr{R}$, is said to be <u>proper</u> if and only if $|P_n| \neq 0$. Show that a proper polynomial matrix is both row proper and column proper, but that a polynomial matrix which is both row proper and column proper need not be proper.

2-22 Show that if $G_R(s)$ is a g.c.r.d. of $P(s)$ and $R(s)$, with $P(s)$ nonsingular, then any other g.c.r.d. $\hat{G}_R(s)$ of $P(s)$ and $R(s)$ is row equivalent to $G_R(s)$. Why do we require that $P(s)$ be nonsingular?

2-23 Show that $G_R(s) = \begin{bmatrix} 1, & s+1 \\ 0, & s^2 \end{bmatrix}$ is a g.c.r.d. of $P(s) = \begin{bmatrix} s^2, & 0 \\ 0, & s^2 \end{bmatrix}$ and $R(s) = [1, \ s+1]$.

2-24 Are the polynomial matrices, $P(s) = \begin{bmatrix} s^2+s, & -s \\ -s^2-1, & s^2 \end{bmatrix}$ and

$R(s) = \begin{bmatrix} s, & 0 \\ -s-1, & 1 \end{bmatrix}$, relatively right prime? relatively left prime?

2-25 Show that $P(s)$ and $R(s)$ are relatively right prime if and only if there exists a pair, $\{M(s),N(s)\}$, of polynomial matrices such that: $M(s)P(s) + N(s)R(s) = I$. (Hint: Use 2.5.20 to establish necessity.)

2-26 It can be shown [R1] that the polynomial matrix pair, $\{R(s),P(s)\}$, with $R(s)$ $(p \times m)$ and $P(s)$ $(m \times m)$ and nonsingular is relatively right prime if and only if $\rho\begin{bmatrix} P(s_o) \\ \hline R(s_o) \end{bmatrix} = m$ for all zeros, s_o, of $|P(s)|$. Use this result to verify that the particular polynomial matrix pair employed in Example 2.5.22 is not relatively right prime but that this same pair is relatively left prime; i.e. that $P^T(s)$ and $R^T(s)$ are relatively right prime.

2-27 Show that if a g.c.r.d. of the polynomial matrix pair, $\{R(s),P(s)\}$, is not square; i.e. that $T_R(s)$ of Theorem 2.5.16 has more columns (q) than rows (m), then one can always find a nonsingular, nonunimodular common right divisor of $R(s)$ and $P(s)$. Extend this result to include the case when $T_R(s)$ is singular.

2-28 Verify that $R(s) = \begin{bmatrix} 1, & s, & 0 \\ s+1, & 0, & 1 \end{bmatrix}$ and $P(s) = \begin{bmatrix} 1, & s^2, & -1 \\ 0, & -1, & 0 \end{bmatrix}$ are relatively right prime, but that $R_1(s) = [1, \; s, \; 0]$ and $P_1(s) = [1, \; s^2, \; -1]$ are not. Find a g.c.r.d. of this latter pair and verify the statement made in Problem 2-27.

2-29 Show that the notion of a "g.c.r.d. (g.c.l.d.) of the rows (columns) of a polynomial matrix" can be defined.

2-30 If the polynomial matrices $P_1(s), P_2(s), \ldots, P_k(s)$, have the same number of columns (rows), show that a g.c.r.d. (g.c.l.d.) of all of the $P_i(s)$ can be defined.

CHAPTER 3
THE STATE SPACE

3.1 INTRODUCTION

Linear systems theory, as the term generally applies today, is not directly concerned with the study of physical systems but rather deals with the study of a class of mathematical representations or models of physical systems which we will call dynamical systems. Rather than present a rigorous abstract definition of this notion of a dynamical system, as is often done, we will employ an intuitive approach more suited to students with engineering backgrounds. In particular, the reader has undoubtedly been introduced to a number of physical laws which are commonly employed to obtain a mathematical description of the behavior of physical systems. For example, Kirchoff's laws can be used to obtain a set of differential equations which, when solved, describe the manner in which various node voltages and loop currents vary with time as functions of driving terms (inputs) as well as various initial conditions. Similarly, Newton's laws and the Euler-Lagrange equations are commonly used for analogous reasons in the study of mechanical systems. In general, therefore, regardless of the particular discipline, well-known physical laws are commonly employed to obtain a set of mathematical relationships which can then be evaluated, simulated, and, in certain cases, explicitly solved in order to provide added insight into the actual physical behavior of the system. It should be noted, however, that not all of the mathematical relationships which result from the application of physical laws fall into the category of what we will term dynamical systems. However, a substantial number do, and we will direct the remainder of our discussions to the study of linear, time-invariant, dynamical systems; i.e. to the study of the dynamical behavior of physical systems which can be mathematically modelled by linear, time-invariant, ordinary differential equations in the differential operator form:

$$P(D)z(t) = Q(D)u(t); \qquad 3.1.1a$$

$$y(t) = R(D)z(t) + W(D)u(t), \qquad 3.1.1b$$

where $z(t)$ is a q-dimensional vector valued function of the time t called the partial state of the dynamical system, $u(t)$ is an m-dimensional vector valued function of time called the input to the dynamical system, $y(t)$ is a p-dimensional vector valued function of time called the output of the dynamical system, and $P(D)$, $Q(D)$, $R(D)$, and $W(D)$ are polynomial matrices of dimensions $q \times q$, $q \times m$, $p \times q$, and $p \times m$ respectively in the differential operator $D = \frac{d}{dt}$ with $P(D)$ nonsingular. The quadruple $\{P(D),Q(D),R(D),W(D)\}$ will sometimes be employed as an alternate to the representation 3.1.1. It might be noted here that the partial state of a dynamical system is usually nothing more than certain selected physical variables of the system which can be measured and controlled; e.g. velocities, accelerations, and positions in mechanical systems or voltages and currents in electrical systems. On the other hand, the (entire) state represents all of the physical variables which characterize the total dynamical behavior of the system. This somewhat heuristic definition of state and the distinction between the partial state and the (entire) state will become more transparent in light of our subsequent discussions.

We will from time to time find it advantageous to distinguish between a scalar system, defined here as any system having only one input and one output, and a multivariable system, defined as any other (multiple input/output) system. It should be pointed out here, at the outset, that although relatively few physical systems are directly representable via 3.1.1, the behavior of a large number of physical systems can be approximated, with surprising accuracy, by the solutions of differential equations of the form 3.1.1. This is particularly true if we consider only small perturbations in the behavior of the physical system about some nominal operating point.[†] Because of this, and the fact that the solutions of 3.1.1 can be readily determined (relatively speaking), a considerable amount of research has focussed on the study of linear, time-invariant dynamical systems.

While the differential operator representation 3.1.1 does represent a general, linear, time-invariant dynamical system, it is not in the more commonly utilized "state-space" form which will be introduced in the next section and then studied in depth in the remainder of this chapter. A detailed analysis of the differential operator representation 3.1.1 will be postponed until Chapter 5.

3.2 STATE REPRESENTATIONS

An important subclass of those dynamical systems which can be represented via the differential operator form 3.1.1 will be considered in this section. In particular, we now define a <u>state representation</u> of a linear, time-invariant dynamical system as the following set of vector-matrix differential equations:

$$\dot{x}(t) = Ax(t) + Bu(t); \qquad 3.2.1a$$

$$y(t) = Cx(t) + Eu(t), \qquad 3.2.1b$$

where $x(t)$ is an n-dimensional vector valued function of time called the <u>state</u> of the system, $u(t)$ and $y(t)$ represent the (m) inputs and (p) outputs of the system respectively as defined in the previous sec-

[†]Here we are speaking of the technique of <u>linearization</u>; i.e. linearly approximating the nonlinear differential equations of the dynamical system about a known operating point by utilizing only the first order terms in a Taylor series expansion of the equations about the solution--a number of texts contain a detailed description of this procedure; e.g. [B1] and [D1].

tion, and A, B, C, and E are $n \times n$, $n \times m$, $p \times n$, and $p \times m$ matrices respectively with elements in the real field $\mathscr{R}$. The quadruple $\{A,B,C,E\}$ will sometimes be employed as an alternate to the state representation 3.2.1. The notation $\dot{x}(t)$ is used to represent the differential operator $D = \frac{d}{dt}$; i.e. $\dot{x}(t) = \frac{dx(t)}{dt} = Dx(t)$. We note that 3.2.1 does indeed represent a subclass of the more general class of dynamical systems represented by 3.1.1 since 3.2.1 can be written as: $(DI-A)x(t) = Bu(t)$; $y(t) = Cx(t) + Eu(t)$, which is of the more general form 3.1.1 with $(DI-A) = P(D)$, $B = Q(D)$, $C = R(D)$, and $E = W(D)$; i.e.

$$\{DI-A,B,C,E\} = \{P(D),Q(D),R(D),W(D)\} \qquad 3.2.2.$$

The state representation for linear, time-invariant dynamical systems is an important and common method of approximating the dynamical behavior of a large class of physical systems, and its popularity can be attributed to a number of factors which will become obvious in our subsequent discussions.

We first note that a state representation can readily be simulated on an analog computer. In particular, consider the general block diagram of a state representation depicted in Figure 3.2.3 below. The block containing the integral sign actually represents a parallel bank of n integrators with (n) outputs $x_1(t),x_2(s),\ldots,x_n(t)$, while the elements of the scalar matrices A, B, C, and E represent various input gains (potentiometer settings) to the appropriate summation amplifiers which are represented by the circled sigmas. It is therefore clear that a total of n integrators (energy storage elements) are required for an analog computer simulation of a dynamical system which is defined via the state representation 3.2.1, and consequently, that n initial conditions can be placed on these n integrators whose outputs represent the state variables $x_1(t),x_2(t)$,

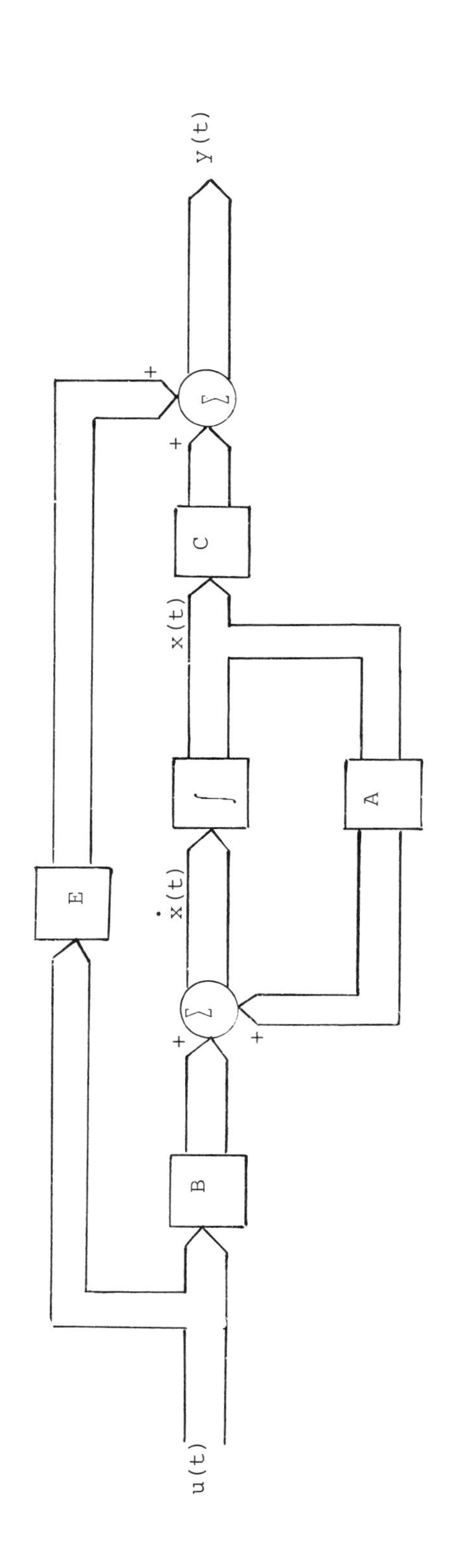

Figure 3.2.3

A STATE REPRESENTATION

$\ldots, x_n(t)$ of the system. We call the integer n the <u>order</u> of the dynamical system and note from our past experiences with analog computer similation that <u>the state of the system 3.2.1 at any initial time</u> t_o <u>along with the input $u(t)$, defined for all</u> $t \varepsilon [t_o, \infty)$, <u>uniquely determines the complete behavior of the system</u> ($x(t)$ and $y(t)$) <u>for all</u> $t \geq t_o$. This fact follows formally from the uniqueness of solutions to linear differential equations of the form considered [C2] and will subsequently be discussed in further detail. Before we present any additional state related results, it might be enlightening to now consider an example of a relatively simple linear dynamical system in order to illustrate the notions of state and state representations.

EXAMPLE 3.2.4: Consider the following electrical network:

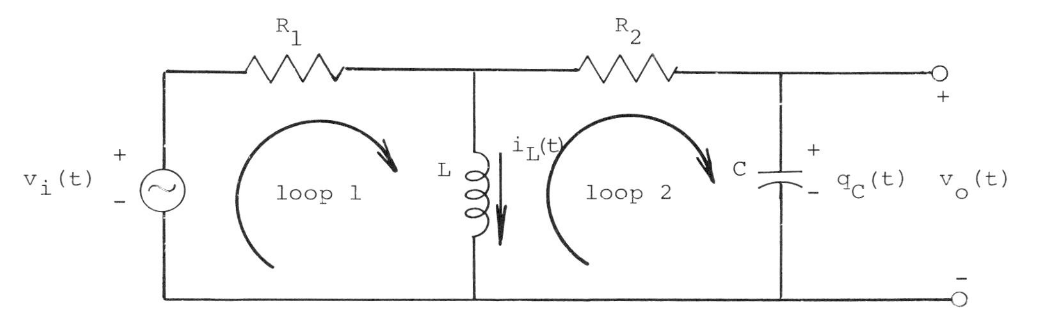

One set of linear differential equations which represents the physical behavior of this system (an RLC network) can readily be found in terms of the inductor current $i_L(t)$ and the capacitor charge $q_C(t)$ by equating to zero the sum of the voltage drops around the two designated loops; i.e. since $q(t) = \int i(t)dt$, or $\frac{dq(t)}{dt} = i(t)$, we determine that for loop 1: $-v_i(t) + \left[i_L(t) + \frac{dq_C(t)}{dt}\right] R_1 + L\frac{di_L(t)}{dt} = 0$, and for loop 2: $-L\frac{di_L(t)}{dt} + \frac{dq_C(t)}{dt} R_2 + \frac{q_C(t)}{C} = 0$. If we now set $v_i(t) = u(t)$ and $v_o(t) = y(t)$ and let $i_L(t) = x_1(t)$ and $q_C(t) =$

$x_2(t)$, where $x_1(t)$ and $x_2(t)$ represent the state of this network; i.e. knowledge of $x_1(t_o) = i_L(t_o)$, $x_2(t_o) = q_C(t_o)$, and $u(t) = v_i(t)$ for all $t \geq t_o$ uniquely determines all loop currents and node voltages for all $t \geq t_o$, then we can rewrite the above equations in the state form 3.2.1. In particular, for this example:

$$\dot{x}(t) = \begin{bmatrix} \dot{x}_1(t) \\ \dot{x}_2(t) \end{bmatrix} = \underbrace{\begin{bmatrix} \dfrac{-R_1R_2}{L(R_1+R_2)}, & \dfrac{R_1}{LC(R_1+R_2)} \\ \dfrac{-R_1}{R_1+R_2}, & \dfrac{-1}{C(R_1+R_2)} \end{bmatrix}}_{A} \begin{bmatrix} x_1(t) \\ x_2(t) \end{bmatrix} + \underbrace{\begin{bmatrix} \dfrac{R_2}{L(R_1+R_2)} \\ \dfrac{1}{R_1+R_2} \end{bmatrix}}_{B} u(t);$$

$$y(t) = \underbrace{\begin{bmatrix} 0, & \dfrac{1}{C} \end{bmatrix}}_{C} \begin{bmatrix} x_1(t) \\ x_2(t) \end{bmatrix} + \underbrace{[\ 0\ ]}_{E} u(t).$$

This example also serves to illustrate an important point, namely that <u>the inductor currents and the capacitor charges represent a complete set of state variables in any nondegenerate RLC network</u>.

One of the primary reasons why the state representation 3.2.1 is so important and useful in the study of linear systems is the relative ease with which the solution of 3.2.1 can be obtained. We will now elaborate this point by first considering the <u>homogeneous equation</u>:

$$\dot{x}(t) = Ax(t) \qquad 3.2.5$$

obtained from 3.2.1 by ignoring, both the input $u(t)$ and the output $y(t)$.

THEOREM 3.2.6: <u>The set of all solutions of</u> $\dot{x}(t) = Ax(t)$ <u>constitutes a vector space of dimension n</u> (called the <u>solution space</u>) <u>over the real field</u> $\mathscr{R}$.

Proof: If $\psi_1(t)$ and $\psi_2(t)$ (both n-dimensional vector valued functions of time) represent any two arbitrary solutions of 3.2.5; i.e. if $\dot{\psi}_1(t) = A\psi_1(t)$ and $\dot{\psi}_2(t) = A\psi_2(t)$, it can easily be verified by direct substitution that all of the following are also solutions of 3.2.5: $\psi_1(t) + \psi_2(t) = \psi_2(t) + \psi_1(t)$, $\psi_1(t) - \psi_1(t) = \psi_2(t) - \psi_2(t) = 0$, $a_1\psi_1(t)$, and $a_2\psi_2(t)$ for any scalars a_1, a_2 in $\mathscr{R}$. Therefore, the set of all solutions of $\dot{x}(t) = Ax(t)$ does constitute a vector space over $\mathscr{R}$, and only the dimension of the space remains to be determined. To show that the dimension is indeed n, as stated, let $\psi_1(t)$, $\psi_2(t),\ldots,\psi_n(t)$ be defined as the particular solutions to 3.2.4 with initial conditions given by $e_1,e_2,\ldots,e_n$ respectively, the standard basis of the set of all n-tuples over $\mathscr{R}$. Thus $\psi_i(t)$ is the n-vector solution $x(t)$ of 3.2.5 obtained by setting $x_i(t_o) = 1$ and $x_j(t_o) = 0$ for all $j \neq i$. We now establish by contradiction that these (n) particular solutions of 3.2.5 are linearly independent; i.e. suppose they are dependent. Then for some set of (n) nonzero scalars $a_1,a_2,\ldots,a_n$ in $\mathscr{R}$, $\sum_1^n a_i\psi_i(t) = 0$ for all $t \geq t_o$. However at $t = t_o$, $\sum_1^n a_i\psi_i(t_o) = \sum_1^n a_ie_i = 0$, which implies that all of the $a_i = 0$ since the e_i are linearly independent, thus establishing, by contradiction, the linear independence of the particular solutions $\psi_1(t)$, $\psi_2(t),\ldots,\psi_n(t)$. We finally show that any solution of 3.2.5 can be expressed (uniquely) as a linear combination of these (n) particular solutions. In particular, let $\psi(t)$ be any solution of 3.2.5 with initial condition e; i.e. $\psi(t_o) = e$, which can be expressed in terms of the standard basis as $\sum_1^n a_ie_i$ for some ordered set of scalars $a_1,a_2,\ldots,a_n$ in $\mathscr{R}$. Since $\sum_1^n a_i\psi_i(t)$ is also a solution of 3.2.5 with initial condition e, we conclude from the uniqueness of solutions of 3.2.5 [C2] that $\psi(t) = \sum_1^n a_i\psi_i(t)$, thus establishing that any solution $\psi(t)$ can be expressed as a linear combination of the basis set

$\psi_1(t), \psi_2(t), \ldots, \psi_n(t)$. Theorem 3.2.6 is thus established. In view of this theorem, we now present:

DEFINITION 3.2.7: An $n \times n$ matrix valued function of time $\Psi(t)$ is called a fundamental matrix of 3.2.5 if the n columns of $\Psi(t)$ are linearly independent solutions of 3.2.5. We thus note that the (n) particular solutions $\psi_1(t), \psi_2(t), \ldots, \psi_n(t)$ employed in the proof of Theorem 3.2.6 qualify as the n column vectors of a suitable $\Psi(t)$.

EXAMPLE 3.2.8: $\Psi(t) = \begin{bmatrix} e^{-t}, & 0 \\ 0, & e^{2t} \end{bmatrix}$ is a fundamental matrix of the differential equation 3.2.5 with $A = \begin{bmatrix} -1 & 0 \\ 0 & 2 \end{bmatrix}$ since $\dot{\Psi}(t) = A\Psi(t)$ and $|\Psi(t)| \neq 0$. $\hat{\Psi}(t) = \begin{bmatrix} e^{-t}, & e^{-t} \\ 2e^{2t}, & e^{2t} \end{bmatrix}$ is another fundamental matrix of 3.2.5 for this A as the reader can readily verify. It should be noted that by definition any fundamental matrix $\Psi(t)$ of $\dot{x}(t) = Ax(t)$ is nonsingular for all $t \varepsilon [t_o, \infty)$.

It should now be clear that once a fundamental matrix of 3.2.5 has been found, $x(t)$, the unique solution of 3.2.5 with initial state $x(t_o) = x_o$, can be expressed as some linear combination of the columns of $\Psi(t)$, which is dependent on the initial state x_o. This is indeed the case, and in view of this observation, we now write:

$$x(t) = \Psi(t)Mx_o, \tag{3.2.9}$$

for some nonsingular real scalar matrix M. To find M we simply evaluate 3.2.9 at the initial time t_o; i.e.

$$x(t_o) = x_o = \Psi(t_o)Mx_o, \tag{3.2.10}$$

and since x_o can be arbitrarily set, M must equal $\Psi^{-1}(t_o)$. Therefore, the unique solution to 3.2.5 is given in terms of any fundamental matrix $\Psi(t)$ of 3.2.5 by $\Psi(t)^{-1}\Psi(t_o)$, and this latter expression is called the state transition matrix of 3.2.5 and denoted by $\Phi(t,t_o)$. In summary, we can now write the solution to 3.2.5 as:

$$x(t) = \Phi(t,t_o)x_o, \tag{3.2.11}$$

where

$$\Phi(t,t_o) = \Psi(t)\Psi^{-1}(t_o) \tag{3.2.12}$$

for any fundamental matrix $\Psi(t)$ of 3.2.5. We can now readily establish that <u>the state transition matrix</u> $\Phi(t,t_o)$ <u>is unique and independent of the particular choice of $\Psi(t)$</u>. In particular, we simply recall from our earlier work in Section 2.2 that any two sets of basis vectors of the same vector space (in this case $\Psi(t)$ and $\hat{\Psi}(t)$ -- any two fundamental matrices of 3.2.5) must satisfy the relation:

$$\Psi(t) = \hat{\Psi}(t)P \tag{3.2.13}$$

for some nonsingular, scalar matrix P with elements in $\mathscr{R}$. Therefore, $\Phi(t,t_o) = \Psi(t)^{-1}\Psi(t_o) = \hat{\Psi}(t)PP^{-1}\hat{\Psi}^{-1}(t_o) = \hat{\Psi}(t)\hat{\Psi}^{-1}(t_o)$ is both unique and independent of the particular choice of $\Psi(t)$. We next note some important properties of the state transition matrix $\Phi(t,t_o)$. In particular, it is now clear in view of 3.2.12 that the following relationships hold:

$$\Phi(t,t) = \Phi(t_o,t_o) = I, \tag{3.2.14}$$

$$\Phi^{-1}(t,t_o) = \Psi(t_o)\Psi^{-1}(t) = \Phi(t_o,t), \tag{3.2.15}$$

$$\Phi(t_2,t_o) = \Psi(t_2)\Psi^{-1}(t_1)\Psi(t_1)\Psi^{-1}(t_o) = \Phi(t_2,t_1)\Phi(t_1,t_o), \tag{3.2.16}$$

and that $\Phi(t,t_o)$ is nonsingular for all t and $t_o \varepsilon (-\infty,\infty)$.

We now consider a method for explicitly determining the state transition matrix which is based on an alternate interpretation of $\Phi(t,t_o)$. In particular, $\Phi(t,t_o)$ is clearly the unique solution of the matrix differential equation:

$$\frac{\partial}{\partial t}\Phi(t,t_o) = A\Phi(t,t_o), \tag{3.2.17}$$

with the initial condition: $\Phi(t_o,t_o) = I$. The solution to the scalar (n=1) version of this elementary differential equation is the matrix exponential $\exp\left[\int_{t_o}^{t} A d\tau\right]$ or $e^{A(t-t_o)}$ which is easily verified to be the solution in the matrix case as well; i.e. if we define $e^{A(t-t_o)}$ by the infinite power series:

$$e^{A(t-t_o)} = I + A(t-t_o) + \frac{A^2(t-t_o)^2}{2!} + \frac{A^3(t-t_o)^3}{3!} + \ldots \qquad 3.2.18$$

$$= \sum_{o}^{\infty} \frac{A^k(t-t_o)^k}{k!}$$

then $e^{A(t_o-t_o)} = I$, the appropriate initial condition, and

$$\frac{\partial}{\partial t} e^{A(t-t_o)} = Ae^{A(t-t_o)} \qquad 3.2.19$$

It is therefore clear that the state transition matrix of $\dot{x}(t) = Ax(t)$ is

$$\Phi(t,t_o) = \Phi(t-t_o) = e^{A(t-t_o)} \qquad 3.2.20$$

In the next section, we will discuss a standard technique used to evaluate $e^{A(t-t_o)}$ in closed form. Before we do so, however, we will first present a physical interpretation of the state transition matrix and, in so doing, we will also define the "state space".

In particular, we recall that from 3.2.11 and 3.2.20

$$x(t_1) = e^{A(t_1-t_o)} x(t_o) \qquad 3.2.21$$

for any $t_1 \geq t_o$ (actually for any t_1 regardless of its position relative to t_o). Furthermore, the state transition matrix evaluated at t_1, $e^{A(t_1-t_o)}$, is a nonsingular $n \times n$ matrix with elements in $\mathscr{R}$. Therefore, it can be interpreted as the matrix representation of a linear operator or transformation which maps the n-tuple $x(t_o)$ in the n-dimensional <u>state space</u>, which is nothing more than the space of all real n-tuples (sometimes denoted by $\mathscr{R}^n$), into the n-tuple $x(t_1) \varepsilon \mathscr{R}^n$; e.g. if we consider a third order system (n=3), then both $x(t_o)$ and $x(t_1)$ can be represented by points in the space of 3-tuples $\mathscr{R}^3$ as depicted in Figure 3.2.22 below. The <u>transition</u> from $x(t_o)$ to $x(t_1)$ as t varies from t_o to t_1 can then be represented by the "smooth" state transition matrix mapping of the initial state $x(t_o)$ to the final state $x(t_1)$ via the relation: $x(t) = e^{A(t-t_o)} x(t_o)$ which is defined at the continuous values of time from t_o to t_1 and visually depicted in Figure 3.2.22.

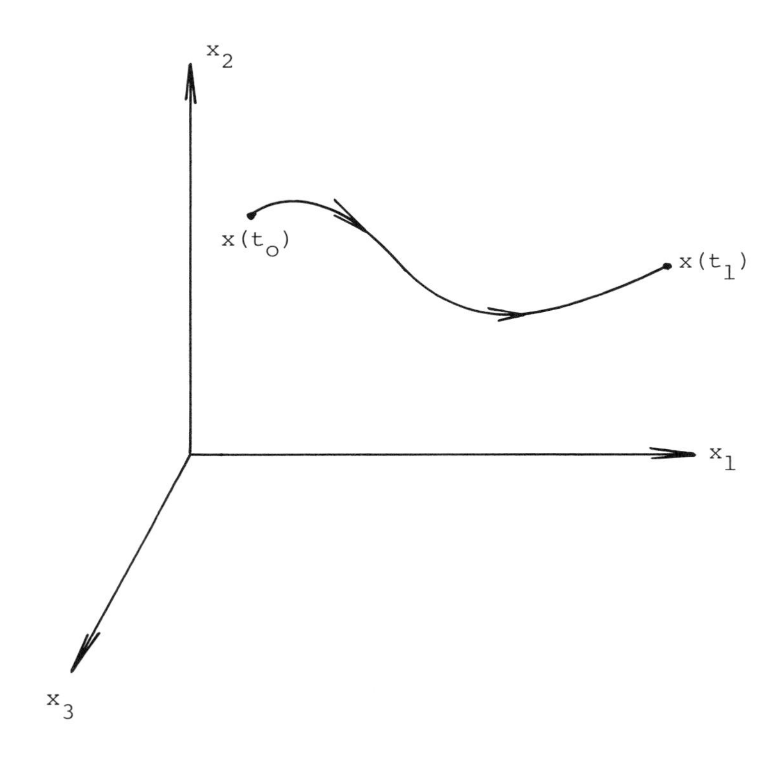

FIGURE 3.2.22

STATE TRANSITION

IN THE STATE SPACE

To illustrate the notions of the state transition matrix as well as state transition in the state space, consider the following:

EXAMPLE 3.2.23: We again employ Example 3.2.8; i.e. $\dot{x}(t) = Ax(t)$, where $A = \begin{bmatrix} -1 & 0 \\ 0 & 2 \end{bmatrix}$. It was shown in Example 3.2.8 that $\Psi(t) = \begin{bmatrix} e^{-t} & 0 \\ 0 & e^{2t} \end{bmatrix}$ is a fundamental matrix for this system. If 3.2.12 is now used, the state transition matrix $\Phi(t,t_o) = e^{A(t-t_o)}$ is readily determined; i.e. $e^{A(t-t_o)} = \Psi(t)\Psi^{-1}(t_o) = \begin{bmatrix} e^{-(t-t_o)}, & 0 \\ 0, & e^{2(t-t_o)} \end{bmatrix}$.

Therefore, if $x(t_o) = \begin{bmatrix} x_1(t_o) \\ x_2(t_o) \end{bmatrix}$, any arbitrary 2-tuple in $\mathscr{R}^2$, then

$x(t_1) = \begin{bmatrix} e^{-(t_1-t_o)} x_1(t_o) \\ e^{2(t_1-t_o)} x_2(t_o) \end{bmatrix}$, another 2-tuple in $\mathscr{R}^2$. We note further that this example also serves to illustrate an important and useful fact--namely <u>if the matrix A in 3.2.5 is diagonal, the state transition matrix $e^{A(t-t_o)}$ is also diagonal and can be evaluated directly, term by term</u>; i.e. if $A = \begin{bmatrix} \lambda_1 & & & 0 \\ & \lambda_2 & & \\ & & \ddots & \\ 0 & & & \lambda_n \end{bmatrix}$ with scalars λ_i in $\mathscr{R}$, then

$$e^{A(t-t_o)} = \begin{bmatrix} e^{\lambda_1(t-t_o)} & & & 0 \\ & e^{\lambda_2(t-t_o)} & & \\ & & \ddots & \\ 0 & & & e^{\lambda_n(t-t_o)} \end{bmatrix}.$$

Unfortunately, this result does not hold in general and an alternate procedure for evaluating $e^{A(t-t_o)}$ must be employed. It might be noted that a general evaluation procedure for the state transition matrix in closed form will be given in the next section.

Once the state transition matrix $\Phi(t,t_o) = e^{A(t-t_o)}$ of the homogeneous part of 3.2.1a has been found, the complete nonhomogeneous (forced) solution can be directly obtained in terms of $e^{A(t-t_o)}$. In particular, we now have:

THEOREM 3.2.24: <u>The complete solution $x(t)$ of the linear time-invariant dynamical system 3.2.1a with initial state</u> $x(t_o) = x_o$ <u>is given by</u>:

$$x(t) = e^{A(t-t_o)}x_o + \int_{t_o}^{t} e^{A(t-\tau)}Bu(\tau)d\tau \tag{3.2.24}$$

<u>Proof</u>: By differentiating this expression with respect to t, we obtain

$$\dot{x}(t) = \frac{\partial}{\partial t} e^{A(t-t_o)}x_o + \frac{\partial}{\partial t}\int_{t_o}^{t} e^{A(t-\tau)}Bu(\tau)d\tau, \tag{3.2.25}$$

and recalling (Leibnitz's rule) that for any "smooth" function $f(t,\tau)$ of two variables, $\frac{\partial}{\partial t}\int_{t_o}^{t} f(t,\tau)d\tau = f(t,t) + \int_{t_o}^{t}\frac{\partial}{\partial t}f(t,\tau)d\tau$, it is clear that

$$\begin{aligned}\dot{x}(t) = Ae^{A(t-t_o)}x_o + Bu(t) + A\int_{t_o}^{t} e^{A(t-\tau)}Bu(\tau)d\tau \\ = Ax(t) + Bu(t),\end{aligned} \tag{3.2.26}$$

or that $x(t)$ as given by 3.2.24 does indeed satisfy Equation 3.2.1a with $x(t_o) = x_o$ and thus represents the unique, complete solution of 3.2.1a. Theorem 3.2.24 is thus established.

In view of 3.2.24 and 3.2.1b, it now follows that the output $y(t)$ is given by:

$$y(t) = Ce^{A(t-t_o)}x_o + C\int_{t_o}^{t} e^{A(t-\tau)}Bu(\tau)d\tau + Eu(t), \qquad 3.2.27$$

the sum of two mutually exclusive quantities, $Ce^{A(t-t_o)}x_o$ and $C\int_{t_o}^{t} e^{A(t-\tau)}Bu(\tau)d\tau + Eu(t)$, which will now be defined, for obvious reasons, as <u>the zero-input response</u> and <u>the zero state response</u> respectively.

If the system 3.2.1 is completely at rest $(x(t) \equiv 0)$ prior to some time t_o when a unit impulse $\delta(t-t_o)$ is applied as the k-<u>th</u> element $u_k(t)$ of the m vector input $u(t) = \begin{bmatrix} u_1(t) \\ u_2(t) \\ \vdots \\ u_m(t) \end{bmatrix}$ to the system with all other inputs identically zero, the resulting p-dimensional output $y(t) = \begin{bmatrix} y_1(t) \\ y_2(t) \\ \vdots \\ y_p(t) \end{bmatrix}$ is equal to the k-<u>th</u> column of what we now define to be the <u>impulse response matrix</u> $T(t-t_o)$ of the system. It readily follows from 3.2.27 that $T(t-t_o) = Ce^{A(t-t_o)}B + E\delta(t-t_o)$ for all $t \geq t_o$ and is identically zero prior to the time t_o. Since t_o is completely arbitrary, it is customarily set equal to 0, with the result that the impulse response matrix of 3.2.1 is dependent on the time t alone; i.e. if $t_o = 0$

$$T(t) = \begin{cases} Ce^{At}B + E\delta(t) & \text{for } t \geq 0 \\ 0 & \text{for } t < 0 \end{cases} \qquad 3.2.28$$

3.3 THE DETERMINATION OF e^{At}

In view of the results presented in the previous section, it is clear that a determination of the state transition matrix $e^{A(t-t_o)}$ must be made before the solution $x(t)$ of the dynamical system 3.2.1 can be found. The infinite series 3.2.18 can be used to evaluate e^{At} if a computer is available and if the series converges within a "reasonable" number of terms, or e^{At} can be determined term by term if A happens to be a diagonal matrix. There are, however, other methods commonly employed to obtain e^{At} in closed form; one of which relies on the well known one-sided Laplace transformation:

$$\mathscr{L}[f(t)] \triangleq \int_o^\infty f(t)e^{-st}dt = f(s) \qquad 3.3.1$$

with $s = \sigma + j\omega$ an (as yet) unspecified complex number. In order to develop this method, we either recall or derive the relation:

$\mathscr{L}\left[\frac{t^k}{k!}\right] = \frac{1}{s^{k+1}}$, which in view of 3.2.18 implies that

$$\mathscr{L}\left[e^{At}\right] = \sum_o^\infty \frac{A^k}{s^{k+1}} = \frac{1}{s}\sum_o^\infty\left(\frac{A}{s}\right)^k \qquad 3.3.2$$

However,

$$\sum_o^\infty\left(\frac{A}{s}\right)^k = I + \frac{A}{s} + \frac{A^2}{s^2} + \ldots = \left(I - \frac{A}{s}\right)^{-1} \qquad 3.3.3$$

since this power series converges whenever the absolute values of all of the eigenvalues of $\frac{A}{s}$ are less than unity [C1]. If s is therefore chosen "large enough" to insure this eigenvalue condition, substituting 3.3.3 into 3.3.2, we have

$$\mathscr{L}\left[e^{At}\right] = \frac{1}{s}\left(I - \frac{A}{s}\right)^{-1} = (sI-A)^{-1} \qquad 3.3.4$$

If we now take the inverse Laplace transform $\mathscr{L}^{-1}[\cdot]$ of both sides of 3.3.4 we obtain the desired result; i.e.

$$e^{At} = \mathscr{L}^{-1}[(sI-A)]^{-1} \qquad 3.3.5$$

A less rigorous derivation of 3.3.5 involves simply calculating the

Laplace transform of 3.2.5; i.e.

$$sx(s) - x(0) = Ax(s), \qquad 3.3.6$$

or $(sI-A)x(s) = x(0)$, which implies that

$$x(s) = (sI-A)^{-1}x(0) \qquad 3.3.7$$

If we now take the inverse Laplace transform of 3.3.7, we find that

$$x(t) = \mathscr{L}^{-1}[(sI-A)]^{-1}x(0), \qquad 3.3.8$$

and by equating 3.3.8 to 3.2.24 with $u(t) \equiv 0$ and $t_o = 0$, it is clear that 3.3.5 holds.

Having established 3.3.5, we now demonstrate its utilization in determining the state transition matrix.

EXAMPLE 3.3.9: Consider the system $\dot{x}(t) = Ax(t)$, where $A = \begin{bmatrix} 0 & 1 & 0 \\ 0 & 0 & 1 \\ 0 & -2 & -3 \end{bmatrix}$. It is clear that $(sI-A) = \begin{bmatrix} s & -1 & 0 \\ 0 & s & -1 \\ 0 & 2 & s+3 \end{bmatrix}$ and, therefore that

$$(sI-A)^{-1} = (sI-A)^{+} \div |sI-A| = \begin{bmatrix} s^2+3s+2, & s+3, & 1 \\ 0, & s^2+3s, & s \\ 0, & -2s, & s^2 \end{bmatrix} \div (s^3+3s^2+2s).$$

We can now evaluate the inverse Laplace transform of $(sI-A)^{-1}$ term by term, noting that $s^3+3s^2+2s = s(s+1)(s+2)$. Beginning with the 1,1 entry, we have: $\mathscr{L}^{-1}\left[\frac{s^2+3s+2}{s^3+3s^2+2s}\right] = \mathscr{L}^{-1}\left[\frac{1}{s}\right] = 1$. The 1,2 entry is given by: $\mathscr{L}^{-1}\left[\frac{s+3}{s(s+1)(s+2)}\right]$ which can be expressed as $\mathscr{L}^{-1}\left[\frac{3/2}{s} - \frac{2}{s+1} + \frac{1/2}{s+2}\right]$ by employing a partial fraction expansion. Thus $e^{At}_{12} = \frac{3}{2} - 2e^{-t} + \frac{1}{2}e^{-2t}$. Continuing in this manner, we can completely determine e^{At} in closed form; i.e. for this example,

$$e^{At} = \begin{bmatrix} 1, & \frac{3}{2} - 2e^{-t} + \frac{1}{2}e^{-2t}, & \frac{1}{2} - e^{-t} + \frac{1}{2}e^{-2t} \\ 0, & 2e^{-t} - e^{-2t}, & e^{-t} - e^{-2t} \\ 0, & -2e^{-t} + 2e^{-2t}, & -e^{-t} + 2e^{-2t} \end{bmatrix}$$

The above example also serves to illustrate a useful point worth noting. In particular, whenever the A matrix is in the indicated companion form with the identity matrix in the upper right block, it follows that $x_{i+1}(t) = \dot{x}_i(t)$ for $i = 1,2,\ldots,n-1$. Since $x(t) = \begin{bmatrix} x_1(t) \\ x_2(t) \\ \vdots \\ x_n(t) \end{bmatrix} = e^{At}x(0)$, it is clear that once $x_1(t)$, the first row of e^{At}, has been found, the remaining n-1 rows of e^{At} can directly be found by successive differentiation of $x_1(t)$; i.e. the reader can verify that in the above example, each succeeding row of e^{At} can be obtained by simply differentiating the previous row.

It should now be clear that the (n) eigenvalues of the state matrix A play a key role in determining the zero-input response of the system (see Section 3.2). In order to determine the relationship between the eigenvalues of A and the zero-input response, we first note that if $u(t) \equiv 0$ for all $t \geq 0$, then, in view of 3.2.24, $x(t) = e^{At}x_o$, or

$$x(s) = \mathscr{L}[x(t)] = (sI-A)^{-1}x_o = \frac{(sI-A)^+}{|sI-A|}x_o. \qquad 3.3.10$$

If we now let $\lambda_1, \lambda_2, \ldots, \lambda_n$ represent the (n) eigenvalues of A which, for convenience, are assumed to be distinct, it follows that $|sI-A| = \prod_1^n (s-\lambda_i)$ and, therefore, that we can express $x(s)$ via the partial fraction expansion:

$$x(s) = \sum_1^n \frac{N_i}{(s-\lambda_i)} x_o, \qquad 3.3.11$$

for some set of matrices, N_i, with complex entries. If we now take the inverse Laplace transform of 3.3.11, we obtain

$$x(t) = \sum_1^n N_i e^{\lambda_i t} x_o, \qquad 3.3.12$$

which implies that the total zero-input response of the system consists of a sum of exponentials of the individual eigenvalues of A times t.

It is thus clear that if the real part of at least one eigenvalue, λ_k, of A is greater than zero, then the zero-input response of the system will increase exponentially with time in response to any arbitrarily small nonzero initial conditions on $x(t)$, including noise. Under these circumstances, we say that the system is unstable. For obvious reasons, we further state that the system is stable or asymptotically stable if $Re(\lambda_i) < 0$ for all i, and if $Re(\lambda_i) \leq 0$, for all i and $= 0$ for at least one λ_k, we say that the system is marginally stable.

3.4 EQUIVALENT SYSTEMS

We now consider the effect on the performance of a dynamical system in the state form 3.2.1 if we alter its state $x(t)$ via the relationship:

$$\hat{x}(t) = Qx(t), \qquad 3.4.1$$

where Q is any $n \times n$ nonsingular matrix with elements in $\mathscr{R}$. In particular, if we substitute 3.4.1 for $x(t) = Q^{-1}\hat{x}(t)$ in 3.2.1, we obtain the altered state representation:

$$\dot{\hat{x}}(t) = \hat{A}\hat{x}(t) + \hat{B}u(t); \qquad 3.4.2a$$

$$y(t) = \hat{C}\hat{x}(t) + Eu(t), \qquad 3.4.2b$$

where $\hat{A} = QAQ^{-1}$, $\hat{B} = QB$, and $\hat{C} = CQ^{-1}$.

DEFINITION 3.4.3: The state representations 3.2.1 and 3.4.2 with states related by 3.4.1 are said to be <u>equivalent</u> and Q is called an <u>equivalence transformation</u>. In other words, the systems $\{A,B,C,E\}$ and $\{\hat{A},\hat{B},\hat{C},\hat{E}\}$ are <u>equivalent</u> if and only if the following relationships hold for some nonsingular real matrix Q:

$$\hat{A} = QAQ^{-1} \qquad 3.4.4a$$

$$\hat{B} = QB \qquad 3.4.4b$$

$$\hat{C} = CQ^{-1} \qquad 3.4.4c$$

$$\hat{E} = E \qquad 3.4.4d$$

The justification for the use of the term "equivalent" in this definition can readily be demonstrated if we note that the solution of either system, $x(t)$ or $\hat{x}(t)$, immediately implies the solution of the other via 3.4.1. We further note that the state transition matrices of the two systems are similar; i.e. 3.2.18 can be used to establish the fact that

$$e^{\hat{A}t} = e^{QAQ^{-1}t} = Qe^{At}Q^{-1}, \qquad 3.4.5$$

which, in turn, implies exact equivalence of the impulse response matrices of the two systems. In particular,

$$\hat{C}e^{\hat{A}t}\hat{B} + E\delta(t) = CQQe^{-1}{}^{At}Q^{-1}QB + E\delta(t) = Ce^{At}B + E\delta(t) \qquad 3.4.6$$

This latter point is an important one in that it establishes the fact that <u>the impulse response matrix of a linear dynamical system is dependent only on the external, input/output behavior of the system and is therefore independent of any particular choice of internal state</u>. The "transfer matrix" of a linear dynamical system, which will be discussed in detail in the next chapter, is another example of an externally (input/output) dependent quantity.

Associated with this notion of equivalence of state representations or systems in state form are the less restrictive notions of "zero-input equivalence" and "zero-state equivalence".

DEFINITION 3.4.7: Two state representations will be called <u>zero-input equivalent</u> if for any initial state of one system there corresponds an initial state of the other such that both systems have the same zero-input response. Two state representations will be called <u>zero-state equivalent</u> if they both have the same impulse response matrix. As noted above, these two definitions represent two less restrictive notions of equivalence; i.e. it can readily be established that equivalence implies both zero-input and zero-state equivalence, but that the converse does not necessarily hold. In particular, consider the equivalent systems represented by 3.2.1, 3.4.1, and 3.4.2. In view of 3.4.6, they are clearly zero-state equivalent. Furthermore, the zero-input response of the first, $Ce^{A(t-t_o)}x(t_o)$, is identical to that of the second provided the initial state $\hat{x}(t_o)$ of the second is set equal to $Qx(t_o)$; i.e.

$$Ce^{A(t-t_o)}x(t_o) = CQQe^{-1\,A(t-t_o)-1}QQx(t_o) = \hat{C}e^{\hat{A}(t-t_o)}\hat{x}(t_o) \qquad 3.4.8$$

It does not necessarily follow, however, that zero-state equivalence, zero-input equivalence, or both together imply equivalence as the following example illustrates:

EXAMPLE 3.4.9: Consider the following electrical networks:

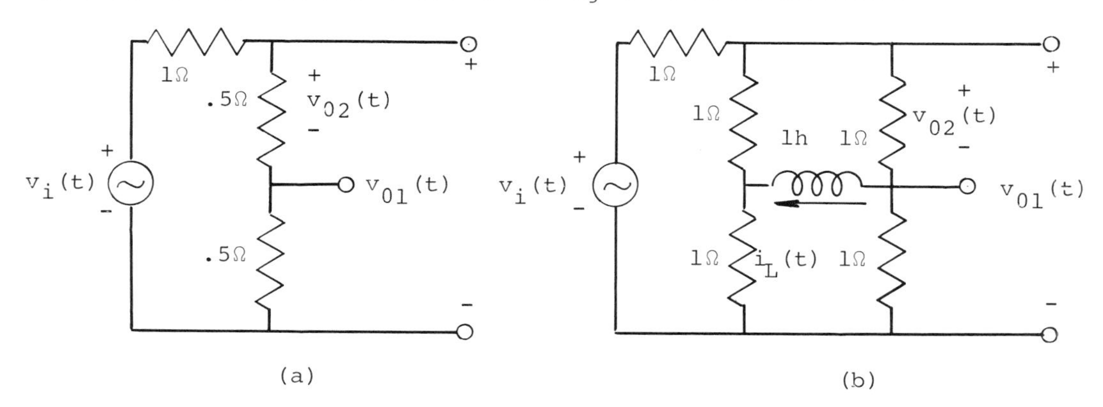

If $v_{01}(t)$ alone is defined as the output $y(t)$ of both systems and $v_i(t)$ as the input $u(t)$ to both systems, then they are zero-state equivalent since $y(t) = \frac{u(t)}{2}$ for both systems for all t. Furthermore, they are also zero-input equivalent since $y(t) = u(t) = 0$ for both systems regardless of $i_L(t)$. However, the two systems are clearly not equivalent since they differ in order. One final point--if $v_{02}(t)$ is now introduced as a second output of both systems with $y(t) = \begin{bmatrix} v_{01}(t) \\ v_{02}(t) \end{bmatrix}$, the two systems will no longer be zero-input equivalent since the output $v_{02}(t)$ of the system (b) will be directly dependent on $i_L(t)$; i.e. $v_{02}(t) = \frac{1}{2} e^{-t} i_L(t_o)$, while $v_{02}(t)$ of system (a) will be identically zero for all t. However, the two systems will remain zero-state equivalent.

3.5 CONTROLLABILITY AND OBSERVABILITY

Up to this point, we have been primarily concerned with the analysis of linear dynamical systems and nothing has been said regarding the synthesis or control of these systems. We now consider some of the fundamental notions associated with the control of linear multivariable systems which are represented in the state form 3.2.1.

DEFINITION 3.5.1: The system 3.2.1a or $\dot{x}(t) = Ax(t) + Bu(t)$ is said to be <u>completely state controllable</u>, or simply <u>controllable</u>, if and only if there exists a control $u(t)$ which transfers any initial state $x(t_o)$ at any time t_o to any arbitrary final state $x(t_1)$ at any time $t_1 > t_o \geq 0$. Otherwise, the system is said to be <u>uncontrollable</u>, although it may be "controllable in part"; i.e. it may be possible to transfer certain states to any desired final states or the entire state to certain regions in the state space. We note that the system output $y(t)$ plays no role in this definition of controllability.

The reader will no doubt agree that this notion of (complete state) controllability is intuitively appealing, since one can "visualize" this transfer from one state $x(t_o)$ to another $x(t_1)$ in the state space via the input or control $u(t)$ (see Figure 3.2.22), and the ability to achieve any desired final state at any $t_1 > t_o$ is particularly attractive. It is for reasons such as this that the state representation 3.2.1 is such an important and useful way to characterize the behavior of physical systems.

As we will soon demonstrate, it is a relatively simple matter not only to determine whether or not 3.2.1a is controllable, but also to explicitly obtain an appropriate control which performs the transfer from $x(t_o)$ to $x(t_1)$. In order to achieve these objectives, we first recall that the complete solution to 3.2.1a is given by 3.2.24; i.e. $x(t) = e^{A(t-t_o)}x(t_o) + \int_{t_o}^{t} e^{A(t-\tau)}Bu(\tau)d\tau$, or at $t = t_1 > t_o$, that

$$e^{A(t_o-t_1)}x(t_1) - x(t_o) = \int_{t_o}^{t_1} e^{A(t_o-\tau)}Bu(\tau)d\tau \qquad 3.5.2$$

The controllability of 3.2.1a therefore implies the existence of a solution $u(t)$ defined for all $t\varepsilon[t_o,t_1]$ which satisfies 3.5.2 for any n-vector $[e^{A(t_o-t_1)}x(t_1) - x(t_o)]$ with elements in $\mathcal{R}$. Loosely speaking, this observation implies that if the system 3.2.1a is controllable, then the right side of 3.5.2 must satisfy some "full rank" condition. This is indeed the case, as we now formally show.

THEOREM 3.5.3: The following statements regarding the linear, time-invariant dynamical system $\dot{x}(t) = Ax(t) + Bu(t)$ are equivalent:

a. The system is (completely state) controllable.

b. $W(t_o,t_1) = \int_{t_o}^{t_1} e^{-A\tau}BB^Te^{-A^T\tau}d\tau$ is nonsingular for all

$t_1 > t_o$.

c. <u>The (n) rows of $e^{-At}B$ (and hence $e^{At}B$) are linearly independent over the real field $\mathscr{R}$ for all $t \varepsilon [0,\infty)$.</u>

d. <u>The rank of the $(n \times nm)$ controllability matrix:</u> $\mathscr{C} \triangleq [B, AB, \ldots, A^{n-1}B]$ <u>is n</u>; i.e. $\rho[\mathscr{C}] = n$.

<u>Proof</u>: We first establish, by contradiction, that 3.4.3c implies 3.5.3b; i.e. assume that the (n) rows $f_i(t)$ of $e^{-At}B = \begin{bmatrix} f_1(t) \\ f_2(t) \\ \vdots \\ f_n(t) \end{bmatrix}$ are linearly independent over $\mathscr{R}$ for all $t \varepsilon [0,\infty)$ but that $|W(t_o,t_1)| = 0$. This latter assumption implies that for some nonzero row vector $\alpha = [\alpha_1, \alpha_2, \ldots, \alpha_n]$ with elements in $\mathscr{R}$, $\alpha \int_{t_o}^{t_1} e^{-A\tau}BB^T e^{-A^T\tau} d\tau = 0$ for all $t \varepsilon [t_o,t_1]$. Postmultiplying this last expression by α^T and placing all terms within the integral sign, we obtain:

$$\int_{t_o}^{t_1} \alpha e^{-A\tau}B(\alpha e^{A\tau}B)^T d\tau = 0, \qquad 3.5.4$$

which clearly implies that $\alpha e^{-At}B = 0$ for all $t \varepsilon [t_o,t_1]$ and some non-zero α, contrary to our assumption that the rows of $e^{-At}B$ are linearly independent for all $t \varepsilon [0,\infty)$. Statement 3.5.3c thus implies 3.5.3b.

We next constructively establish that 3.5.3b implies 3.5.3a. In particular, if $W(t_o,t_1)$ is nonsingular, then the input:

$$u(t) = B^T e^{-A^T t} W^{-1}(t_o,t_1)\left[e^{-At_1}x(t_1) - e^{-At_o}x(t_o)\right] \qquad 3.5.5$$

satisfies the required relationship 3.5.2 regardless of t_1, t_o, $x(t_o)$, and $x(t_1)$. Statement 3.5.3b thus implies 3.5.3a.

The fact that 3.5.3a implies 3.5.3c will now be established by contradiction. In particular, we assume that the system 3.2.1a is controllable but that the (n) rows of $e^{-At}B$ are linearly dependent for all $t \varepsilon [0,\infty)$; i.e. that

$$\alpha e^{-At}B = 0 \quad \text{for all} \quad t\varepsilon[0,\infty) \qquad 3.5.6$$

for some nonzero real vector α. Since the system is assumed controllable, 3.5.2 holds for some input $u(t)$. Premultiplying both sides of 3.5.2 by αe^{-At_o}, we obtain:

$$\alpha e^{-At_1}x(t_1) - \alpha e^{-At_o}x(t_o) = \alpha\int_{t_o}^{t_1} e^{-A\tau}Bu(\tau)d\tau, \qquad 3.5.7$$

or in view of 3.5.6, that

$$\alpha e^{-At_1}x(t_1) = \alpha e^{-At_o}x(t_o) \qquad 3.5.8$$

If we now choose $x(t_o) = e^{At_o}\alpha^T$, which is nonzero since e^{At_o} is nonsingular and $\alpha \neq 0$, then

$$\alpha e^{-At_1}x(t_1) = \alpha\alpha^T \neq 0 \qquad 3.5.9$$

Therefore $x(t_1)$ can never be driven to the zero state via $u(t)$ which contradicts our assumption of (complete state) controllability. Statement 3.5.3a thus implies 3.5.3c. In summary, therefore, statements 3.5.3a, b, and c imply and are implied by one another, and we need only establish the equivalence of 3.5.3d to establish the theorem. We do this by first establishing:

LEMMA 3.5.10: <u>The (n) rows of $e^{-At}B$ are linearly independent over $\mathscr{R}$ for all $t\varepsilon[0,\infty)$ if and only if the (n) rows of the $(n \times \infty)$ matrix valued function of time</u> $[e^{-At}B, -e^{-At}AB, \ldots, (-1)^{n-1}e^{-At}A^{n-1}B, \ldots]$ <u>are linearly independent over $\mathscr{R}$ for all $t\varepsilon[0,\infty)$; i.e. if and only if</u>

$$\rho[e^{-At}B, -e^{-At}AB, \ldots, (-1)^{n-1}e^{-At}A^{n-1}B, \ldots] = n \qquad 3.5.11$$

Proof: We establish necessity first; i.e. assume that the rows of $e^{-At}B$ are linearly independent but that $\rho[e^{-At}B, -e^{-At}AB, \ldots] < n$ at some time $t_o \varepsilon [0,\infty)$, which implies that for some nonzero real vector $\alpha = [\alpha_1, \alpha_2, \ldots, \alpha_n]$,

$$\alpha \left[e^{-At_o}B, -e^{-At_o}AB, \ldots, (-1)^{n-1} e^{-At_o}A^{n-1}B, \ldots \right] = 0 \quad 3.5.12$$

However, since the entries of $e^{-At}B$ are analytic everywhere (for all $t\varepsilon[0,\infty)$ including t_o), $e^{-At}B$ can be expanded via a Taylor series about t_o; i.e.

$$e^{-At}B = \sum_{o}^{\infty} \frac{(t-t_o)^k}{k!} (-1)^k e^{-At_o}A^k B, \quad 3.5.13$$

an expression which holds wherever $e^{-At}B$ is analytic. If we now premultiply both sides of 3.5.13 by α, we see, in view of 3.5.12, that

$$\alpha e^{-At}B = \sum_{o}^{\infty} \frac{(t-t_o)^k}{k!} (-1)^k \alpha e^{-At_o}A^k B = 0 \quad 3.5.14$$

for all t where 3.5.13 holds and therefore, for all $t\varepsilon[0,\infty)$, which is contrary to the assumption that the rows of $e^{-At}B$ are linearly independent. Necessity is therefore established by contradiction. We note, in view of this proof--3.5.14 in particular, that if the rows of $e^{-At}B$ are linearly dependent (independent) at any time t_o, then they are linearly dependent (independent) for all $t\varepsilon[0,\infty)$.

Sufficiency is also readily established by contradiction. In particular, assume that 3.5.11 holds, but that for some nonzero real vector α,

$$\alpha e^{-At}B = 0. \quad 3.5.15$$

for all $t\varepsilon[0,\infty)$. By successively differentiating 3.5.15 with respect to time, we find that $-\alpha e^{-At}AB = \alpha e^{-At}A^2B = \ldots = (-1)^{n-1}\alpha e^{-At}A^{n-1}B = \ldots = 0$, or that

$$\alpha\left[e^{-At}B,-e^{-At}AB,\ldots,(-1)^{n-1}e^{-At}A^{n-1}B,\ldots\right] = 0, \qquad 3.5.16$$

since all derivatives of the identically zero, analytic, vector-valued function $\alpha e^{-At}B$ are also zero. Since 3.5.16 is contrary to the assumption that 3.5.11 holds, sufficiency is established by contradiction and the proof of Lemma 3.5.10 is complete.

In view of Lemma 3.5.10 and, in particular, 3.5.11 evaluated at $t = 0$, it is clear that 3.5.3c holds if and only if

$$\rho[B,-AB,\ldots,(-1)^{n-1}A^{n-1}B,\ldots] = n \qquad 3.5.17$$

This latter relationship can now be simplified if we employ the Cayley-Hamilton Theorem (Section 2.4) which enables us to express all terms of the form A^kB with $k \geq n$ as some linear combination of the first (n) terms: $B,AB,A^2B,\ldots,A^{n-1}B$ which comprise the matrix defined in 3.5.17. Furthermore, since the rank of a matrix does not change if we multiply any column by -1, it follows that 3.5.3c holds if and only if

$$\rho[B,AB,\ldots,A^{n-1}B] = n \qquad 3.5.18$$

Therefore, 3.5.3d implies and is implied by 3.5.3c, as are 3.5.3a and 3.5.3b. Theorem 3.5.3 is thus established.

It should be pointed out that 3.5.3d, the determination of the rank of the controllability matrix $\mathscr{C}$, is the test usually employed to ascertain whether or not a system in the state form 3.2.1a is controllable; i.e. <u>the system $\dot{x}(t) = Ax(t) + Bu(t)$ or, equivalently, the pair $\{A,B\}$ is controllable if and only if $\rho[\mathscr{C}] = n$</u>. Also, whenever a state form system is controllable, the relation 3.5.5 yields an input which transfers any $x(t_o)$ to any $x(t_1)$ at any $t_1 > t_o$. It is interesting to note that the particular input given by 3.5.5 performs the required transfer while simultaneously minimizing

$\int_{t_o}^{t_1} u^T(t)u(t)dt$, which represents a measure of the total energy expended. This fact is established in a number of other texts (e.g. [Al]) and will not be formally established here. To illustrate certain of the preceeding results, we now present:

EXAMPLE 3.5.19: Consider a system in the state form 3.2.1a with $A = \begin{bmatrix} -1 & 0 \\ 0 & 2 \end{bmatrix}$. In view of Example 3.2.23, $e^{At} = e^{A^T t} = \begin{bmatrix} e^{-t} & 0 \\ 0 & e^{2t} \end{bmatrix}$. If there is only one input, $B = \begin{bmatrix} b_1 \\ b_2 \end{bmatrix}$, then by 3.5.3d the system is controllable if and only if $\rho[\mathscr{C}] = 2$, where $\mathscr{C} = \begin{bmatrix} b_1 & -b_1 \\ b_2 & 2b_2 \end{bmatrix}$. Since $\mathscr{C}$ has full rank 2 if and only if its determinant $3b_1b_2$ is nonzero, it is clear that this scalar system is controllable if and only if both b_1 and b_2 are nonzero. The reader can readily verify that in the more general multivariable case, where $B = \begin{bmatrix} b_{11}b_{12}\cdots b_{1m} \\ b_{21}b_{22}\cdots b_{2m} \end{bmatrix}$, the system with A matrix given above would be controllable if and only if both the first and second rows of B are nonzero.

Suppose there is only one input and $B = \begin{bmatrix} 1 \\ 1 \end{bmatrix}$, which implies complete state controllability as we have just shown. One control $u(t)$ which transfers the system from $x(0) = \begin{bmatrix} 1 \\ 4 \end{bmatrix}$ to the origin of the state space at $t = 1$, $x(1) = \begin{bmatrix} 0 \\ 0 \end{bmatrix}$, is given by 3.5.5; i.e.

$$W(t_o,t_1) = W(0,1) = \int_o^1 \begin{bmatrix} e^{\tau} & 0 \\ 0 & e^{-2\tau} \end{bmatrix} \begin{bmatrix} 1 & 1 \\ 1 & 1 \end{bmatrix} \begin{bmatrix} e^{\tau} & 0 \\ 0 & e^{-2\tau} \end{bmatrix} d\tau = \begin{bmatrix} \frac{e}{2} - \frac{1}{2}, & 1 - e^{-1} \\ 1 - e^{-1}, & \frac{1}{4} - \frac{e^{-4}}{4} \end{bmatrix}$$

$= \begin{bmatrix} .859, & .432 \\ .432, & .246 \end{bmatrix}$, and $u(t) = [e^t, \ e^{-2t}] \begin{bmatrix} .859, & .432 \\ .432, & .246 \end{bmatrix}^{-1} \begin{bmatrix} -1 \\ -4 \end{bmatrix} =$ $52.9e^t - 107.3e^{-2t}$.

It is often more important to drive the output $y(t) = Cx(t) + Eu(t)$, rather than the state $x(t)$, from some initial value $y(t_o)$ to a desired value $y(t_1)$ at some time $t_1 > t_o$. For this reason, we now have:

DEFINITION 3.5.20: The system 3.2.1 is said to be output controllable if and only if there exists a control $u(t)$ which transfers any initial output $y(t_o)$ at time t_o to any arbitrary final output $y(t_1)$ at any $t_1 > t_o$. As in the case of complete state controllability, nothing is implied or stated in this definition regarding the path which $y(t)$ must follow from $y(t_o)$ to $y(t_1)$ since only the two end points and the times t_o and t_1 are specified. It might be noted here, however, that the ability to transfer $y(t)$ along any "sufficiently smooth" prespecified path via $u(t)$ is also important in certain applications, and any system possessing this ability is said to be output function controllable, or the output is said to be functionally reproducible. A good deal more will be said regarding this notion in Chapter 5 and Section 5.5 in particular.

In view of the controllability results presented thus far and, in particular, Definition 3.5.20 and statement d of Theorem 3.5.3, we now have:

THEOREM 3.5.21: The system $\dot{x}(t) = Ax(t) + Bu(t)$; $y(t) = Cx(t) + Eu(t)$ is output controllable if and only if the $p \times (nm + m)$ matrix $[CB, CAB, \ldots, CA^{n-1}B, E]$ has rank p.

The proof of this theorem follows virtually the same reasoning as applied to the proof of 3.5.3 and will therefore not be given here.

Closely linked to the notion of controllability of linear dynamical systems is the "dual" notion of "observability". As we have shown, complete state controllability is an input $u(t)$ and state $x(t)$ dependent notion exclusively, and is completely unrelated to and independent of the defined system output $y(t)$. Furthermore, system controllability implies the ability to completely control the entire state of the system. Suppose, however, that the complete state is not directly accessible at the output terminals $y(t)$ of the system (see Figure 3.2.3). How can we "observe" the dynamical behavior of the entire n-dimensional state if the only available measurements are the $p(< n)$ output variables $y(t)$? The answer to this question is supplied, in large part, by the following:

DEFINITION 3.5.22: The system 3.2.1 or $\dot{x}(t) = Ax(t) + Bu(t)$; $y(t) = Cx(t) + Eu(t)$ is said to be <u>completely state observable</u>, or simply <u>observable</u>, if and only if the entire state $x(t)$ of the system can be determined over any finite time interval $[t_o,t_1]$ from complete knowledge of the system input and output over the time interval $[t_o,t_1]$ with $t_1 > t_o \geq 0$. Otherwise, the system is said to be <u>unobservable</u> although it may be "observable in part"; i.e. it may be possible to determine some portion of the entire state over the time interval. This notion of observability thus implies the ability to reconstruct the <u>entire</u> n-dimensional state of a system from knowledge of $y(t)$ and $u(t)$ alone.

As in the case of controllability, it is a relatively simple matter not only to ascertain whether or not a system in the state form 3.2.1 is observable, but also to constructively determine $x(t)$ from knowledge of $y(t)$ and $u(t)$ over any finite time interval $[t_o,t_1]$. In particular, we recall that the solution $y(t)$ of 3.2.1 is given by 3.2.27; i.e. $y(t) = Ce^{A(t-t_o)}x(t_o) + C\int_{t_o}^{t} e^{A(t-\tau)}Bu(\tau)d\tau + Eu(t)$, where we assume complete knowledge of $y(t)$, $u(t)$, and the state transition matrix e^{At} over the time interval $[t_o,t_1]$. The problem is therefore to determine $x(t)$ for all $t\varepsilon[t_o,t_1]$ and we begin by subtracting $C\int_{t_o}^{t} e^{A(t-\tau)}Bu(\tau)d\tau + Eu(t)$ from both sides of 3.2.27, in order to obtain an equation involving $x(t_o)$ and a known p-dimensional vector valued function of time $f(t) = y(t) - C\int_{t_o}^{t} e^{A(t-\tau)}Bu(\tau)d\tau - Eu(t)$; i.e.

$$Ce^{A(t-t_o)}x(t_o) = f(t) \qquad 3.5.23$$

If we now premultiply both sides of 3.5.23 by $e^{A^T(t-t_o)}C^T$ and integrate the resulting expression between the limits t_o and t_1, we obtain:

$$\int_{t_o}^{t_1} e^{A^T(t-t_o)}C^TCe^{A(t-t_o)}dt\ x(t_o) = \int_{t_o}^{t_1} e^{A^T(t-t_o)}C^Tf(t)dt \qquad 3.5.24$$

Clearly, if $\int_{t_o}^{t_1} e^{A^T(t-t_o)}C^TCe^{A(t-t_o)}dt$ is invertible, the initial state $x(t_o)$ can readily be found; i.e.

$$x(t_o) = e^{At_o}V(t_o,t_1)^{-1}e^{A^Tt_o}\int_{t_o}^{t_1} e^{A^T(t-t_o)}C^Tf(t)dt, \qquad 3.5.25$$

where $V(t_o,t_1) = \int_{t_o}^{t_1} e^{A^T\tau}C^TC\,e^{A\tau}d\tau$. Furthermore, knowledge of $x(t_o) = x_o$ now enables us to reconstruct the entire state $x(t)$ for all $t\varepsilon[t_o,t_1]$ via 3.2.24.

It is interesting to note the similarity between this constructive procedure for determining $x(t)$ for all $t\varepsilon[t_o,t_1]$, and the constructive procedure employed in the case of controllability to determine a control $u(t)$ which drives any controllable system from one state $x(t_o)$ to any other state $x(t_1)$ at any time $t_1 > t_o$. In particular, both constructive procedures involve the nonsingularity of a matrix obtained by integrating a certain product involving either B or C and the state transition matrix e^{At}. As we have shown via Theorem 3.5.3 in the case of controllability, the nonsingularity of the matrix $W(t_o,t_1)$ is equivalent to the linear independence of the rows of $e^{-At}B$ or $e^{At}B$ over $\mathscr{R}$ as well as the full rank of the controllability matrix $\mathscr{C}$. Furthermore, each of these (3) conditions are both necessary and sufficient for complete state controllability. In view of these observations, it follows that by repeating essentially the same arguments used to establish the controllability Theorem 3.5.3, we can also establish its "dual", namely:

THEOREM 3.5.26: <u>The following statements regarding the linear, time-invariant dynamical system $\dot{x}(t) = Ax(t) + Bu(t);\ y(t) = Cx(t) + Eu(t)$ are equivalent</u>:

a. <u>The system is (completely state) observable</u>.

b. $V(t_o,t_1) = \int_{t_o}^{t_1} e^{A^T\tau}C^TCe^{A\tau}d\tau$ <u>is nonsingular for all</u> $t_1 > t_o$.

c. <u>The (n) columns of Ce^{At} are linearly independent for all $t\varepsilon[0,\infty)$ over the real field $\mathscr{R}$</u>.

d. <u>The rank of the $(np \times n)$ observability matrix</u> $\mathscr{O} = \begin{bmatrix} C \\ \hline CA \\ \hline \vdots \\ \hline CA^{n-1} \end{bmatrix}$

<u>is n</u>; i.e. $\rho[\mathscr{O}] = n$.

It should be noted that as in the case of controllability, the determination of the rank of the observability matrix $\mathscr{O}$ is the test usually employed to ascertain whether or not the system 3.2.1 is observable; i.e. <u>the system 3.2.1 or, equivalently, the pair $\{A,C\}$ is observable if and only if $\rho[\mathscr{O}] = n$</u>. In view of the above, it is clear that observability is an output $y(t)$ and state $x(t)$ dependent notion exclusively, and is completely independent of the particular input $u(t)$ applied to the system, in direct contrast to controllability, which is independent of the system output $y(t)$ but is dependent on both the input $u(t)$ and state $x(t)$.

We finally note that both system controllability and observability are "unaffected" by a transformation of state (an equivalence transformation Q), an observation which will be further clarified in the next section. At this point, we merely remark that the rank of the controllability (observability) matrix $\mathscr{C}(\mathscr{O})$ of the state representation $\{A,B,C,E\}$ is identical to the rank of the controllability (observability) matrix $\hat{\mathscr{C}}(\hat{\mathscr{O}})$ of the equivalent system $\{\hat{A},\hat{B},\hat{C},\hat{E}\} = \{QAQ^{-1},QB,CQ^{-1},E\}$, a fact which we leave as an exercise for the reader to formally verify (see Problem 3-4).

The notion of "duality" has been mentioned repeatedly throughout this discussion of observability and will be employed in certain subsequent discussions in the remainder of this text. It therefore warrants an explanation and some degree of formalism which we now provide. Generally speaking, a "dual" result or notion in linear systems theory is ultimately tied to the "dual" concepts of controllability and observability which have just been defined. More specifically, given any mathematical representation for a linear dynamical system and a result involving controllability (observability), an analogous or "dual" result involving observability (controllability) can usually be established directly, without repeating a proof. The usual procedure involves simply transposing all matrices involved in the mathematical representation for the system and then interchanging the transposed input and output matrices, thereby obtaining the p input, m output "dual system or representation". Loosely speaking, this procedure interchanges the input and outputs of the system, thereby interchanging the notions of controllability and observability.

To formalize the above, consider the state representation 3.2.1 for a linear dynamical system or, equivalently, the quadruple $\{A,B,C,E\}$. The "dual system" is obtained by transposing all (four) matrices in this representation and then interchanging B^T and C^T, the transposed input and output matrices respectively. The resulting quadruple $\{A^T,C^T,B^T,E^T\}$ or, equivalently, the state representation:

$$\dot{\bar{x}}(t) = A^T\bar{x}(t) + C^T\bar{u}(t); \qquad 3.5.27a$$

$$\bar{y}(t) = B^T\bar{x}(t) + E^T\bar{u}(t) \qquad 3.5.27b$$

is defined as the _dual system_, or simply the _dual_ of 3.2.1. If we now compare Theorems 3.5.3 and 3.5.26 in light of this definition, it is readily apparent that _the state representation 3.2.1 is controllable (observable) if and only if its dual representation 3.5.27 is observable (controllable) and vice versa_. The notions of controllability and observability are therefore defined as _dual notions_, while Theorems 3.5.3 and 3.5.26 are _dual results_. This somewhat heuristic notion of duality is a most useful one as we will further illustrate in our subsequent discussions.

3.6 CONTROLLABLE AND OBSERVABLE COMPANION FORMS

If a system $\{A,B,C,E\}$ in the state form 3.2.1 is (completely state) controllable, it can be reduced via a nonsingular transformation Q to an equivalent controllable system (see Section 3.4) in a certain structured form which we will call a "controllable companion form". As we will demonstrate, it is often advantageous to work with a structure equivalent system rather than the original system for a number of reasons which will become apparent in our subsequent discussions. For the most part, however, the purpose of this section will be to simply introduce the dual notions of "controllable and observable companion forms" and to present a constructive procedure for reducing systems to these forms.

We begin by considering the simplest class of controllable systems, namely single-input or scalar systems which are expressible via the state representation

$$\dot{x}(t) = Ax(t) + bu(t) \qquad 3.6.1$$

where the lower case b is used instead of B to signify a vector input gain matrix. We assume that the system 3.6.1, or the pair $\{A,b\}$, is controllable which, in view of 3.5.3d, implies that the rank of the $n \times n$ controllability matrix $\mathscr{C} = [b, Ab, \ldots, A^{n-1}b]$ is n, or that $|\mathscr{C}| \neq 0$. We now consider the $n \times n$ matrix Q obtained from $\mathscr{C}$ by setting q_1, the first row of Q, equal to the last (n-th) row of $\mathscr{C}^{-1}$, and recursively computing the remaining rows of Q by successive postmultiplication of each preceding row of Q by A. In particular,

$$Q = \begin{bmatrix} q_1 \\ q_1 A \\ \vdots \\ q_1 A^{n-1} \end{bmatrix}, \qquad 3.6.2$$

where q_1 is the n-th row of $\mathscr{C}^{-1}$. It is thus readily apparent that $q_1 b = q_1 Ab = \ldots = q_1 A^{n-2} b = 0$, but that $q_1 A^{n-1} b = 1$, which immediately implies the relation: $Qb = \begin{bmatrix} 0 \\ 0 \\ \vdots \\ 0 \\ 1 \end{bmatrix}$. These observations also imply the nonsingularity of Q, since

$$Q\mathscr{C} = \begin{bmatrix} 0 & 0 & \cdots & 0 & 1 \\ 0 & 0 & \cdots & 1 & x \\ \vdots & & \cdot^{\cdot^{\cdot}} & & \vdots \\ 1 & x & \cdot & \cdot & x \end{bmatrix}, \qquad 3.6.3$$

where the x's denote unspecified real scalars. It is thus clear in view of 3.6.3 that $|Q| \times |\mathscr{C}| = \pm 1$, or that $|Q| = \frac{\pm 1}{|\mathscr{C}|}$. If $\hat{x}(t)$ is now defined as $Qx(t)$ we see that the first element $\hat{x}_1(t)$ of $\hat{x}(t)$, when differentiated with respect to time, yields the relation: $\dot{\hat{x}}_1(t) = q_1Ax(t) + q_1bu(t)$ which, in turn, equals $q_2x(t) = \hat{x}_2(t)$. Furthermore, $\dot{\hat{x}}_2(t) = q_1A^2x(t) + q_1Abu(t) = \hat{x}_3(t)$ and so forth, or in general, $\dot{\hat{x}}_i(t) = \hat{x}_{i+1}(t)$ for $i = 1,2,\ldots,n-1$. It therefore follows that the equivalent scalar state representation $\{\hat{A},\hat{b}\}$ or $\dot{\hat{x}}(t) = \hat{A}\hat{x}(t) + \hat{b}u(t)$, where $\hat{A} = QAQ^{-1}$ and $\hat{b} = Qb$ is in a particular structural form which we call the scalar controllable companion form; i.e.

$$\hat{A} = QAQ^{-1} = \begin{bmatrix} 0 & 1 & 0 & \cdots & 0 \\ 0 & 0 & 1 & \cdots & 0 \\ \vdots & & & \ddots & \\ & & & & 1 \\ -a_o & -a_1 & \cdot & \cdots & -a_{n-1} \end{bmatrix}, \text{ and } \hat{b} = Qb = \begin{bmatrix} 0 \\ 0 \\ \vdots \\ 0 \\ 1 \end{bmatrix} \qquad 3.6.4$$

where $\hat{A}$ is an n-dimensional companion matrix with the identity matrix in the upper right block, and $\hat{b}$ is identically zero with the exception of the nonzero entry 1 in the n-th row. We note that as in the case of controllability, the output $y(t)$ plays no role in the determination of this scalar controllable companion form.

Some immediate benefits are derived from the reduction of $\{A,b\}$ to controllable companion form. In particular, the characteristic polynomial $|\lambda I-A|$ of A is immediately apparent by inspection of the last row of $\hat{A}$ since $|\lambda I-A| = |Q^{-1}\lambda IQ - Q^{-1}\hat{A}Q| = |Q^{-1}|$ $|\lambda I-\hat{A}||Q| = |\lambda I-\hat{A}|$, and if we evaluate the determinant of $\lambda I-\hat{A} =$

$$\begin{bmatrix} \lambda & -1 & 0 & 0.. & 0 \\ 0 & \lambda & -1 & 0.. & 0 \\ \\ a_o & a_1 & \cdot & \cdot & \lambda+a_{n-1} \end{bmatrix}$$

by last row minors we immediately obtain:

$$|\lambda I-\hat{A}| = |\lambda I-A| = a_o + a_1\lambda + \ldots + a_{n-1}\lambda^{n-1} + \lambda^n \quad 3.6.5$$

Furthermore, the state transition matrix e^{At} associated with A can also be readily determined in light of Example 3.3.9 and Equation 3.4.5. In particular, since $\hat{A}$ is a companion matrix, once the first row of $e^{\hat{A}t}$ has been found, the remaining rows can be obtained by successive differentiation of the preceding rows as in Example 3.3.9. Equation 3.4.5 can then be employed to obtain e^{At} since $e^{At} =$ $Q^{-1}e^{\hat{A}t}Q$. There are other benefits to be derived from this reduction of the system to controllable companion form as we will demonstrate in our subsequent discussions.

The notion of a "controllable companion form" is not confined solely to scalar systems, and we can now extend it to the more general multivariable case. In particular, consider any (completely state) controllable system of the form 3.2.1a; i.e. $\dot{x}(t) = Ax(t) + Bu(t)$, with B assumed to be of full rank $m \leq n$. This latter assumption, which can be removed later on, implies that all m available inputs are mutually independent, which is usually the case in practice. We now define $\bar{\mathscr{C}}$ as the $n \times \bar{n}$ matrix obtained by selecting from left to right as many $(\bar{n})$ linearly independent columns of the controllability matrix $\mathscr{C} = [B,AB,\ldots,A^{n-1}B]$ as possible. Since the system was assumed to be controllable, it follows from 3.5.3d that $\mathscr{C}$ has full rank n and hence that $\bar{n} = n$. Therefore, $\bar{\mathscr{C}}$ has full rank n and $|\bar{\mathscr{C}}| \neq 0$. We now construct the nonsingular $n \times \bar{n}$ matrix L

by simply reordering the $\bar{n}$ (=n) columns of $\bar{\mathscr{C}}$, beginning with a "power ordering" of those first (d_1) columns of $\bar{\mathscr{C}}$ which involve b_1, the first column of B, and then employing those (d_2) columns of $\bar{\mathscr{C}}$ which involve b_2 next and so forth. In particular,

$$L = [b_1, Ab_1, \ldots, A^{d_1-1}b_1, b_2, \ldots, A^{d_2-1}b_2, \ldots, A^{d_m-1}b_m] \qquad 3.6.6$$

We now define the m integers d_i as the <u>controllability indices</u> of the system and denote by the Greek letter μ, max d_i for $i = 1,2,\ldots,m$, which we further define as the <u>controllability index</u> of the system; i.e. max $d_i = \mu$. It should now be noted that all m columns of B are present in L since we assumed that B was of full rank $m \leq n$. Therefore, the first m vectors of $\bar{\mathscr{C}}$ to be selected were the m columns $b_1, b_2, \ldots, b_m$ of B. Furthermore, if $A^k b_j$ (for $1 \leq k \leq n-m$ and $1 \leq j \leq m$) is present in $\bar{\mathscr{C}}$ or L, then $A^{k-1}b_j$ is also present; i.e. if it were not, it would have been eliminated due to its linear dependence on the previous columns of $\mathscr{C}$, and consequently, so would $A^k b_j$. We now set

$$\sigma_k = \sum_{1}^{k} d_i \quad \text{for} \quad k = 1,2,\ldots,m, \qquad 3.6.7$$

which implies that $\sigma_1 = d_1$, $\sigma_2 = d_1 + d_2, \ldots,$ and $\sigma_m = d_1+d_2+\ldots+d_m = n$.

We can now enlarge the algorithm employed in the scalar case to determine an appropriate equivalence transformation matrix Q in the multivariable case; i.e. we set q_k equal to the σ_k-<u>th</u> row of L^{-1} for $k = 1,2,\ldots,m$, and consider the following $(n \times n)$ matrix Q:

$$Q = \begin{bmatrix} q_1 \\ q_1 A \\ \vdots \\ q_1 A^{d_1 - 1} \\ q_2 \\ \vdots \\ q_2 A^{d_2 - 1} \\ \vdots \\ q_m A^{d_m - 1} \end{bmatrix} \qquad 3.6.8$$

If Q, thus defined, is postmultiplied by L, it can be shown that $|QL| = 1$ in absolute value (see Problem 3-19) and, therefore, that Q is nonsingular. By now using the same reasoning which applied in the development of 3.6.4 in the scalar case, it follows that Q represents an equivalence transformation which reduces the given system to an equivalent state representation: $\dot{\hat{x}}(t) = \hat{A}\hat{x}(t) + \hat{B}u(t)$, where the pair $\{\hat{A},\hat{B}\}$ assumes a particularly useful structured form, namely a multi-input or multivariable controllable <u>companion form</u>; i.e.

$$\hat{A} = QAQ^{-1} = \begin{bmatrix} \hat{A}_{11} & \hat{A}_{12} & \cdots & \hat{A}_{1m} \\ \hat{A}_{21} & \hat{A}_{22} & \cdots & \hat{A}_{2m} \\ \vdots & & \ddots & \vdots \\ \hat{A}_{m1} & \cdots & & \hat{A}_{mm} \end{bmatrix} =$$

$$
= \left[\begin{array}{c|c|c|c}
\begin{matrix} 0 & 1 & 0 & \ldots & 0 \\ 0 & 0 & 1 & 0\ldots & 0 \\ & & & \ddots & \\ & & & & 1 \\ x & x & \cdot & \cdot\;\cdot & x \end{matrix} &
\begin{matrix} 0 & \ldots & & & 0 \\ 0 & \ldots & & & 0 \\ & & & & \\ & & & & \\ x & x & \cdot & \cdot\;\cdot & x \end{matrix} &
\cdots &
\begin{matrix} 0 & 0\ldots & & & 0 \\ \vdots & \vdots & & & \vdots \\ & & & & \\ & & & & \\ x & x & \cdot & \cdot\;\cdot & x \end{matrix} \\
\hline
\begin{matrix} 0 & 0 & \cdot & \cdot\;\cdot & 0 \\ \vdots & \vdots & & & \vdots \\ & & & & \\ & & & & \\ x & x & \cdot & \cdot\;\cdot & x \end{matrix} &
\begin{matrix} 0 & 1 & 0 & \ldots & 0 \\ 0 & 0 & 1 & 0\ldots & 0 \\ \vdots & & & \ddots & \\ & & & & 1 \\ x & x & \cdot & \cdot\;\cdot & x \end{matrix} &
\cdots &
\begin{matrix} 0 & 0\ldots & & & 0 \\ \vdots & \vdots & & & \vdots \\ & & & & \\ & & & & \\ x & x & \cdot & \cdot\;\cdot & x \end{matrix} \\
\hline
\vdots & \vdots & \ddots & \vdots \\
\hline
\begin{matrix} 0 & 0\ldots & & & 0 \\ \vdots & \vdots & & & \vdots \\ & & & & \\ & & & & \\ x & x & \cdot & \cdot\;\cdot & x \end{matrix} &
\begin{matrix} 0 & 0\ldots & & & 0 \\ \vdots & \vdots & & & \vdots \\ & & & & \\ & & & & \\ x & x & \cdot & \cdot\;\cdot & x \end{matrix} &
\cdots &
\begin{matrix} 0 & 1 & 0 & \ldots & 0 \\ 0 & 0 & 1 & 0\ldots & 0 \\ & & & \ddots & \\ & & & & 1 \\ x & x & \cdot & \cdot\;\cdot & x \end{matrix}
\end{array}\right] ; \qquad 3.6.9a
$$

and

$$
\hat{B} = QB = \left[\begin{array}{cccc}
0 & 0 & \ldots & 0 \\
0 & 0 & \ldots & 0 \\
\vdots & \vdots & & \vdots \\
1 & x & \ldots & x \\
\hline
0 & 0 & \ldots & 0 \\
0 & 0 & \ldots & 0 \\
\vdots & \vdots & & \vdots \\
0 & 1 & x\ldots & x \\
\hline
\vdots & \vdots & & \vdots \\
 & & & \\
\vdots & \vdots & & \\
0 & 0 & \ldots & 1
\end{array}\right] , \qquad 3.6.9b
$$

where the (m) diagonal blocks $\hat{A}_{ii}$ of $\hat{A}$ are each an upper right identity companion matrix of dimension d_i, while the off-diagonal blocks, A_{ij} for $i \neq j$ are each identically zero except for their respective final rows. We therefore note that <u>all information regarding the equivalent state matrix $\hat{A}$ can be derived from knowledge of the m ordered controllability indices d_i and the m ordered σ_k rows of $\hat{A}$</u>. The same can also be said of $\hat{B}$, since we note that <u>only these same ordered σ_k rows of $\hat{B}$ are nonzero</u>. This particular structured form for the controllable pair $\{\hat{A},\hat{B}\}$ will play an important and most useful role in much of what follows.

To illustrate this algorithm for reducing a multi-input controllable system to multivariable controllable companion form, we now present:

EXAMPLE 3.6.10: Consider the system 3.2.1a where $A = \begin{bmatrix} 0 & 0 & 1 & 0 \\ 3 & 0 & -3 & 1 \\ -1 & 1 & 4 & -1 \\ 1 & 0 & -1 & 0 \end{bmatrix}$

$B = \begin{bmatrix} 0 & 0 \\ 1 & 0 \\ 0 & 1 \\ 0 & 0 \end{bmatrix}$. We readily verify that the system is controllable; i.e. $\bar{\mathscr{C}} = [b_1, b_2, Ab_2, A^2b_2]$, the matrix consisting of the first $n(= 4)$ linearly independent columns of $\mathscr{C}$. Therefore, for this example, no reordering of the columns of $\mathscr{C}$ is required and $L = \bar{\mathscr{C}} = \begin{bmatrix} 0 & 0 & 1 & 4 \\ 1 & 0 & -3 & -10 \\ 0 & 1 & 4 & 13 \\ 0 & 0 & -1 & -3 \end{bmatrix}$,

$d_1 = 1$, $d_2 = 3$, $\sigma_1 = d_1 = 1$, and $\sigma_2 = d_1 + d_2 = n = 4$. The transformation matrix Q is computed next by first calculating q_1 and q_2, the first and fourth (corresponding to σ_1 and σ_2) rows of L^{-1}. For this example, $q_1 = [1\ 1\ 0\ -2]$ and $q_2 = [1\ 0\ 0\ 1]$, which implies that $Q =$

$$\begin{bmatrix} q_1 \\ q_2 \\ q_2A \\ q_2A^2 \end{bmatrix} = \begin{bmatrix} 1 & 1 & 0 & -2 \\ 1 & 0 & 0 & 1 \\ 1 & 0 & 0 & 0 \\ 0 & 0 & 1 & 0 \end{bmatrix} \quad \text{and} \quad Q^{-1} = \begin{bmatrix} 0 & 0 & 1 & 0 \\ 1 & 2 & -3 & 0 \\ 0 & 0 & 0 & 1 \\ 0 & 1 & -1 & 0 \end{bmatrix}. \quad \text{Therefore,}$$

$$\hat{A} = QAQ^{-1} = \left[\begin{array}{c|ccc} 0 & 1 & 0 & 0 \\ \hline 0 & 0 & 1 & 0 \\ 0 & 0 & 0 & 1 \\ 1 & 1 & -3 & 4 \end{array}\right], \text{ while } \hat{B} = QB = \left[\begin{array}{cc} 1 & 0 \\ \hline 0 & 0 \\ 0 & 0 \\ 0 & 1 \end{array}\right]. \text{ We now note that}$$

the pair $\{\hat{A},\hat{B}\}$ is indeed in controllable companion form. In particular, $\hat{A}_{11}$ is a companion matrix of dimension $d_1 = 1$, and $\hat{A}_{22}$ is a companion matrix of dimension $d_2 = 3$. Consequently, only the first and fourth (corresponding to σ_1 and σ_2) rows of $\hat{A}$ and $\hat{B}$ are "nontrivial" and thus contain all the pertinent information regarding the equivalent state matrix $\hat{A}$. Also, the controllability index $\mu = \max d_i$ is clearly equal to $d_2 = 3$ in this example.

We can now consider certain implications and extensions of the preceding results when the multivariable system 3.2.1a is not completely state controllable. In particular, we will still assume that $\rho[B] = m \leq n$, but we now consider the case when the rank of the controllability matrix $\mathscr{C}$ is $\bar{n} < n$. Note that it is still possible to define the $n \times \bar{n}$ matrix $\bar{\mathscr{C}}$, consisting of the first $\bar{n}$ linearly independent columns of $\mathscr{C}$, as well as the $n \times \bar{n}$ matrix L as given by 3.6.6, but with $\sigma_m = \sum_1^m d_i = \bar{n}$ instead of n. The $\bar{n}$ linearly independent columns of L clearly form a basis of some subspace $\mathscr{W}$ of $\mathscr{R}^n$, the space of all real n-tuples. If we define $\mathscr{W}^\perp$ as the <u>orthogonal complement</u> of $\mathscr{W}$; i.e. the subspace of $\mathscr{R}^n$ consisting of all vectors in $\mathscr{R}^n$ perpendicular to $\mathscr{W}$ in the sense of a zero inner product (see Section 2.2), it follows that any vector v in $\mathscr{R}^n$ can be expressed as a linear combination of some vector w in $\mathscr{W}$ and some vector $w^\perp$ in $\mathscr{W}^\perp$. In particular, $v = \alpha w + \beta w^\perp$ for all v in $\mathscr{R}^n$, which implies that $\mathscr{R}^n$ can be defined as the direct sum of $\mathscr{W}$ and $\mathscr{W}^{\perp}$; i.e. $\mathscr{R}^n = \mathscr{W} \oplus \mathscr{W}^{\perp}$. It is thus clear that the dimension q of $\mathscr{W}^{\perp}$ is $n - \bar{n}$, since $\mathscr{R}^n$ has dimension n. We now let $\beta_1, \beta_2, \ldots, \beta_q$ be any basis of $\mathscr{W}^{\perp}$ and consider the "extended" state representation:

$$\dot{x}(t) = Ax(t) + B_e u_e(t), \qquad 3.6.11$$

where B_e is the $n \times (m + q)$ matrix obtained by appending to B the q basis vectors of $\mathscr{W}^{\perp}$; i.e. $B_e = [B, \beta_1, \beta_2, \ldots, \beta_q]$, while $u_e(t)$ is an $(m + q)$-dimensional input obtained by appending to $u(t)$, q additional input elements; i.e. $u(t) = [u_1(t), u_2(t), \ldots, u_m(t), u_{m+1}(t), \ldots, u_{m+q}(t)]^T$. The extended system 3.6.11, thus defined, is clearly a controllable one, and it is therefore possible to employ the algorithm presented earlier in this section to obtain an n-dimensional equivalence transformation which reduces the extended system to controllable companion form; i.e. we denote the appropriate transformation matrix as Q_e and utilize it to reduce the original system 3.2.1a to the equivalent representation:

$$\dot{\hat{x}}(t) = \hat{A}\hat{x}(t) + \hat{B}u(t), \qquad 3.6.12$$

where $\hat{A} = Q_e A Q_e^{-1}$ and $\hat{B} = Q_e B$. Due to this specific choice for Q_e, it follows that the equivalent pair $\{\hat{A}, \hat{B}\}$ "partially resembles" the multivariable controllable companion form 3.6.9. In particular,

$$\hat{A} = \left[\begin{array}{c|c} \hat{A}_c & \hat{A}_{c\bar{c}} \\ \hline 0 & \hat{A}_{\bar{c}} \end{array}\right]; \quad \hat{B} = \left[\begin{array}{c} \hat{B}_c \\ \hline 0 \end{array}\right], \qquad 3.6.13$$

where the pair $\{\hat{A}_c, \hat{B}_c\}$ is in an $\bar{n}$-dimensional controllable companion form; i.e. the pair $\{\hat{A}_c, \hat{B}_c\}$ assumes the structure indicated by 3.6.9 with $\sum_1^m d_i = \bar{n}$. Furthermore, the lower left $q \times \bar{n}$ block of $\hat{A}$ as well as the final q rows of $\hat{B}$ are identically zero. On closer inspection, it becomes apparent that the controllable and completely uncontrollable "portions" of the system have been separated via the form 3.6.13. More specifically, the $\bar{n}$-dimensional subsystem defined by the first $\bar{n}$ rows of 3.6.12, namely $\dot{\hat{x}}_c(t) = \hat{A}_c\hat{x}_c(t) + \hat{A}_{c\bar{c}}\hat{x}_{\bar{c}}(t) + \hat{B}_c u(t)$ is clearly controllable since $\hat{A}_{c\bar{c}}x_{\bar{c}}(t)$ can be treated as a known disturbance. Furthermore, the q-dimensional subsystem defined by the remaining rows of 3.6.12, namely $\dot{\hat{x}}_{\bar{c}}(t) = \hat{A}_{\bar{c}}\hat{x}_{\bar{c}}(t)$ is completely independent of $u(t)$ and, therefore, is <u>completely uncontrollable</u>. We further note that in view of 3.6.5 and 3.6.13, the characteristic polynomial $|\lambda I-A|$ of A (and hence $\hat{A} = Q_e^{-1}AQ_e$) can be written as the product of the characteristic polynomials of the controllable and completely uncontrollable "portions" of the system: i.e.

$$|\lambda I-A| = |\lambda I-\hat{A}| = |\lambda I-\hat{A}_c| \times |\lambda I-\hat{A}_{\bar{c}}| \qquad 3.6.14$$

In view of the above, we call the zeros of $|\lambda I-\hat{A}_c|$ $(|\lambda I-\hat{A}_{\bar{c}}|)$ the <u>controllable (uncontrollable) modes or poles</u> of the system.

Now that an algorithm for reducing controllable systems to controllable companion form has been presented, we can employ duality (see Section 3.5) to directly achieve an analogous result for observable systems. In particular, we now consider the (completely state) observable system:

$$\dot{x}(t) = Ax(t); \; y(t) = Cx(t), \qquad 3.6.15$$

or, equivalently, the observable pair $\{A,C\}$ with C of full rank $p \leq n$. The input $u(t)$ has been omitted for convenience, since it plays no role in the discussion which follows. The dual of system 3.6.15 is readily determined to be the (completely state) controllable system:

$$\dot{\bar{x}}(t) = A^T\bar{x}(t) + C^T\bar{u}(t) \qquad 3.6.16$$

Since this controllable system is in the state form 3.2.1a with C^T of full rank $p \leq n$, it can be reduced to controllable companion form via an equivalence transformation $\overline{Q}$, which can be found by applying the algorithm to the pair $\{A^T, C^T\}$. We thus obtain the equivalent system:

$$\dot{\hat{\overline{x}}}(t) = \hat{A}^T\hat{\overline{x}}(t) + \hat{C}^T\overline{u}(t) \qquad 3.6.17$$

where $\hat{A}^T = \overline{Q}A^T\overline{Q}^{-1}$, $\hat{C}^T = \overline{Q}C^T$, and both matrices are in the controllable companion form 3.6.9. We can now revert back to the original system by employing the dual of 3.6.17; i.e.

$$\dot{\hat{x}}(t) = \hat{A}\hat{x}(t);\; y(t) = \hat{C}\hat{x}(t), \qquad 3.6.18$$

where $\hat{A} = \overline{Q}^{-T}A\overline{Q}^T$ and $\hat{C} = C\overline{Q}^T$. It is thus clear that the equivalence transformation $\overline{Q}^{-T}$ reduces the original observable system 3.6.15 to a particular structure form which we now define as a multi-output or multivariable observable companion form. This form "closely resembles" the controllable companion form 3.6.9 and is, in particular, a transposed version of that form. More specifically, $\hat{A} = \begin{bmatrix} A_{11} A_{12} \cdots A_{1p} \\ A_{21} A_{22} \cdots A_{2p} \\ \vdots \quad \ddots \\ A_{p1} \cdot\cdot\cdot A_{pp} \end{bmatrix}$, where each diagonal submatrix $\hat{A}_{ii}$ of $\hat{A}$ is a $\overline{d}_i$-dimensional companion matrix with lower left rather than upper right identity, while the off-diagonal blocks, $\hat{A}_{ij}$ for $i \neq j$ are each identically zero except for their respective final columns; i.e.

$$\hat{A}_{ii} = \begin{bmatrix} 0 & 0 & \cdots & & x \\ 1 & 0 & & & \\ 0 & 1 & & & \\ \vdots & \vdots & \ddots & & \vdots \\ 0 & 0 & \cdots & 1 & x \end{bmatrix}, \text{ and } \hat{A}_{ij} = \begin{bmatrix} 0 & 0 & \cdots & x \\ 0 & 0 & \cdots & x \\ \vdots & \vdots & & \vdots \\ 0 & 0 & \cdots & x \end{bmatrix} \qquad 3.6.19a$$

Furthermore,

$$\hat{C} = \begin{bmatrix} 0 & 0 & \cdots & 1 & 0 & \cdots & 0 & \cdots & 0 \\ 0 & 0 & \cdots & x & 0 & \cdots & 1 & \cdots & 0 \\ \vdots & \vdots & & \vdots & \vdots & & \vdots & & \vdots \\ 0 & 0 & \cdots & x & 0 & \cdots & x & \cdots & 1 \end{bmatrix} \qquad 3.6.19b$$

and has p nonzero $\bar{\sigma}_k = \sum_1^k \bar{d}_i$ columns for $k = 1,2,\ldots,p$.

As in the controllable case, we now define the p integers $\bar{d}_i$ as the observability indices of the system and denote by the Greek letter ν, max $\bar{d}_i$ for $i = 1,2,\ldots,p$, which we further define as the observability index of the system; i.e. max $\bar{d}_i = \nu$. It can now be noted that, in comparison to the controllable case, all information regarding the equivalent state matrix $\hat{A}$ can be derived from knowledge of the p ordered observability indices $\bar{d}_i$ and the p ordered $\bar{\sigma}_k$ columns of $\hat{A}$. The same can also be said of $\hat{C}$, since only these same ordered σ_k columns of $\hat{C}$ are nonzero.

We finally note that if the system 3.6.15 is unobservable, it can be reduced to an equivalent system which "partially resembles" the observable companion form 3.6.19 by employing a technique which is the dual of that presented in the uncontrollable case, a point which we will illustrate via Example 3.6.20. Before doing so, however, it should be perhaps emphasized that whenever we affect a system reduction to either a controllable or observable companion form, three system matrices are suitably altered, and not just two (see 3.4.2). We have been somewhat remiss in noting this fact and, for convenience, have omitted C(B) in the discussion of controllable (observable) reduction. Thus C is reduced to $\hat{C} = CQ^{-1}$ in the case of the controllable companion form, while B is reduced to $\hat{B} = \bar{Q}^{-T}B$ in the case of the observable companion form. However, neither of these two matrices assumes any particular structured form.

EXAMPLE 3.6.20: To illustrate the reduction of an unobservable system of the form 3.6.15 to a structured form which "partially resembles" the observable companion form 3.6.19, consider the pair $\{A,C\}$, where $A = \begin{bmatrix} -1 & 1 & 0 & 1 \\ 1 & 1 & 0 & 1 \\ 0 & 0 & 4 & -1 \\ 1 & 0 & 0 & 2 \end{bmatrix}$ and $C = \begin{bmatrix} 1 & 0 & 0 & 0 \\ -1 & 0 & 0 & 1 \end{bmatrix}$. The reader can readily verify that this system is not completely state observable, since the rank of the observability matrix is $3 < n = 4$. Therefore, although we cannot reduce this system to observable companion form, we can obtain an equivalent representation in which the completely unobservable "portion" is separate and distinct from the observable "portion" of the system, which is in observable companion form. In particular, we can first define the dual system: $\dot{\bar{x}}(t) = \bar{A}\bar{x}(t) + \bar{B}\bar{u}(t)$, where $\bar{A} = A^T$ and $\bar{B} = C^T$ and then apply the extended algorithm for systems which are not completely state controllable to this dual pair $\{\bar{A},\bar{B}\}$. The first three linearly independent columns of the controllability matrix $\bar{\mathscr{C}} = [\bar{b}_1, \bar{b}_2, \bar{A}\bar{b}_1] = \begin{bmatrix} 1 & -1 & -1 \\ 0 & 0 & 1 \\ 0 & 0 & 0 \\ 0 & 1 & 1 \end{bmatrix}$, and also represent the transpose of the first three linearly independent rows of the observability matrix $\mathscr{O}$ of the original system. To employ the algorithm, we now append to $\bar{B}$ an additional column $\bar{\beta}_1$ which is orthogonal to the three columns of $\bar{\mathscr{C}}$. The obvious choice for $\bar{\beta}_1$ is the vector $\begin{bmatrix} 0 \\ 0 \\ 1 \\ 0 \end{bmatrix}$, since the third row of $\bar{\mathscr{C}}$ is identically zero. We thus obtain the extended system $\{A,B_e\}$, which is completely controllable. We can now directly apply the algorithm for completely state controllable systems to the pair $\bar{A},\bar{B}_e$; i.e. $\bar{L}_e = \begin{bmatrix} 1 & -1 & -1 & 0 \\ 0 & 1 & 0 & 0 \\ 0 & 0 & 0 & 1 \\ 0 & 1 & 1 & 0 \end{bmatrix} =$

$[\bar{b}_1, \bar{A}\bar{b}_1, \bar{b}_2, \bar{\beta}_1]$, and $\bar{Q}_e = \begin{bmatrix} \bar{q}_1 \\ \bar{q}_1\bar{A} \\ \bar{q}_2 \\ \bar{q}_{e1} \end{bmatrix} = \begin{bmatrix} 0 & 1 & 0 & 0 \\ 1 & 1 & 0 & 0 \\ 0 & -1 & 0 & 1 \\ 0 & 0 & 1 & 0 \end{bmatrix}$ while $\bar{Q}_e^{-1} =$

$\begin{bmatrix} -1 & 1 & 0 & 0 \\ 1 & 0 & 0 & 0 \\ 0 & 0 & 0 & 1 \\ 1 & 0 & 1 & 0 \end{bmatrix}$. Their respective transposes, $\bar{Q}_e^T = \begin{bmatrix} 0 & 1 & 0 & 0 \\ 1 & 1 & -1 & 0 \\ 0 & 0 & 0 & 1 \\ 0 & 0 & 1 & 0 \end{bmatrix}$

and $\bar{Q}_e^{-T} = \begin{bmatrix} -1 & 1 & 0 & 1 \\ 1 & 0 & 0 & 0 \\ 0 & 0 & 0 & 1 \\ 0 & 0 & 1 & 0 \end{bmatrix}$ are immediately apparent, the latter

matrix $\bar{Q}_e^{-T}$ representing an appropriate transformation matrix (see 3.6.18) which reduces the original system $\{A,C\}$ to the appropriate structured form; i.e. $\bar{d}_1 = 2$, $\bar{d}_2 = 1$, $\bar{\sigma}_1 = 2$, $\bar{\sigma}_2 = 3$, $\hat{A} = \bar{Q}_e^{-T} A \bar{Q}_e^T =$

$\left[\begin{array}{cc:c:c} 0 & 3 & 2 & 0 \\ 1 & 0 & 0 & 0 \\ \hdashline 0 & 1 & 2 & 0 \\ \hdashline 0 & 0 & -1 & 4 \end{array}\right]$, and $\hat{C} = C\bar{Q}_e^T = \begin{bmatrix} 0 & 1 & 0 & 0 \\ 0 & -1 & 1 & 0 \end{bmatrix}$. Note that the upper left 3×3 submatrix of $\hat{A}$ and the first three columns of $\hat{C}$ are in multivariable observable companion form, while the last (fourth) columns of $\hat{A}$ and $\hat{C}$ clearly represent the completely unobservable "portion" of the system; i.e. $\hat{x}_4(t)$ does not affect the output of the system. We further note that $|\lambda I - A| = |\lambda I - \hat{A}| = |\lambda I - \hat{A}_o| \times |\lambda I - \hat{A}_{\bar{o}}|$. The <u>observable modes or poles</u> of the system are therefore the zeros of

$$|\lambda I - \hat{A}_o| = \begin{bmatrix} \lambda & -3 & -2 \\ -1 & \lambda & 0 \\ 0 & -1 & \lambda-2 \end{bmatrix} = \lambda^3 - 2\lambda^2 - 3\lambda + 4 \text{ or } 1 \text{ and } \frac{1 \pm \sqrt{17}}{2},$$ while

the <u>unobservable mode or pole</u> of this system is the zero of

$|\lambda I - A_{\bar{o}}| = (\lambda-4)$ or 4.

3.7 CONCLUDING REMARKS AND REFERENCES

The notion of state and a state representation for modeling the dynamical behavior of physical systems, which were introduced in this chapter, represent the foundation of what is commonly referred to as the "modern approach" to control system analysis and design. The time domain notion of state is highly dependent on and interrelated with the study of ordinary differential equations, and is clearly distinguishable from the "classical" frequency domain techniques such as Nyquist, Bode, and Nichols diagrams, and the root locus which were developed almost three decades ago.

The trend from frequency domain to time domain techniques for theoretical investigations has been quite evident over the past ten years, and was first popularized in this country by a number of authors in the early 1960's, who were primarily concerned with the then popular and somewhat exotic notions of optimal control [P1] and Liapunov stability [L1]. The formalism of the notions of controllability and observability can be attributed to Kalman [K1][K2][K3], although the rank condition on the controllability matrix $\mathscr{C}$ appears in some earlier reports; e.g. [P1] and [L2]. Various extensions of controllability, including the notions of output and output function controllability can be found in [K4] and [B2]. The controllable and observable companion forms for linear multivariable systems were first presented by Luenberger [L3], and have since been rediscovered and improved on by a number of other investigators, including the author [W4]. In particular, the extension of the algorithm to include systems which are not completely state controllable first appeared in [W4], motivated, in part, by the "canonical decomposition theorem" [K2][K3][Z1].

One of the first rigorous texts devoted exclusively to state-space ideas is due to Zadeh and Desoer [Z1], and a very thorough and detailed exposition of most of the results presented in this chapter can be found there. Since then (1963), a number of other state oriented texts have appeared. The state representation for dynamical systems is perhaps the most commonly accepted method employed in the current control literature for mathematically describing the dynamical behavior of physical systems, although there is some evidence that frequency domain procedures are once again gaining some degree of popularity, perhaps due in part to the fact that the majority of physical systems are still designed via the traditional frequency domain methods. One goal of this text, which is in accord with this latter observation, is to display, whenever possible, the similarities between the modern and classical approaches to control system analysis and synthesis, in order to offer system designers a choice of design techniques.

PROBLEMS - CHAPTER 3

3-1 Find both a differential operator representation and a state representation for the RLC network depicted below. What can you say regarding the controllability and observability of this network?

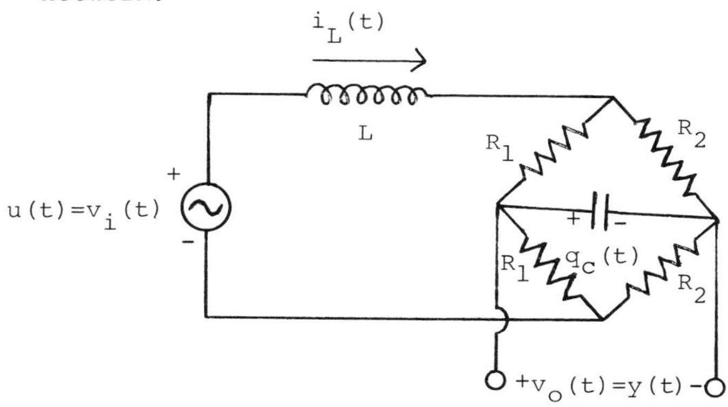

3-2 Draw an analog computer diagram which can be used to simulate the above network.

3-3 Determine $\Phi(t-t_o) = e^{A(t-t_o)}$ for each of the following system matrices. (a) $A = \begin{bmatrix} 1 & 0 & 0 \\ 0 & 2 & 0 \\ 0 & 0 & -1 \end{bmatrix}$; (b) $A = \begin{bmatrix} -2 & 1 \\ -2 & 0 \end{bmatrix}$; (c) $A = \begin{bmatrix} 0 & 1 & 0 \\ 0 & 0 & 1 \\ 0 & -1 & -2 \end{bmatrix}$.

3-4 Show that the rank of both the controllability and observability matrices of a state form system is unaltered by any equivalence transformation of the state $x(t)$.

3-5 Which of the systems with A matrices given in Problem 3-3 are (a) stable (b) marginally stable (c) unstable?

3-6 Which of the following system pairs are equivalent?

(a)

S 1	S 2
$A = \begin{bmatrix} 0 & 1 \\ -2 & -3 \end{bmatrix}; \quad B = \begin{bmatrix} 0 \\ 1 \end{bmatrix}$	$\hat{A} = \begin{bmatrix} -1 & 1 \\ 0 & -2 \end{bmatrix}; \quad B = \begin{bmatrix} 0 \\ 1 \end{bmatrix}$
$C = \begin{bmatrix} 1 & 1 \\ 0 & 1 \end{bmatrix}; \quad E = \begin{bmatrix} 1 \\ 0 \end{bmatrix}$	$\hat{C} = \begin{bmatrix} 0 & 1 \\ -1 & 1 \end{bmatrix}; \quad E = \begin{bmatrix} 1 \\ 0 \end{bmatrix}$

(b)

$$A = \begin{bmatrix} 0 & 1 & 3 \\ 1 & 0 & 1 \\ 0 & -2 & 0 \end{bmatrix}; \quad B = \begin{bmatrix} 1 & 2 \\ 0 & 0 \\ 1 & 0 \end{bmatrix} \qquad \hat{A} = \begin{bmatrix} 1 & 0 & 1 \\ 0 & 1 & 3 \\ 0 & -2 & 0 \end{bmatrix}; \quad \hat{B} = \begin{bmatrix} 1 & 2 \\ 1 & 0 \\ 0 & 0 \end{bmatrix}$$

$$C = \begin{bmatrix} 1 & 0 & 0 \\ 0 & 0 & 1 \end{bmatrix}; \quad E = \begin{bmatrix} 1 & -1 \\ 0 & 2 \end{bmatrix} \qquad \hat{C} = \begin{bmatrix} 1 & 0 & 0 \\ 0 & 0 & 1 \end{bmatrix}; \quad \hat{E} = \begin{bmatrix} 0 & 2 \\ 1 & -1 \end{bmatrix}$$

3-7 Show that two state form systems which have the same impulse response matrix are not necessarily equivalent.

3-8 If two systems are zero state equivalent, do they necessarily have the same impulse response matrix? Need they be of equal order? Explain your answers.

3-9 If two systems are zero input equivalent, do they necessarily have the same impulse response matrix? Explain.

3-10 Determine any conditions under which the system employed in Example 3.2.4 is either uncontrollable or unobservable.

3-11 Prove that the state form system 3.2.1 is controllable if and only if it is possible to drive any arbitrary initial state to the origin of the state space in any arbitrary finite time.

3-12 For each of the three system matrices given in Problem 3-3, find a column vector b such that the pair $\{A,b\}$ is controllable. Verify, by example, that it is not always possible to do this.

3-13 Show that a system can be output controllable but not state controllable. Show that a system can be state controllable but not output controllable.

3-14 If $A = \begin{bmatrix} 1 & 0 & 0 \\ 0 & 1 & 0 \\ 0 & 0 & -1 \end{bmatrix}$ and $B = \begin{bmatrix} 0 & 0 \\ 1 & 0 \\ 0 & 1 \end{bmatrix}$, are the rows of $e^{At}B$ independent over $\mathscr{R}$ for all finite t?

3-15 Reduce the following pair to controllable companion form:

$$A = \begin{bmatrix} 0 & 1 & 1 & 1 \\ 0 & 0 & 0 & 1 \\ 0 & 1 & 0 & 0 \\ 0 & 0 & 1 & 1 \end{bmatrix}; \qquad B = \begin{bmatrix} 1 & 0 \\ 0 & 0 \\ 0 & 1 \\ 1 & 0 \end{bmatrix}$$

3-16 Reduce the following pair to observable companion form:

$$A = \begin{bmatrix} 0 & 2 & 0 & 0 \\ 0 & 1 & -2 & 0 \\ 0 & 0 & 3 & 1 \\ 1 & 0 & 0 & 0 \end{bmatrix}; \qquad C = \begin{bmatrix} 0 & 1 & 0 & 0 \\ 0 & 0 & 1 & 0 \end{bmatrix}$$

3-17 If in evaluating $\rho[\mathscr{C}]$ for some state pair $\{A,B\}$, all of the columns of $A^k B$ are linearly dependent on the previous (left-most) columns of $\mathscr{C}$, show that it is unnecessary to continue the evaluation since $\rho[\mathscr{C}]$ will not increase.

3-18 Reduce the following uncontrollable $\{A,B\}$ pair to a form which "partially resembles" the controllable companion form and determine the uncontrollable modes of each.

(a) $A = \begin{bmatrix} 1 & 0 & 0 \\ 0 & -1 & 0 \\ 0 & 0 & 2 \end{bmatrix}$; $B = \begin{bmatrix} 1 & 0 \\ 0 & 1 \\ 0 & 0 \end{bmatrix}$ (b) $A = \begin{bmatrix} 0 & 0 & 1 & 0 \\ 0 & 0 & 1 & 0 \\ 0 & 0 & 0 & 0 \\ 0 & 0 & 0 & -1 \end{bmatrix}$;

$$B = \begin{bmatrix} 0 & 1 \\ 0 & 0 \\ 1 & 0 \\ 0 & 0 \end{bmatrix}.$$

3-19 Show that if Q and L are given by 3.6.8 and 3.6.6 respectively, then the absolute value of the determinant of QL is 1.

3-20 Show that the state representation $\{A,I\}$ is already in controllable companion form and furthermore, that the controllability index $\mu = 1$.

3-21 Show that two controllable fourth order systems, the first with $d_1 = d_2 = 2$ and the second with $d_1 = 1$ and $d_2 = 3$ cannot be equivalent. What can you conclude in general regarding the controllability indices of equivalent systems?

3-22 Verify, by example, that in general, the multivariable controllable companion form 3.6.9 is not uniquely specified by a controllable pair $\{A,B\}$. Does this result hold in the scalar case?

3-23 The linear differential equation, $\dot{x}(t) = Ax(t) + Bu(t)$, where

$$A = \begin{bmatrix} 0 & 1 & 0 & 0 \\ 3w^2 & 0 & 0 & 2w \\ 0 & 0 & 0 & 1 \\ 0 & -2w & 0 & 0 \end{bmatrix} \qquad B = \begin{bmatrix} 0 & 0 \\ 1 & 0 \\ 0 & 0 \\ 0 & 1 \end{bmatrix}$$

is obtained as the result of "linearizing" the non-linear equations of motion of an orbiting satellite about a steady-state solution. The state $x_1(t)$ is the differential radius, $r_\Delta(t)$, while $x_3(t)$ is the differential angle, $\theta_\Delta(t)$. The input $u_1(t)$ is the radial thrust and $u_2(t)$ is the tangential thrust.

(a) Is the system stable?

(b) Is the system controllable?

(c) If the radial thruster fails, can the system be completely controlled?

(d) If the tangential thruster fails, can the system be completely controlled?

(e) If $y(t) = \begin{bmatrix} x_1(t) \\ x_3(t) \end{bmatrix}$, determine the transfer matrix of the system.

CHAPTER 4

FREQUENCY DOMAIN REPRESENTATIONS

4.1 INTRODUCTION

The reader is undoubtedly familiar with the well known frequency domain methods such as the Nyquist and Bode diagrams, the Nichols chart, and the root locus for analyzing the dynamical behavior of scalar systems. These methods are dependent only on the external (input/output) frequency response characteristics of a system, or equivalently, on the "transfer function" $T(s)$ of a system which is usually expressible as the ratio of two polynomials $r(s)$ and $p(s)$ in the Laplace operator s with real coefficients; i.e. in the scalar case, the transfer function can usually be expressed as:

$$T(s) = \frac{r(s)}{p(s)}, \qquad 4.1.1$$

where the zeros of $r(s)$ and $p(s)$ represent, respectively, the <u>zeros</u> and <u>poles</u> of the system. Frequency domain methods are so named because an evaluation of $T(s)$ at $s = j\omega$ for any positive real value of ω, yields a complex number $T(j\omega) = \alpha(\omega) + j\beta(\omega)$, the magnitude of which $\sqrt{\alpha^2(\omega) + \beta^2(\omega)}$ represents the ratio of the output to the input amplitude response in steady-state due to a sinusoidal input signal of frequency ω, while $\theta = \tan^{-1}\frac{\beta(\omega)}{\alpha(\omega)}$ represents the difference in phase between the two waveforms. Frequency domain techniques represent what is commonly referred to as the "classical" approach (in contrast to the "modern" or state-space approach) to control system analysis and design, since they were the first to be developed and extensively employed in the field of control systems engineering. Frequency domain tachniques are still somewhat heuristic in nature, and often rely quite heavily on the personal whims and experiences of the design engineer. They have, nevertheless, been quite successfully employed far longer than the now popular state dependent methods and still retain their dominance from the point of view of practical control system design. It should be noted, however, that the classical frequency domain techniques are primarily suited to scalar system analysis and design and usually cannot be extended to include the multivariable case. This fact, along with the ever-increasing need to more efficiently control increasingly complex systems, spurred the trend towards the more precise and mathematically rigorous time domain methods which have dominated the control literature for the past decade.

As noted earlier, one of the primary goals of this text will be to merge frequency domain with time domain techniques for linear system analysis and design. This will be done by displaying a number of similarities between the two approaches, thus providing the control system designer with a greater degree of flexibility than he would have if confined to either domain alone. Part of this objective will be accomplished in this chapter by first extending the notion of a scalar "transfer function" to the matrix case, and then presenting a most useful result, called the "structure theorem", which establishes a direct relationship between the "transfer matrix" of a multivariable system and the controllable and observable companion forms introduced in Section 3.6. We will later employ the "structure theorem" to resolve certain questions related to the transfer between time and frequency domain representations for dynamical systems.

4.2 THE TRANSFER MATRIX

The state method 3.2.1 for representing the dynamical behavior of linear systems was formally introduced and discussed in some detail in the previous chapter. We now introduce a second method which is also used quite often to represent the behavior of dynamical systems, one which can be obtained by simply extending, to the matrix case, the well known notion of a scalar "transfer function".

DEFINITION 4.2.1: The <u>transfer matrix</u> $T(s)$ of a linear multivariable system is the Laplace transform of its impulse response matrix $T(t)$, and the <u>transfer function</u> is simply the name given to the transfer matrix of a scalar system. If we recall that the impulse response matrix $T(t)$ of any state representation $\{A,B,C,E\}$ is uniquely defined via 3.2.28, it is clear that

$$\begin{aligned} T(s) = \mathscr{L}[T(t)] = \int_0^\infty T(t)e^{-st}dt &= \int_0^\infty [Ce^{At}B + E\delta(t)]e^{-st}dt \\ &= C(sI-A)^{-1}B + E \end{aligned} \qquad 4.2.2$$

Some observations are in order at this point. In particular, we first recall that the state representation 3.2.1 for a linear dynamical system specifies the internal (state) as well as the external (input/output) behavior of the system. Furthermore, since two equivalent systems share the identical impulse response matrix, they also have the same transfer matrix although their internal states are, in general, quite different. In particular,

$$T(s) = C(sI-A)^{-1}B + E = CQQ^{-1}(sI-A)^{-1}QQ^{-1}B + E = \hat{C}(sI-\hat{A})^{-1}\hat{B} + E \qquad 4.2.3$$

We therefore can expect that certain information which is present in a state representation is "lost" when a conversion to the transfer matrix via 4.2.2 is made. This is indeed the case and, moreover, the information which is "lost" involves the initial conditions on the state variables as we now show; i.e. if we take the Laplace transform of the state representation 3.2.1, and solve for the Laplace transform of the output $y(s)$ in terms of the Laplace transform of the input $u(s)$, we obtain:

$$y(s) = C(sI-A)^{-1}x(0) + \underbrace{[C(sI-A)^{-1}B + E]}_{T(s)}u(s), \qquad 4.2.4$$

which we recognize as the sum of a term which is dependent on the initial state, $x(0)$, and the transfer matrix $T(s)$ premultiplying $u(s)$. It is thus clear that information regarding the initial state of the system is "lost" when $T(s)$ alone is used to represent the dynamical behavior of the system. More specifically, the transfer matrix $T(s)$ describes only the external (input/output) behavior of a system and is independent of any particular choice of the internal state. We also expect that unlike the unique transfer from any time domain representation $\{A,B,C,E\}$ to the frequency domain transfer matrix $T(s)$ via 4.2.2, the transfer from any transfer matrix $T(s)$ to a state-space quadruple $\{A,B,C,E\}$ which satisfies 4.2.2 is not unique. This is indeed the case, and more will be said regarding this point in Section 4.4.

As in the case of transfer functions, each of the entries of the $p \times m$ transfer matrix $T(s)$ is itself a transfer function which provides the "same information" as in the scalar case. In particular, the i,j-*th* element, $T_{ij}(j\omega)$, of $T(j\omega)$ represents the steady-state frequency response (amplitude ratio and phase difference) between the i-*th* output and the j-*th* input, with all other inputs identically zero. This observation follows directly from the definition of an impulse response matrix and 4.2.1.

We further note that the transfer matrix $T(s) = C(sI-A)^{-1}B + E$ can always be written as:

$$T(s) = C(sI-A)^{+}B \div |sI-A| + E = [C(sI-A)^{+}B + E|sI-A|] \div |sI-A|, \qquad 4.2.5$$

where $|sI-A| = \Delta(s)$, an n-*th* degree polynomial called the characteristic polynomial of A (see Section 2.4) whose (n) zeros represent the poles of the system; i.e. the (n) eigenvalues of A are equivalent to the poles of the dynamical system. Since $(sI-A)^{+}$ is an $n \times n$ polynomial matrix whose entries are polynomials of degree no greater than $n-1$, it is clear that whenever $E = 0$, the numerator degree of each entry, $T_{ij}(s)$, of $T(s) = C(sI-A)^{+}B \div \Delta(s)$ will be strictly less than the denominator degree, and $T(s)$ will be called a <u>strictly proper</u> transfer matrix. However, if $E \neq 0$, the numerator degrees will be less than or equal to the corresponding denominator degrees, and $T(s)$ will be called a <u>proper</u> transfer matrix. We note that strictly proper transfer matrices form a subclass of proper transfer matrices and that for any proper transfer matrix $T(s)$ obtained via 4.2.2,

$$E = \lim_{s \to \infty} T(s) \qquad 4.2.6$$

It should finally be noted that we have chosen to introduce the notion of a transfer matrix here by first employing the notion of a state representation, although the concept of a transfer matrix, or its time domain equivalent--the impulse response matrix, could have been introduced with no reference whatsoever to state-space ideas. Thus, as in the scalar case, the transfer matrix of a multivariable system can be obtained experimentally by taking a total of pm frequency response measurements at the various (p) output and (m) input terminals of the physical plant, assuming, of course, that the plant is stable.

EXAMPLE 4.2.7: To illustrate the notion of a transfer matrix as well as a transfer from the time domain state representation to the frequency domain transfer matrix via 4.2.2, consider a system of the form 3.2.1 with $A = \begin{bmatrix} 1 & 2 & 0 \\ 4 & -1 & 0 \\ 0 & 0 & 1 \end{bmatrix}$, $B = \begin{bmatrix} 1 \\ 0 \\ 1 \end{bmatrix}$, $C = \begin{bmatrix} 0 & 1 & -1 \\ 0 & 0 & 1 \end{bmatrix}$, and $E = \begin{bmatrix} 0 \\ 1 \end{bmatrix}$. To determine the transfer matrix of this system via 4.2.2, we first calculate $(sI-A)$ and its inverse $(sI-A)^+ \div |sI-A|$; i.e.

$$(sI-A) = \begin{bmatrix} s-1 & -2 & 0 \\ -4 & s+1 & 0 \\ 0 & 0 & s-1 \end{bmatrix}, \quad (sI-A)^+ = \begin{bmatrix} s^2-1 & 2s-2 & 0 \\ 4s-4 & s^2-2s+1 & 0 \\ 0 & 0 & s^2-9 \end{bmatrix}, \text{ and}$$

$|sI-A| = \Delta(s) = (s-1)(s+3)(s-3)$, which implies that the poles of the system are equal to 1, -3, and 3. We now determine that $(sI-A)^{-1} =$

$$(sI-A)^+ \div \Delta(s) = \frac{\begin{bmatrix} (s+1)(s-1) & 2(s-1) & 0 \\ 4(s-1) & (s-1)^2 & 0 \\ 0 & 0 & (s+3)(s-3) \end{bmatrix}}{(s-1)(s+3)(s-3)},$$

and reducing each entry of $(sI-A)^{-1}$ to prime factors to simplify the remaining computations, we find that

$$(sI-A)^{-1} = \begin{bmatrix} \frac{s+1}{s^2-9} & \frac{2}{s^2-9} & 0 \\ \frac{4}{s^2-9} & \frac{s-1}{s^2-9} & 0 \\ 0 & 0 & \frac{1}{s-1} \end{bmatrix}.$$

If we now premultiply this expression by C and then postmultiply it by B, we obtain:

$$C(sI-A)^{-1}B = \begin{bmatrix} \frac{(s+1)(-s+5)}{(s-1)(s+3)(s-3)} \\ \frac{1}{s-1} \end{bmatrix},$$

which is clearly a strictly proper transfer matrix, consistent with our earlier discussions. The transfer matrix $T(s)$ for this system can now be completely determined by simply adding $E = \begin{bmatrix} 0 \\ 1 \end{bmatrix}$ to this latter expression. In particular

$$T(s) = C(sI-A)^{-1}B + E = \begin{bmatrix} \frac{(s+1)(-s+5)}{(s-1)(s+3)(s-3)} \\ \frac{s}{s-1} \end{bmatrix},$$

which is a proper transfer matrix as expected; i.e. the 2,1 entry, $T_{21}(s)$, of $T(s)$, corresponding to the nonzero 2,1 entry of E, is a proper rational function.

4.3 THE STRUCTURE THEOREM

We are now in a position to establish a result which displays a fundamental structure of dynamical systems and provides a most useful relationship between time and frequency domain representations for linear multivariable systems. In order to establish this result, we will employ the material presented in Sections 3.6 and 4.2. In particular, from Section 3.6, we recall that if a state representation $\{A,B,C,E\}$ is controllable with B of full rank $m \leq n$, then it can be reduced via a nonsingular transformation Q to a controllable companion form $\{\hat{A},\hat{B},\hat{C},E\} = \{QAQ^{-1},QB,CQ^{-1},E\}$ as given by 3.6.9. Furthermore, associated with the structured form are the (m) controllability indices d_i, for $i = 1,2,...,m$, which specify the dimensions of the various diagonal companion-form submatrices of $\hat{A}$, as well as the (m) ordered integers $\sigma_k = \sum_1^k d_i$, for $k = 1,2,...,m$, which denote the "nontrivial" rows of $\hat{A}$ and $\hat{B}$.

We now define $\hat{A}_m$ as the $(m \times n)$ matrix consisting of the (m) ordered σ_k rows of $\hat{A}$, and $\hat{B}_m$ as the $(m \times m)$ matrix consisting of the (m) ordered σ_k rows of $\hat{B}$. By inspection of 3.6.9b, it is clear that $\hat{B}_m$, thus defined, is an upper right triangular matrix with ones along the diagonal; i.e.

$$\hat{B}_m = \begin{bmatrix} 1 & x & x & \cdots & x \\ 0 & 1 & x & \cdots & x \\ \vdots & & & \ddots & \vdots \\ \vdots & & & & x \\ 0 & 0 & & \cdots & 1 \end{bmatrix} \tag{4.3.1}$$

and is clearly nonsingular since $|\hat{B}_m| = 1$. $\hat{A}_m$ assumes no particular special form. We also define $S(s)$ as the following $(n \times m)$ polynomial matrix with n nonzero, monic, single-term entries:

$$S(s) = \begin{bmatrix} 1 & 0 & \cdots & 0 \\ s & 0 & \cdots & 0 \\ \vdots & \vdots & & \vdots \\ s^{d_1-1} & 0 & \cdots & 0 \\ 0 & 1 & & \vdots \\ 0 & s & & \\ \vdots & \vdots & & \\ 0 & s^{d_2-1} & \cdots & 0 \\ 0 & 0 & & \vdots \\ \vdots & \vdots & & \\ & & & 0 \\ & & & 1 \\ & & & s \\ & & & \vdots \\ 0 & 0 & \cdots & s^{d_m-1} \end{bmatrix} \qquad 4.3.2$$

In view of these definitions, we can now state and establish:

THEOREM 4.3.3 (THE STRUCTURE THEOREM): <u>If a state representation $\{A,B,C,E\}$ is controllable with B of full rank</u> $m \leq n$, <u>its transfer matrix</u> $T(s)$ <u>given by</u> $C(sI-A)^{-1}B + E$, <u>can be expressed as</u>:

$$T(s) = \hat{C}S(s)\delta^{-1}(s)\hat{B}_m + E = [\hat{C}S(s) + E\hat{B}_m^{-1}\delta(s)][\hat{B}_m^{-1}\delta(s)]^{-1}, \qquad 4.3.3$$

where

$$\delta(s) = \begin{bmatrix} s^{d_1} & & & 0 \\ & s^{d_2} & & \\ & & \ddots & \\ 0 & & & s^{d_m} \end{bmatrix} - \hat{A}_m S(s) \qquad 4.3.4$$

<u>Proof</u>: We first recall that, in view of 4,2,3, the transfer matrices of two equivalent systems are identical; i.e. $T(s) = C(sI-A)^{-1}B + E = \hat{C}(sI-\hat{A})^{-1}\hat{B} + E$, and we need only show that $\hat{C}(sI-\hat{A})^{-1}\hat{B} = \hat{C}S(s)\delta^{-1}(s)\hat{B}_m$. It is therefore sufficient to show that

$$(sI-\hat{A})^{-1}\hat{B} = S(s)[\hat{B}_m^{-1}\delta(s)]^{-1}, \qquad 4.3.5$$

or, equivalently, that

$$(sI-\hat{A})S(s) = \hat{B}\hat{B}_m^{-1}\delta(s) \qquad 4.3.6$$

Note, however, that 4.3.6 is an immediate consequence of the definition of $S(s)$, $\hat{B}_m$, and $\delta(s)$, and can be established directly by substitution. The basic structure theorem 4.3.3 is thus established.

Some important observations regarding this theorem can now be made. In particular, by equating the n-th degree denominator polynomials of 4.3.5, we readily obtain the following relationship:

$$|sI-\hat{A}| = |sI-A| = \Delta(s) = |\delta(s)| \qquad 4.3.7$$

Perhaps the most important aspect of the structure theorem, however, is that it enables us to express the transfer matrix $T(s)$ of a time domain dynamical system as the product of a $(p \times m)$ polynomial matrix

$$R(s) = \hat{C}S(s) + E\hat{B}_m^{-1}\delta(s), \qquad 4.3.8$$

and the inverse of another $(m \times m)$ polynomial matrix

$$P(s) = \hat{B}_m^{-1}\delta(s), \qquad 4.3.9$$

i.e., in view of 4.2.3 and the above, it is clear that

$$T(s) = R(s)P(s)^{-1} \qquad 4.3.10$$

We further note that these two polynomial matrices possess certain important properties. In particular, P(s) is column proper since $\Gamma_c[P(s)] = \hat{B}_m^{-1}$ (Section 2.5), while the degree of each column $R_j(s)$ of $R(s)$ is less than or equal to the degree (d_j) of the corresponding column $P_j(s)$ of $P(s)$. This degree relationship is immediately obvious by inspection of 4.3.8 and 4.3.9 and can be more succinctly expressed as $\partial_{cj}[R(s)] \leq \partial_{cj}[P(s)]$ for $j = 1,2,\ldots,m$, where $\partial_{cj}[(\cdot)]$ denotes the degree of the j-th column of $(\cdot)$, or simply as

$$\partial_c[R(s)] \leq \partial_c[P(s)] \qquad 4.3.11$$

We can now establish that the results just presented also hold, with slight modification, whenever a state representation $\{A,B,C,E\}$ is not completely state controllable, but B is of full rank $m \leq n$. In particular, we recall from Section 3.6 that an uncontrollable (not completely state controllable) system can also be reduced via a nonsingular transformation Q_e to a particular structured form $\{\hat{A},\hat{B},\hat{C},E\} = \{Q_eAQ_e^{-1},Q_eB,CQ_e^{-1},E\}$, where $\hat{A}$ and $\hat{B}$ are given by 3.6.13. More specifically, $\hat{A} = \left[\begin{array}{c|c} \hat{A}_c & \hat{A}_{c\bar{c}} \\ \hline 0 & \hat{A}_{\bar{c}} \end{array}\right]$, $\hat{B} = \left[\begin{array}{c} \hat{B}_c \\ \hline 0 \end{array}\right]$, $\hat{C} = [\hat{C}_c \mid \hat{C}_{\bar{c}}]$, and $E = E$, where the $\bar{n} \times \bar{n}$ submatrix $\hat{A}_c$ of $\hat{A}$ and the $\bar{n} \times m$ submatrix $\hat{B}_c$ of $\hat{B}$ are in controllable companion form and $\hat{A}_{\bar{c}}$ represents the completely uncontrollable "portion" of the state matrix $\hat{A}$. If we now evaluate $T(s)$ via 4.2.3, we obtain the relation:

$$T(s) = [\hat{C}_c \mid \hat{C}_{\bar{c}}] \left[\begin{array}{c|c} sI-\hat{A}_c & -\hat{A}_{c\bar{c}} \\ \hline 0 & sI-\hat{A}_{\bar{c}} \end{array}\right]^{-1} \left[\begin{array}{c} \hat{B}_c \\ \hline 0 \end{array}\right] + E, \qquad 4.3.12$$

or, since

$$\left[\begin{array}{c|c} sI-\hat{A}_c & -\hat{A}_{c\bar{c}} \\ \hline 0 & sI-\hat{A}_{\bar{c}} \end{array}\right]^{-1} = \frac{\left[\begin{array}{c|c} (sI-\hat{A}_c)^+|sI-\hat{A}_{\bar{c}}| & (sI-\hat{A}_c)^+\hat{A}_{c\bar{c}}(sI-\hat{A}_{\bar{c}})^+ \\ \hline 0 & (sI-\hat{A}_{\bar{c}})^+|sI-\hat{A}_c| \end{array}\right]}{|sI-\hat{A}_c| \times |sI-\hat{A}_{\bar{c}}|}, \qquad 4.3.13$$

that

$$T(s) = \frac{\hat{C}_c(sI-\hat{A}_c)^+|sI-\hat{A}_{\bar{c}}|\hat{B}_c}{|sI-\hat{A}_c| \times |sI-\hat{A}_{\bar{c}}|} + E. \qquad 4.3.14$$

With the exception of the "cancellable" term $|sI-\hat{A}_{\bar{c}}| = \Delta_{\bar{c}}(s)$ which appears in both the numerator and denominator of $T(s)$ and represents the uncontrollable poles or modes of the system, this latter expression for $T(s)$ can be expressed via the structure theorem by the following equivalent one; i.e. since $\{\hat{A}_c, \hat{B}_c\}$ is in controllable companion form,

$$T(s) = \hat{C}_c S_c(s) \delta_c^{-1}(s) \hat{B}_{cm} \frac{\Delta_{\bar{c}}(s)}{\Delta_{\bar{c}}(s)} + E \qquad 4.3.15$$

$$= [\hat{C}_c S_c(s) + E\hat{B}_{cm}^{-1} \delta_c(s)] \Delta_{\bar{c}}(s) [\hat{B}_{cm}^{-1} \delta_c(s) \Delta_{\bar{c}}(s)]^{-1}$$

where $S_c(s)$ is given by 4.3.2 with $\sum_1^m d_i = \bar{n}$ instead of n,

$$\delta_c(s) = \begin{bmatrix} s^{d_1} & & & \\ & s^{d_1} & & 0 \\ & & \ddots & \\ 0 & & & s^{d_m} \end{bmatrix} - \hat{A}_{cm} S_c(s),$$

and $\hat{A}_{cm}$ and $\hat{B}_{cm}$ are given by the (m) ordered $\sigma_k = \sum_1^k d_i$ rows of the controllable companion pair $\hat{A}_c$ and $\hat{B}_c$ respectively. We thus note that 4.3.15 is entirely analogous to 4.3.3 with the exception of the fact that <u>pole-zero cancellation of all $(n - \bar{n})$ of the completely uncontrollable poles or modes of the system, as represented by the zeros of</u> $\Delta_{\bar{c}}(s) = |sI-\hat{A}_{\bar{c}}|$, <u>will occur in the transfer matrix</u>. It might be noted that the controllable poles of the system are equal to the zeros of $\Delta_c(s) = |sI-\hat{A}_c|$, and in view of 3.6.14, that $\Delta_{\bar{c}}(s)\Delta_c(s) = \Delta(s)$.

We can now employ duality to establish an entirely analogous result from the point of view of system observability rather than controllability. In particular, we first recall from Section 3.6 that if the state representation $\{A,B,C,E\}$ is completely observable with C of full rank $p \leq n$, then the system can be reduced via a nonsingular transformation $\bar{Q}^{-T}$ to an observable companion form $\{\hat{A},\hat{B},\hat{C},E\} = \{\bar{Q}^{-T}A\bar{Q}^{T},\bar{Q}^{-T}B,C\bar{Q}^{T},E\}$ with $\hat{A}$ and $\hat{C}$ given by 3.6.19. As in the case of controllability, we associate with this form (p) ordered observability indices $\bar{d}_i$ for $i = 1,2,\ldots,p$, which specify the dimensions of the various diagonal companion-form submatrices of $\hat{A}$, as well as (p) ordered integers $\bar{\sigma}_k = \sum_1^k \bar{d}_i$, for $k = 1,2,\ldots,p$, which denote the "nontrivial" columns of $\hat{A}$ and $\hat{C}$.

If we now define the $(n \times p)$ matrix $\hat{A}_p$ and the $(p \times p)$ matrix $\hat{C}_p$ as the (p) ordered "nontrivial" $\bar{\sigma}_k$ columns of $\hat{A}$ and $\hat{C}$ respectively and $\bar{S}(s)$ as the following $(p \times n)$ polymomial matrix of n nonzero, monic single-term entries: i.e.

$$\bar{S}(s) = \begin{bmatrix} 1 & s & \ldots & s^{\bar{d}_1-1} & 0 & 0 & \ldots & 0 & 0 & \ldots & & & 0 \\ 0 & 0 & \ldots & 0 & 1 & s & \ldots & s^{\bar{d}_2-1} & 0 & \ldots & & & 0 \\ \vdots & \vdots & & \vdots & \vdots & & & \vdots & \vdots & & & & \\ 0 & 0 & & 0 & 0 & \ldots & & 0 & 0 & \ldots & 1 & s & \ldots s^{\bar{d}_p-1} \end{bmatrix}, \qquad 4.3.16$$

we can readily establish, by duality, the following "observable version" of the structure theorem; i.e.

$$T(s) = \hat{C}_p\bar{\delta}(s)^{-1}\bar{S}(s)\hat{B}+E = [\bar{\delta}(s)\hat{C}_p^{-1}]^{-1}[\bar{S}(s)\hat{B} + \bar{\delta}(s)\hat{C}_p^{-1}E], \qquad 4.3.17$$

where

$$\bar{\delta}(s) = \begin{bmatrix} s^{\bar{d}_1} & & & 0 \\ & s^{\bar{d}_2} & & \\ & & \ddots & \\ 0 & & & s^{\bar{d}_p} \end{bmatrix} - \bar{S}(s)\hat{A}_p \qquad 4.3.19$$

We note that $\hat{C}_p$ assumes a special form here as $\hat{B}_m$ did in the controllable version. In particular, $\hat{C}_p$ is lower left triangular with ones along the diagonal; i.e.

$$\hat{C}_p = \begin{bmatrix} 1 & 0 & \cdots & 0 \\ x & 1 & \cdots & 0 \\ \vdots & & \ddots & \vdots \\ \vdots & & & 0 \\ x & \cdots & x & 1 \end{bmatrix} \qquad 4.3.19$$

and is clearly nonsingular since $|\hat{C}_p| = 1$.

In view of the above, we can now make some observations analogous to those which were made in the controllable case. In particular, it is immediately apparent, by equating the monic n-th degree denominator polynomials of $C(sI-A)^{-1}B + E$ and $T(s)$, as given by 4.3.17, that

$$|sI-\hat{A}| = |sI-A| = \Delta(s) = |\bar{\delta}(s)| \qquad 4.3.20$$

It is also clear that, as in the controllable case, the transfer matrix $T(s)$ can be expressed as a product; i.e. in view of 4.3.17

$$T(s) = \bar{P}^{-1}(s)Q(s), \qquad 4.3.21$$

where

$$\bar{P}(s) = \bar{\delta}(s)\hat{C}_p^{-1}, \qquad 4.3.22$$

and

$$Q(s) = \bar{S}(s)\hat{B} + \bar{\delta}(s)\hat{C}_p^{-1}E \qquad 4.3.23$$

Furthermore, $\underline{\bar{P}(s)}$ is row proper since $\Gamma_r[\bar{P}(s)] = \hat{C}_p^{-1}$ (Section 2.5), while the degree of each row $Q_i(s)$ of $Q(s)$ is less than or equal to the degree $(\bar{d}_i)$ of the corresponding row $\bar{P}_i(s)$ of $\bar{P}(s)$. We express this row degree relationship as

$$\partial_r[Q(s)] \leq \partial_r[\bar{P}(s)], \qquad 4.3.24$$

the dual of 4.3.11.

It should also be clear that if a state representation $\{A,B,C,E\}$ is not completely state observable, it can be reduced via an equivalence transformation $\bar{Q}_e^{-T}$ to a form which separates the observable and completely unobservable "portions" of the system while placing the observable "portion" in companion form; i.e. $\{\hat{A},\hat{B},\hat{C},E\} = \{\overline{Q}_e^{-T} A\overline{Q}_e^{T}, \overline{Q}_e^{-T} B, C\overline{Q}_e^{T}, E\}$, where $\hat{A} = \left[\begin{array}{c|c} \hat{A}_o & 0 \\ \hline \hat{A}_{o\bar{o}} & \hat{A}_{\bar{o}} \end{array}\right]$, $\hat{B} = \left[\begin{array}{c} \hat{B}_o \\ \hline \hat{B}_{\bar{o}} \end{array}\right]$, and $\hat{C} = [\hat{C}_o \mid 0]$ with $\hat{A}_o$ and $\hat{C}_o$ in observable companion form. By duality with 4.3.15, it then follows that

$$T(s) = \hat{C}_{op}\overline{\delta}_o^{-1}(s)\overline{S}_o(s)\hat{B}_o \frac{\Delta_{\bar{o}}(s)}{\Delta_{\bar{o}}(s)} + E \qquad 4.3.25$$

$$= [\overline{\delta}_o(s)\hat{C}_{op}^{-1}\Delta_{\bar{o}}(s)]\Delta_{\bar{o}}^{-1}(s)[\overline{S}_o(s)\hat{B}_o + \overline{\delta}_o(s)\hat{C}_{op}^{-1}E],$$

where the zeros of $\Delta_{\bar{o}}(s) = |sI-\hat{A}_{\bar{o}}|$ represent the completely unobservable poles of the system. It should perhaps be noted that the observable poles of the system are equal to the zeros of $\Delta_o(s) = |sI-\hat{A}_o|$, and that $\Delta_{\bar{o}}(s)\Delta_o(s) = \Delta(s)$, which is analogous to the dual controllable relationship, $\Delta_{\bar{c}}(s)\Delta_c(s) = \Delta(s)$. We leave the detailed development of 4.3.25 as an exercise for the reader, noting that, as in the controllable case, <u>pole-zero cancellation of all of the unobservable poles or modes of the system, as represented by the zeros of $\Delta_{\bar{o}}(s) = |sI-\hat{A}_{\bar{o}}|$, will occur in the transfer matrix</u>.

EXAMPLE 4.3.26: To illustrate the derivation of $T(s)$ via the structure theorem, we again employ the state representation $\{A,B,C,E\}$ of Example 4.2.7. In particular, $A = \begin{bmatrix} 1 & 2 & 0 \\ 4 & -1 & 0 \\ 0 & 0 & 1 \end{bmatrix}$, $B = \begin{bmatrix} 1 \\ 0 \\ 1 \end{bmatrix}$, $C = \begin{bmatrix} 0 & 1 & -1 \\ 0 & 0 & 1 \end{bmatrix}$, and $E = \begin{bmatrix} 0 \\ 1 \end{bmatrix}$.

<u>Part a - the controllable version</u>: The reader can readily verify that this system is controllable, and can therefore be reduced via the nonsingular transformation

$Q = \begin{bmatrix} 1 & 0 & -1 \\ 1 & 2 & -1 \\ 9 & 0 & -1 \end{bmatrix} \div 8$ (see Section 3.6) to controllable companion form;

i.e. $QAQ^{-1} = \hat{A} = \begin{bmatrix} 0 & 1 & 0 \\ 0 & 0 & 1 \\ -9 & 9 & 1 \end{bmatrix}$, $QB = \hat{B} = \begin{bmatrix} 0 \\ 0 \\ 1 \end{bmatrix}$, $CQ^{-1} = \hat{C} = \begin{bmatrix} 5 & 4 & -1 \\ -9 & 0 & 1 \end{bmatrix}$,

and $E = \begin{bmatrix} 0 \\ 1 \end{bmatrix}$. Since there is only one input, $m = 1$, $d_1 = 3 = \sigma_1 = n$.

Therefore, $S(s) = \begin{bmatrix} 1 \\ s \\ s^2 \end{bmatrix}$, $\hat{A}_m = [-9 \quad 9 \quad 1]$, $\hat{B}_m = 1$, and $\delta(s) = s^3 - \hat{A}_m S(s) = s^3 - s^2 - 9s + 9 = (s-1)(s+3)(s-3)$. The transfer matrix of this system can now be directly determined via 4.3.3; i.e.

$$T(s) = \hat{C}S(s)\delta(s)^{-1}\hat{B}_m + E = \begin{bmatrix} -s^2 + 4s + 5 \\ s^2 - 9 \end{bmatrix} \frac{1}{(s-1)(s+3)(s-3)} + \begin{bmatrix} 0 \\ 1 \end{bmatrix},$$

which, after all possible pole-zero cancellations have been made, yields the same transfer matrix expression as in Example 4.2.7, namely

$T(s) = \begin{bmatrix} \dfrac{(s+1)(-s+5)}{(s-1)(s+3)(s-3)} \\ \\ \dfrac{s}{s-1} \end{bmatrix}$. The "factored" representation, $R(s)P^{-1}(s)$, for $T(s)$, as given by 4.3.10, can also be readily determined. In

particular, $R(s) = \hat{C}S(s) + E\hat{B}_m^{-1}\delta(s) = \begin{bmatrix} -s^2 + 4s + 5 \\ s^3 - 9s \end{bmatrix} = \begin{bmatrix} (s+1)(-s+5) \\ s(s+3)(s-3) \end{bmatrix}$,

while $P(s) = \hat{B}_m^{-1}\delta(s) = s^3 - s^2 - 9s + 9 = (s-1)(s+3)(s-3)$. The product $R(s)P(s)^{-1}$ also produces the same transfer matrix as above after all possible pole-zero cancellations have been made.

<u>Part b - the observable version</u>: The reader can now verify that this system is also observable, and can therefore be reduced via the non-

singular transformation $\bar{Q}^{-T} = \begin{bmatrix} 4 & -1 & -1 \\ 0 & 1 & -1 \\ 0 & 0 & 1 \end{bmatrix}$ (see Section 3.6) to observable companion form; i.e. $\bar{Q}^{-T}A\bar{Q}^{T} = \hat{A} = \left[\begin{array}{cc|c} 0 & 9 & 8 \\ 1 & 0 & 0 \\ \hline 0 & 0 & 1 \end{array}\right]$, $\bar{Q}^{-T}B = \hat{B} = \begin{bmatrix} 3 \\ -1 \\ 1 \end{bmatrix}$, $C\bar{Q}^{T} = \hat{C} = \begin{bmatrix} 0 & 1 & 0 \\ 0 & 0 & 1 \end{bmatrix}$, and $E = \begin{bmatrix} 0 \\ 1 \end{bmatrix}$. It is thus clear that $p = 2$, $\bar{d}_1 = 2$, $\bar{d}_2 = 1$, $\bar{\sigma}_1 = 2$, and $\bar{\sigma}_2 = 3 = n$. Furthermore, $\bar{S}(s) = \begin{bmatrix} 1 & s & 0 \\ 0 & 0 & 1 \end{bmatrix}$, $\hat{A}_p = \begin{bmatrix} 9 & 8 \\ 0 & 0 \\ 0 & 1 \end{bmatrix}$, $\hat{C}_p = I_2$, and $\bar{\delta}(s) = \begin{bmatrix} s^2 & 0 \\ 0 & s \end{bmatrix} - \bar{S}(s)\hat{A}_p = \begin{bmatrix} s^2-9, & -8 \\ 0, & s-1 \end{bmatrix}$. The transfer matrix of the system can now be directly determined via 4.3.17; i.e. $T(s) = \hat{C}_p\bar{\delta}^{-1}(s)\bar{S}(s)\hat{B} + E =$

$$\frac{\begin{bmatrix} s-1, & 8 \\ 0, & s^2-9 \end{bmatrix}\begin{bmatrix} -s+3 \\ 1 \end{bmatrix}}{s^3-s^2-9s+9} + \begin{bmatrix} 0 \\ 1 \end{bmatrix},$$

which yields the same transfer matrix as obtained in Part a and Example 4.2.7; i.e. $T(s) = \begin{bmatrix} \dfrac{(s+1)(-s+5)}{(s-1)(s+3)(s-3)} \\[2ex] \dfrac{s}{s-1} \end{bmatrix}$.

The "factored" representation, $\bar{P}^{-1}(s)Q(s)$, for $T(s)$, as given by 4.3.21 can also be readily determined. In particular $\bar{P}(s) = \bar{\delta}(s)\hat{C}_p^{-1} = \begin{bmatrix} s^2-9, & -8 \\ 0, & s-1 \end{bmatrix}$, and $Q(s) = \bar{S}(s)\hat{B} + \bar{\delta}(s)\hat{C}_p^{-1}E = \begin{bmatrix} -s-5 \\ s \end{bmatrix}$.

4.4 REALIZATION THEORY (TIME DOMAIN REDUCTION)

In the previous two sections, we developed two techniques which can be used to obtain the transfer matrix of a system whose dynamics are expressed in the state form 3.2.1, namely (i) the direct employment of 4.2.2 and (ii) the utilization of the structure theorem after an initial reduction of the quadruple $\{A,B,C,E\}$ to an appropriate companion form. We also noted that the transfer matrix of a system is an external system description which can be determined experimentally via steady-state frequency response measurements. Regardless of how the transfer matrix of a system is obtained, however, it is often desirable to obtain a time domain representation for the dynamical behavior of the system, given its transfer matrix, a fact which motivates the following:

DEFINITION 4.4.1: A <u>realization</u> of a proper, rational transfer matrix $T(s)$ is any state representation $\{A,B,C,E\}$ whose transfer matrix is $T(s)$; i.e. $\{A,B,C,E\}$ is a <u>realization</u> of $T(s)$ if and only if

$$C(sI-A)^{-1}B + E = T(s) \qquad 4.4.2$$

A realization $\{A_c,B_c,C_c,E\}(\{A_o,B_o,C_o,E\})$ of $T(s)$ which is controllable (observable) is called a <u>controllable (observable) realization</u> of $T(s)$, while a realization $\{\tilde{A},\tilde{B},\tilde{C},E\}$ of $T(s)$ of least order $\tilde{n}$ is called a <u>minimal or irreducible realization</u> of $T(s)$. The primary purpose of this section will be to develop constructive algorithms for obtaining various types of realizations, as well as to display a fundamental relationship between minimality of a realization and system controllability and observability.

To begin, we recall from our earlier discussion in this chapter, that since the transfer matrix of a system describes only the external (input/output) behavior of the system, certain internal (state) information is either "lost" or simply ignored when a conversion from a state representation to the transfer matrix via 4.2.2 is made. In particular, the initial conditions on the state of the system as well as the actual number, n, of state variables are quantities which are not always apparent from knowledge of $T(s)$ alone. We further noted that equivalent state representations have the identical transfer matrix. We therefore expect a lack of uniqueness, both in the definition of state as well as the specification of the state dimension n, whenever a determination of a realization of $T(s)$ is made. This is indeed the case, as we will now illustrate using a simple scalar example.

EXAMPLE 4.4.3: Consider the system whose transfer function $T(s) = \frac{s+1}{s^2+3s+2} = \frac{s+1}{(s+1)(s+2)} = \frac{1}{s+2}$. The reader can easily verify that all four of the following state representations realize $T(s)$:

(i) $A = \begin{bmatrix} 0 & 1 \\ -2 & -3 \end{bmatrix}$; $B = \begin{bmatrix} 0 \\ 1 \end{bmatrix}$; $C = [1 \quad 1]$, which is controllable but unobservable.

(ii) $A = \begin{bmatrix} 0 & -2 \\ 1 & -2 \end{bmatrix}$; $B = \begin{bmatrix} 1 \\ 1 \end{bmatrix}$; $C = [0 \quad 1]$, which is observable but uncontrollable.

(iii) $A = \begin{bmatrix} -1 & 0 \\ 0 & -2 \end{bmatrix}$; $B = \begin{bmatrix} 0 \\ 1 \end{bmatrix}$; $C = [0 \quad 1]$, which is neither controllable nor observable.

(iv) $A = [-2]$; $B = [1]$; $C = [1]$, which is both controllable and observable.

In view of Problem 3-4, it is clear that no two of these realizations can represent equivalent dynamical systems. Note further that a cancellation of the common pole-zero factor $(s+1)$ does not occur in the fourth realization. Therefore, although we did not explicitly state so in our definition of a realization of $T(s)$, we now note that <u>a realization of $T(s)$ need exactly reproduce only the uncancellable portion of $T(s)$</u>. This latter observation is important in establishing minimality and, therefore, some degree of uniqueness of a realization, as we will later show. In view of this observation, the following would also qualify as realizations of the $T(s)$ of Example 4.4.3:

(v) $A = \begin{bmatrix} -1 & 0 & 0 \\ 0 & -2 & 0 \\ 0 & 0 & -3 \end{bmatrix}$; $B = \begin{bmatrix} 0 \\ 1 \\ 0 \end{bmatrix}$; $C = [1 \quad 1 \quad 1]$, which is observable but uncontrollable.

(vi) $A = \begin{bmatrix} -2 & 0 \\ 0 & -3 \end{bmatrix}$; $B = \begin{bmatrix} 1 \\ 1 \end{bmatrix}$; $C = [1 \quad 0]$, which is controllable but unobservable.

Based on the results which we have developed thus far, we are now in a position to present an algorithm for obtaining either a controllable or an observable realization of any proper, rational transfer matrix $T(s)$. It might be noted that the algorithm is based on a direct employment of the structure theorem. In particular, let $T(s) = [t_{ij}(s)]$ be any $p \times m$ matrix whose entries $t_{ij}(s) = \frac{r_{ij}(s)}{p_{ij}(s)}$ are proper, rational functions of s. If we now let $g_j(s)$, for $j = 1,2,\ldots,m$, be the least common (monic) multiple of the (p) denominator polynomials, $p_{1j}(s), p_{2j}(s), \ldots, p_{pj}(s)$, which appear in each (j-th) column of $T(s)$[†]; i.e. the least common denominator of the j-th column entries of $T(s)$, we can represent $T(s)$ by the following expression:

[†] The algorithm which now follows is based on the assumption that $\partial[g_j(s)] > 0$ for all j. If for some j, however, $\partial[g_j(s)] = 0$, the j-th column of $T(s)$ will be a constant real vector, independent of s, and will therefore equal E_j, the j-th column of E. B_j can consequently be set equal to zero, and the algorithm modified by simply ignoring the j-th column of $T(s)$ while determining a realization of the resulting "column reduced" transfer matrix. Example 4.4.14 and Problems 4-11 and 4-13 serve to illustrate this procedure.

$$T(s) = \begin{bmatrix} \frac{r^*_{11}(s)}{g_1(s)}, & \frac{r^*_{12}(s)}{g_2(s)}, & \cdots, & \frac{r^*_{1m}(s)}{g_m(s)} \\ \frac{r^*_{21}(s)}{g_1(s)}, & \frac{r^*_{22}(s)}{g_2(s)}, & \cdots, & \vdots \\ \vdots & \vdots & \ddots & \\ \frac{r^*_{p1}(s)}{g_1(s)}, & \cdots & , & \frac{r^*_{pm}(s)}{g_m(s)} \end{bmatrix} = [r^*_{ij}(s)]\{\text{diag}[g_j(s)]\}^{-1}, \quad 4.4.4$$

where $\text{diag}[g_j(s)] = \begin{bmatrix} g_1(s) & & 0 \\ & g_2(s) & \\ & & \ddots \\ 0 & & & g_m(s) \end{bmatrix}$, and $r^*_{ij}(s)$ is equal to $r_{ij}(s)$ multiplied by the polynomial obtained by dividing $g_j(s)$ by $p_{ij}(s)$. $T(s)$, as given by 4.4.4, can now be directly related to the controllable structure theorem representation 4.3.3 for a transfer matrix, namely $[\hat{C}S(s) + E\hat{B}_m^{-1}\delta(s)][\hat{B}_m^{-1}\delta(s)]^{-1} = R(s)P(s)^{-1}$, and, consequently, to an appropriate controllable realization $\{A_c, B_c, C_c, E\}$ of $T(s)$, as we now show.

In particular, we first note that 4.4.4 is an expression for $T(s)$ in terms of the product of one polynomial matrix $[r^*_{ij}(s)]$, and the inverse of another, $\text{diag}[g_j(s)]$, which is column proper with $\Gamma_c[\text{diag}[g_j(s)]] = I_m$. In view of the results given in Section 4.3, $\text{diag}[g_j(s)]$ can therefore be directly equated to $P(s)$ and $[r^*_{ij}(s)]$ to $R(s)$. More specifically, if we set $\text{diag}[g_j(s)] = P(s) = \hat{B}_m^{-1}\delta(s)$, it is clear that $I_m = \Gamma_c[\hat{B}_m^{-1}\delta(s)] = \hat{B}_m^{-1}$ or that $\hat{B}_m = I_m$. Furthermore, if d_j is used to denote the degree of $g_j(s)$, i.e. if $d_j = \partial[g_j(s)]$, then $g_j(s)$ can be written as

$$g_j(s) = s^{d_j} + g_{j1}s^{d_j-1} + \dots + g_{jd_j}, \qquad 4.4.5$$

or as

$$g_j(s) = s^{d_j} + [g_{jd_j}, g_{jd_j-1}, \dots, g_{j1}] \begin{bmatrix} 1 \\ s \\ \vdots \\ s^{d_j-1} \end{bmatrix} = s^{d_j} + \bar{g}_j \begin{bmatrix} 1 \\ s \\ \vdots \\ s^{d_j-1} \end{bmatrix}, \qquad 4.4.6$$

where $\bar{g}_j = [g_{jd_j}, g_{jd_j-1}, \dots, g_{j1}]$. Therefore,

$$\underbrace{\begin{bmatrix} g_1(s) & & & 0 \\ & g_2(s) & & \\ & & \ddots & \\ 0 & & & g_m(s) \end{bmatrix}}_{\mathrm{diag}[g_j(s)]} = \underbrace{\begin{bmatrix} s^{d_1} & & & 0 \\ & s^{d_2} & & \\ & & \ddots & \\ 0 & & & s^{d_m} \end{bmatrix}}_{\mathrm{diag}[s^{d_i}]} - \underbrace{\begin{bmatrix} -\bar{g}_1 & & & 0 \\ & -\bar{g}_2 & & \\ & & \ddots & \\ 0 & & & -\bar{g}_m \end{bmatrix}}_{\hat{A}_m} S(s) = \delta(s) \qquad 4.4.7$$

The $p \times m$ polynomial matrix $[r^*_{ij}(s)]$, with elements given by

$$r^*_{ij}(s) = r^*_{ijo}s^{d_j} + r^*_{ij1}s^{d_j-1} + \dots + r^*_{ijd_j} \qquad 4.4.8$$

can now be equated to $R(s) = \hat{C}S(s) + E\delta(s)$ with $\hat{B}_m^{-1} = I_m$. In particular, since $T(s)$ is proper by assumption, $\partial_c[r^*_{ij}(s)] \leq \partial_c[\mathrm{diag}[g_j(s)]]$ and

$$\lim_{s \to \infty} [r^*_{ij}(s)]\{\mathrm{diag}[g_j(s)]\}^{-1} = E = [r^*_{ijo}] \qquad 4.4.9$$

Therefore, $[r^*_{ij}(s)] = \hat{C}S(s) + E\delta(s)$, or in view of 4.4.7 and 4.4.9,

$$[r^*_{ij}(s)] - [r^*_{ijo}]\ \mathrm{diag}[g_j(s)] = \hat{C}S(s) \qquad 4.4.10$$

for some real $p \times n$ scalar matrix $\hat{C}$. If we now define A_c as the controllable companion form matrix obtained by replacing the (m) ordered $\sigma_k(=\sum_1^k d_i)$ rows of the $n(=\sum_1^m d_i)$ dimensional companion matrix with upper right identity by the (m) ordered rows of $\hat{A}_m = \begin{bmatrix} -\bar{g}_1 & & & 0 \\ & -\bar{g}_2 & & \\ & & \ddots & \\ 0 & & & -\bar{g}_m \end{bmatrix}$, B_c as the matrix obtained by replacing the (m) σ_k rows of the $n \times m$ null matrix (all elements $\equiv 0$) by the (m) ordered rows of I_m, and let E and $C_c = \hat{C}$ be given by 4.4.9 and 4.4.10, respectively, it follows immediately from the structure theorem that the resulting controllable companion form quadruple $\{A_c, B_c, C_c, E\}$ realizes $T(s)$. To illustrate the above procedure, consider the following:

EXAMPLE 4.4.11: Consider the system whose transfer matrix

$T(s) = \begin{bmatrix} \frac{1}{s+1} & \frac{s}{s-2} \\ 2 & 0 \\ \frac{2}{s-2} & 1 \end{bmatrix}$. To find a controllable realization of this $T(s)$

we first determine the least common denominator of each column of $T(s)$ and then express $T(s)$ in the form 4.4.4; i.e. $g_1(s) = (s+1)(s-2) = s^2 - s - 2$, $g_2(s) = s-2$, $[r_{ij}^*(s)] = \begin{bmatrix} s-2, & s \\ 2s^2-2s-4, & 0 \\ 2s+2, & s-2 \end{bmatrix}$, and $\text{diag}[g_j(s)] = \begin{bmatrix} s^2-s-2, & 0 \\ 0, & s-2 \end{bmatrix}$. $T(s)$ can clearly be represented as $[r_{ij}^*(s)]\{\text{diag}[g_j(s)]\}^{-1}$. $\text{Diag}[g_j(s)]$ can now be equated to $\delta(s)$ via 4.4.7; i.e. since $d_1 = 2$ and $d_2 = 1$,

$$\begin{bmatrix} s^2-s-2, & 0 \\ 0, & s-2 \end{bmatrix} = \begin{bmatrix} s^2 & 0 \\ 0 & s \end{bmatrix} - \begin{bmatrix} 2 & 1 & 0 \\ 0 & 0 & 2 \end{bmatrix} \begin{bmatrix} 1 & 0 \\ s & 0 \\ 0 & 1 \end{bmatrix} = \text{diag}[s^{d_i}] - \hat{A}_m S(s),$$

where $\hat{A}_m = \begin{bmatrix} 2 & 1 & 0 \\ 0 & 0 & 2 \end{bmatrix}$. By inspection of $[r^*_{ij}(s)]$, it is also clear that $[r^*_{ijo}] = \begin{bmatrix} 0 & 1 \\ 2 & 0 \\ 0 & 1 \end{bmatrix} = E = \lim_{s \to \infty} T(s)$. We can now determine $\hat{C}S(s)$ using 4.4.10. In particular, $\hat{C}S(s) = [r^*_{ij}(s)] - [r^*_{ijo}]\ \text{diag}[g_j(s)] = \begin{bmatrix} s-2 & 2 \\ 0 & 0 \\ 2s+2 & 0 \end{bmatrix}$, which, in view of the fact that $S(s) = \begin{bmatrix} 1 & 0 \\ s & 0 \\ 0 & 1 \end{bmatrix}$, immediately implies that $\hat{C} = \begin{bmatrix} -2 & 1 & 2 \\ 0 & 0 & 0 \\ 2 & 2 & 0 \end{bmatrix} = C_c$. $\hat{A}_m$ and the d_i now directly specify $A_c = \begin{bmatrix} 0 & 1 & 0 \\ 2 & 1 & 0 \\ 0 & 0 & 2 \end{bmatrix}$ and $B_c = \begin{bmatrix} 0 & 0 \\ 1 & 0 \\ 0 & 1 \end{bmatrix}$. In view of Theorem 4.3.3, it follows that this controllable, companion form quadruple $\{A_c, B_c, C_c, E\}$ realizes $T(s)$.

By duality, we can now readily outline a procedure for obtaining an observable realization of any proper, rational transfer matrix $T(s)$. More precisely, using dual operations, we can modify the algorithm just presented for finding a controllable realization, or we can apply the exact same algorithm to the transpose $T^T(s)$ of $T(s)$ and, as the final step, simply calculate the dual of the controllable realization of $T^T(s)$ thus obtained. If we employ this latter procedure, the dual of the controllable realization $\{A_{cT}, B_{cT}, C_{cT}, E_T\}$ of $T^T(s)$, namely

$$\{A^T_{cT}, C^T_{cT}, B^T_{cT}, E^T_T\} = \{A_o, B_o, C_o, E\} \tag{4.4.12}$$

represents an observable realization of $T(s)$, since

$$\begin{aligned}[T^T(s)]^T = T(s) = [C_{cT}(sI-A_{cT})^{-1}B_{cT}+E_T]^T &= B^T_{cT}(sI-A^T_{cT})^{-1}C^T_{cT}+E^T_T \\ &= C_o(sI-A_o)^{-1}B_o + E,\end{aligned} \tag{4.4.13}$$

with the pair $\{A_{cT}, B_{cT}\}$ or $\{A_o^T, C_o^T\}(\{A_o, C_o\})$ controllable (observable). The following example illustrates this procedure.

EXAMPLE 4.4.14: To find an observable realization of the transfer matrix $T(s)$ of Example 4.4.11, we need only find a controllable realization of $T^T(s) = \begin{bmatrix} \frac{1}{s+1} & 2 & \frac{2}{s-2} \\ \frac{s}{s-2} & 0 & 1 \end{bmatrix}$. If we again use the algorithm for obtaining controllable realizations, we determine that the controllable quadruple: $A_{cT} = \begin{bmatrix} 0 & 1 & 0 \\ 2 & 1 & 0 \\ 0 & 0 & 2 \end{bmatrix}$, $B_{cT} = \begin{bmatrix} 0 & 0 & 0 \\ 1 & 0 & 0 \\ 0 & 0 & 1 \end{bmatrix}$, $C_{cT} = \begin{bmatrix} -2 & 1 & 2 \\ 2 & 2 & 0 \end{bmatrix}$, $E_T = \begin{bmatrix} 0 & 2 & 0 \\ 1 & 0 & 1 \end{bmatrix}$ realizes $T^T(s)$, and therefore, that the observable realization $\{A_o, B_o, C_o, E\} = \{A_{cT}^T, C_{cT}^T, B_{cT}^T, E_T^T\}$ realizes $T(s)$, a fact which the reader can readily verify. It should now also be clear that the dual $\{A_c^T, C_c^T, B_c^T, E^T\}$ of the controllable realization $\{A_c, B_c, C_c, E\}$ of $T(s)$ given in Example 4.4.11, represents an observable realization of $T^T(s)$, a result which, of course, holds in general as well.

We have now illustrated procedures for obtaining both controllable and observable realizations of any proper, rational transfer matrix $T(s)$. Perhaps the most important class of realizations, however, are those which have least order $\tilde{n}$, namely minimal or irreducible realizations, since they contain no extraneous or unnecessary information and require a minimal number $(\tilde{n})$ of energy storage elements for physical implementation (e.g. on an analog computer). The question of whether or not a particular realization of $T(s)$ is minimal is closely tied to the notions of controllability and observability, a point we now formally illustrate and establish.

THEOREM 4.4.15: <u>An n-dimensional realization</u> $\{A,B,C,E\}$ <u>of</u> $T(s)$ <u>is minimal (irreducible) if and only if it is both controllable and ob-</u>

servable; i.e. if and only if both $\rho[\mathscr{C}]$ and $\rho[\mathscr{O}]$ equal n.

Proof: We first recall from our earlier work in Section 4.3 that whenever a state representation $\{A,B,C,E\}$ is uncontrollable we can separate the part which is completely uncontrollable from the controllable part, and thus obtain a lower dimension ($\hat{n}$) quadruple $\{\hat{A}_c,\hat{B}_c,\hat{C}_c,E\}$ which yields the same transfer matrix $T(s)$ as the original representation. By duality, this same result holds from the point of view of system observability. In view of these observations, we therefore conclude that a realization of $T(s)$ cannot be minimal unless it is both controllable and observable, or equivalently, that minimality implies both controllability and observability of a realization.

We therefore need only establish the converse, or that controllability and observability together imply minimality of any realization of $T(s)$. We do this by contradiction. In particular, suppose that some n-dimensional realization $\{A,B,C,E\}$ of $T(s)$ is both controllable and observable, but that there exists a realization $\{\tilde{A},\tilde{B},\tilde{C},E\}$ of $T(s)$ of order $\tilde{n}$, which is lower than n. Since both realizations have the same transfer matrix $T(s)$, they also have the same impulse response matrix $T(t)$ as given by 3.2.28; i.e.

$$Ce^{At}B + E\delta(t) = \tilde{C}e^{\tilde{A}t}\tilde{B} + E\delta(t) \qquad 4.4.16$$

for all $t\varepsilon[0,\infty)$. Equating both sides of 4.4.16 and their k-th derivatives with respect to time at $t = 0$, we obtain the relations:

$$CA^kB = \tilde{C}\tilde{A}^k\tilde{B}, \qquad 4.4.17$$

for $k = 0,1,2,\ldots$. If we now consider the product of the observability matrix $\mathscr{O}$ and controllability matrix $\mathscr{C}$ of the n-dimensional controllable and observable realization $\{A,B,C,E\}$, it is clear that

$$\mathcal{O}\mathcal{C} = \begin{bmatrix} C \\ CA \\ \vdots \\ CA^{n-1} \end{bmatrix} [B, AB, \ldots, A^{n-1}B] = \begin{bmatrix} CB, & CAB, & \ldots, & CA^{n-1}B \\ CAB, & CA^2B, & \ldots, & CA^nB \\ \vdots & \vdots & & \vdots \\ CA^{n-1}B, & CA^nB, & \ldots, & CA^{2(n-1)}B \end{bmatrix}, \quad 4.4.18$$

where $\rho[\mathcal{O}] = \rho[\mathcal{C}] = n$. We next recall that if $\mathcal{O}$ and $\mathcal{C}$ are $q \times n$ and $n \times r$ matrices respectively with elements in the same field, then in view of Sylvester's Inequality (see Problem 2-16):

$$\rho[\mathcal{O}] + \rho[\mathcal{C}] - n \le \rho[\mathcal{O}\mathcal{C}] \le \min(\rho[\mathcal{O}], \rho[\mathcal{C}]), \quad 4.4.19$$

which clearly implies in this case that $\rho[\mathcal{O}\mathcal{C}] = n$. If we now replace each submatrix CA^kB of 4.4.18 by its equivalent $\tilde{C}\tilde{A}^k\tilde{B}$, as given by 4.4.17, and represent the result by the product:

$$\begin{bmatrix} \tilde{C} \\ \tilde{C}\tilde{A} \\ \vdots \\ \tilde{C}\tilde{A}^{n-1} \end{bmatrix} [\tilde{B}, \tilde{A}\tilde{B}, \ldots, \tilde{A}^{n-1}\tilde{B}] = \tilde{\mathcal{O}}_n \tilde{\mathcal{C}}_n,$$

it then follows by Sylvester's Inequality, 4.4.19, and the assumption that $\tilde{n} < n$ that

$$\rho[\tilde{\mathcal{O}}_n \tilde{\mathcal{C}}_n] \le \tilde{n} < n = \rho[\mathcal{O}\mathcal{C}], \quad 4.4.20$$

since both $\tilde{\mathcal{O}}_n$ and $\tilde{\mathcal{C}}_n$ can have rank no greater than $\tilde{n}$. Since $\mathcal{O}\mathcal{C} = \tilde{\mathcal{O}}_n \tilde{\mathcal{C}}_n$, however, 4.4.20 represents a contradiction to the initial assumption that a realization of lower order than one which is both controllable and observable can exist, thus establishing the theorem.

An immediate consequence of this theorem is the fact that we already have at hand a technique for obtaining a minimal realization $\{\tilde{A}, \tilde{B}, \tilde{C}, E\}$ of any rational, proper transfer matrix $T(s)$. In particular, we already know how to (i) obtain either a controllable or observable realization of $T(s)$ and (ii) separate the completely uncontrollable (unobservable) "portion" of a realization from the controllable (observable) part. By combining these two steps we can therefore construct minimal realizations; i.e. the controllable (observable) part of an observable (controllable) realization represents a realization which is both controllable and observable and therefore minimal.

The reader can readily verify that the controllable realization of $T(s)$, as given in Example 4.4.11, is also observable, and the observable realization of $T(s)$, as given in Example 4.4.14, is also controllable. In view of Theorem 4.4.15, it therefore follows that both realizations are minimal, although they are in different companion forms. This latter observation serves to motivate the following result which clarifies the relationship between any two minimal realizations of the same transfer matrix.

THEOREM 4.4.21: If $\{A,B,C,E\}$ is a minimal realization of $T(s)$, then $\{\hat{A},\hat{B},\hat{C},E\}$ is also a minimal realization of $T(s)$ if and only if the two realizations are equivalent; i.e. if and only if $\{\hat{A},\hat{B},\hat{C},E\} = \{QAQ^{-1},QB,CQ^{-1},E\}$ for some nonsingular matrix Q.

Proof: In view of 4.2.3 and the fact that controllability and observability are unaffected by an equivalence transformation Q (see Problem 3-4), sufficiency is readily established, and we need only establish (necessity) that any two minimal realizations of $T(s)$ are equivalent. We do this constructively by first noting, in light of Theorem 4.4.15, that both realizations must have the same order n. In view of 4.4.17, it also follows that $CA^kB = \hat{C}\hat{A}^k\hat{B}$ for $k = 0,1,2,\ldots$, or that

$$\mathcal{O}\mathcal{C} = \hat{\mathcal{O}}\ \hat{\mathcal{C}}, \tag{4.4.22}$$

where $\mathcal{O}$ and $\mathcal{C}$ ($\hat{\mathcal{O}}$ and $\hat{\mathcal{C}}$) represent the observability and controllability matrices respectively, of the state representation $\{A,B,C,E\}(\{\hat{A},\hat{B},\hat{C},E\})$. Furthermore, in view of Theorem 4.4.15,

$\rho[\mathscr{C}] = \rho[\mathscr{O}] = \rho[\hat{\mathscr{C}}] = \rho[\hat{\mathscr{O}}] = n$, the dimension of the state matrix A or $\hat{A}$. Since the $(pn \times n)$ matrix $\hat{\mathscr{O}}$ has full rank n, it follows by Sylvester's Inequality that the $(n \times n)$ matrix $\hat{\mathscr{O}}^T\hat{\mathscr{O}}$ is nonsingular. Therefore, from 4.4.22, we can now derive the following relation:

$$Q\mathscr{C} = [\hat{\mathscr{O}}^T\hat{\mathscr{O}}]^{-1}\hat{\mathscr{O}}^T\mathscr{O}\mathscr{C} = \hat{\mathscr{C}}, \tag{4.4.23}$$

where

$$Q = [\hat{\mathscr{O}}^T\hat{\mathscr{O}}]^{-1}\hat{\mathscr{O}}^T\mathscr{O}. \tag{4.4.24}$$

Since both $\mathscr{C}$ and $\hat{\mathscr{C}}$ have full rank n, the $(n \times n)$ matrix Q, thus defined, is nonsingular and therefore qualifies as an equivalence transformation. From 4.4.23, we now obtain the relation:

$$Q = \hat{\mathscr{C}}\mathscr{C}^T[\mathscr{C}\mathscr{C}^T]^{-1}, \tag{4.4.25}$$

or, in view of 4.4.22, that

$$\mathscr{O} = \hat{\mathscr{O}}Q \tag{4.4.26}$$

Finally, as a direct consequence of 4.4.22, $\mathscr{O}A\mathscr{C} = \hat{\mathscr{O}}\hat{A}\hat{\mathscr{C}}$ which, in view of 4.4.24 and 4.4.25, directly implies that

$$QA = \hat{A}Q \tag{4.4.27}$$

Equations 4.4.23, 4.4.26, and 4.4.27 immediately yield the desired relations; i.e. $QB = \hat{B}$, $CQ^{-1} = \hat{C}$, and $QAQ^{-1} = \hat{A}$, and thus constructively establish the theorem.

EXAMPLE 4.4.28: To illustrate the constructive proof of Theorem 4.4.21, we now consider the two minimal realizations of the same transfer matrix as derived in Examples 4.4.11 and 4.4.14. In particular, we have shown that both the controllable companion form state representation (Example 4.4.11):

$$A = \begin{bmatrix} 0 & 1 & 0 \\ 2 & 1 & 0 \\ 0 & 0 & 2 \end{bmatrix}, \quad B = \begin{bmatrix} 0 & 0 \\ 1 & 0 \\ 0 & 1 \end{bmatrix}, \quad C = \begin{bmatrix} -2 & 1 & 2 \\ 0 & 0 & 0 \\ 2 & 2 & 0 \end{bmatrix}, \text{ and } \quad E = \begin{bmatrix} 0 & 1 \\ 2 & 0 \\ 0 & 1 \end{bmatrix}$$ and the

observable companion form state representation (Example 4.4.14):

$$\hat{A} = \begin{bmatrix} 0 & 2 & 0 \\ 1 & 1 & 0 \\ 0 & 0 & 2 \end{bmatrix}, \quad \hat{B} = \begin{bmatrix} -2 & 2 \\ 1 & 2 \\ 2 & 0 \end{bmatrix}, \quad \hat{C} = \begin{bmatrix} 0 & 1 & 0 \\ 0 & 0 & 0 \\ 0 & 0 & 1 \end{bmatrix}, \text{ and } E = \begin{bmatrix} 0 & 1 \\ 2 & 0 \\ 0 & 1 \end{bmatrix}$$ realize

the same $T(s)$. Furthermore, they are both minimal realizations of $T(s)$, since the pair $\{A,C\}$ is observable and the pair $\{\hat{A},\hat{B}\}$ is controllable. By Theorem 4.4.21, these two realizations are therefore equivalent, and the equivalence transformation Q which relates them is given by either 4.4.24 or 4.4.25. Using 4.4.25 for example, we determine that $\mathscr{C} = \begin{bmatrix} 0 & 0 & 1 & 0 & 1 & 0 \\ 1 & 0 & 1 & 0 & 3 & 0 \\ 0 & 1 & 0 & 2 & 0 & 4 \end{bmatrix}$ and, therefore, that $\mathscr{C}\mathscr{C}^T =$

$\begin{bmatrix} 2 & 4 & 0 \\ 4 & 11 & 0 \\ 0 & 0 & 21 \end{bmatrix}$. Furthermore, $\hat{\mathscr{C}} = \begin{bmatrix} -2 & 2 & 2 & 4 & -2 & 8 \\ 1 & 2 & -1 & 4 & 1 & 8 \\ 2 & 0 & 4 & 0 & 8 & 0 \end{bmatrix}$ and

$\hat{\mathscr{C}}\mathscr{C}^T = \begin{bmatrix} 0 & -6 & 42 \\ 0 & 3 & 42 \\ 12 & 30 & 0 \end{bmatrix}$. Since $[\mathscr{C}\mathscr{C}^T]^{-1} = \begin{bmatrix} \frac{11}{6} & -\frac{2}{3} & 0 \\ -\frac{2}{3} & \frac{1}{3} & 0 \\ 0 & 0 & \frac{1}{21} \end{bmatrix}$, the product $\hat{\mathscr{C}}\mathscr{C}^T[\mathscr{C}\mathscr{C}^T]^{-1} = Q = \begin{bmatrix} 4 & -2 & 2 \\ -2 & 1 & 2 \\ 2 & 2 & 0 \end{bmatrix}$ with

$Q^{-1} = \begin{bmatrix} 2 & -2 & 3 \\ -2 & 2 & 6 \\ 3 & 6 & 0 \end{bmatrix} \div 18$. The reader can now verify that $QAQ^{-1} = \hat{A}$, $QB = \hat{B}$, and $CQ^{-1} = \hat{C}$, which thus establishes Q as the appropriate equivalence transformation.

4.5 CONCLUDING REMARKS AND REFERENCES

We have now formally introduced the notion of a transfer matrix of a linear, time-invariant, multivariable system and discussed certain aspects of this very common method for representing the dynamical behavior of linear systems. In particular, we have shown that the transfer matrix represents only the external or input/output behavior of a system and is, consequently, a more general system representation than a state representation; i.e. while the state representation {A,B,C,E} yields a unique T(s) via 4.2.2, the converse does not hold, and any proper, rational T(s) can be realized by any number of state representations, including those of differing state dimension. We have restricted our discussion in this chapter to proper transfer matrices, since this is the only class which arises from the state representation {A,B,C,E}. This restriction will be dropped in the next chapter.

The structure theorem, which first appeared in [W4], was introduced in Section 4.3 and employed to directly obtain the transfer matrix of a system in either controllable or observable companion form. The procedure was then generalized to include systems which are not completely state controllable or observable. We showed that in this latter case, all uncontrollable and unobservable modes of the system appear as cancellable pole-zero factors in the transfer matrix T(s), a result which has long been recognized in the scalar case [G2]. The application of the structure theorem to realization theory first appeared in [W4], although other comparable methods for obtaining realizations and minimal realizations do exist [G2][H2][M2][B1]. It might be noted that an alternate method for obtaining minimal realizations based on the structure theorem, but employing frequency domain reduction rather than time domain reduction, will be presented in the next chapter.

We finally note that the fundamental relationship (Theorem 4.4.15) linking minimality of a transfer matrix realization to system controllability and observability was first established by Kalman [K5] and Youla and Tissi [Y1]. However, the particular approach employed here to establish this result, along with the constructive procedure (Theorem 4.4.21) for relating two equivalent minimal realizations of the same transfer matrix, was motivated by a similar development in Chen [C1].

PROBLEMS - CHAPTER 4

4-1 Does the system $\{A,B,C,E\}$, where $A = \begin{bmatrix} 1 & -1 \\ 2 & 0 \end{bmatrix}$; $B = \begin{bmatrix} 1 & 2 \\ 1 & 0 \end{bmatrix}$;

$C = [1,\ 0]$; and $E = [2\ \ 0]$ realize the transfer matrix

$$T(s) = \frac{[2s^2-s+3, \quad 2s]}{s^2-s+2}\ ?$$

4-2 Verify by direct substitution that Equation 4.3.6 does hold.

4-3 Establish "Theorem" 4.3.17 directly without employing duality.

4-4 Find a fourth order controllable realization of the transfer function $T(s)$ employed in Example 4.4.3. Is the realization observable?

4-5 If the McMillan degree of a proper rational transfer matrix $T(s)$ is defined as the order, n, of a realization of $T(s)$ which has the property that no lower order realization of $T(s)$ exists, show that the McMillan degree of $T(s)$ is equivalent to the order of a minimal realization of $T(s)$.

4-6 Show that the equivalence transformation Q employed in Theorem 4.4.21 is unique and can be found directly (and perhaps most easily) via either 4.4.23 or 4.4.26.

4-7 Employ the controllable version of the structure theorem to obtain the transfer matrix of the system defined in Problem 3-15 if $C = I$ and $E = 0$.

4-8 Employ the observable version of the structure theorem to obtain the transfer matrix of the system defined in Problem 3-16 if $B = I$ and $E = 0$.

4-9 How might the controllable (observable) version of the structure theorem be modified when $\{A,B\}$ is controllable ($\{C,A\}$ is observable) but $\rho[B] < m$ ($\rho[C] < p$)?

4-10 Consider the transfer function $T(s) = \dfrac{2s-4}{s^3-7s+6}$. Find (a) a controllable realization of $T(s)$, (b) an observable realization of $T(s)$, and (c) a minimal realization of $T(s)$.

4-11 Find an observable realization of the transfer matrix

$$T(s) = \begin{bmatrix} \frac{s-1}{s+1} & \frac{1}{s^2-1} \\ 1 & 0 \end{bmatrix}$$

using the dual of the algorithm detailed in Section 4.4. Is the realization minimal?

4-12 Consider the transfer matrix $T(s) = \left[\frac{s-1}{s}, \ \frac{1}{s^2-1}\right]$. Find a realization of $T(s)$ in controllable companion form such that $|sI-A_c| = s^3-s$. Find a realization of $T(s)$ in observable companion form such that $|sI-A_o| = s^3-s$. Verify that both realizations are minimal and determine the unique equivalence transformation Q which relates them.

4-13 Find a minimal realization of $T(s) = \begin{bmatrix} 1 & \frac{1}{s} & \frac{s-1}{s} \\ 0 & \frac{s+1}{s^2} & 0 \end{bmatrix}$.

4-14 Determine the transfer matrix of the system defined in Problem 3-18(b) if $C = [1 \ \ 0 \ \ 1 \ \ 0]$ and $E = [1 \ \ -1]$ via the structure theorem, and verify that all uncontrollable modes of the system do appear as cancellable pole-zero factors in $T(s)$.

4-15 Show that if $\{A,b,c,e\}$ realizes $T(s) = \dfrac{1}{s^2+4s-2}$ and $|sI-A| = s^2+4s-2$, then, $\{A,b\}$ must be controllable and $\{A,c\}$ must be observable.

4-16 Faddeev has shown [F2] that given any $n \times n$ constant matrix A, the characteristic polynomial $\Delta(s) = s^n + a_{n-1}s^{n-1} + \ldots + a_1 s + a_o$ of A, and the adjoint of $(sI-A)$; i.e. $(sI-A)^+ = Is^{n-1} + \tilde{A}_1 s^{n-2} + \ldots + \tilde{A}_{n-1}$ can be found via the following recursive relations which define Faddeev's Algorithm:

Let $A_1 \triangleq A$, then $a_{n-1} = -\text{trace } A \triangleq -\sum_1^n a_{ii}$, and $\tilde{A}_1 = A_1 + a_{n-1}I$

Furthermore,

$$A_2 = A\tilde{A}_1 \qquad a_{n-2} = -\frac{1}{2}\text{ trace } A_2 \qquad \tilde{A}_2 = A_2 + a_{n-2}I$$

$$\vdots \qquad\qquad \vdots \qquad\qquad \vdots$$

$$A_{n-1} = A\tilde{A}_{n-2} \qquad a_1 = -\frac{1}{n-1}\text{ trace } A_{n-1} \qquad \tilde{A}_{n-1} = A_{n-1} + a_1 I$$

$$A_n = A\tilde{A}_{n-1} \qquad a_o = -\frac{1}{n}\text{ trace } A_n \qquad \tilde{A}_n = A_n + a_o I = 0$$

For each of the following A matrices, determine $(sI-A)^{-1}$ in the "usual" manner and then verify that Faddeev's Algorithm does hold.

(a) $A = \begin{bmatrix} 2 & -1 \\ 0 & -1 \end{bmatrix}$ (b) $A = \begin{bmatrix} 0 & 1 & 0 \\ 0 & 0 & 1 \\ 0 & -2 & -3 \end{bmatrix}$ (c) $A = \begin{bmatrix} -1 & 0 & 0 & 0 \\ 0 & -2 & 0 & 0 \\ 0 & 0 & 0 & 3 \\ 0 & 0 & -3 & 0 \end{bmatrix}$

4-17 If $T^T(s) = T(s)$, what can be said regarding the dual $\{A_c^T, C_c^T, B_c^T, E_c^T\}$ of any controllable realization $\{A_c, B_c, C_c, E_c\}$ of $T(s)$?

4-18 Show that if $\{A,B,C,E\}$ is a realization of $T(s) = \left[\frac{r_{ij}(s)}{p_{ij}(s)}\right]$, where $r_{ij}(s)$ and $p_{ij}(s)$ are relatively prime for all i,j, and $|sI-A| = \Delta(s)$ is the least common (monic) multiple of all of the $p_{ij}(s)$, then $\{A,B,C,E\}$ is a minimal realization of $T(s)$. Verify, by example, that one cannot always find a realization of $T(s)$ which satisfies this property.

4-19 If $\{A,B,C,E\}$ is a minimal realization of $T(s)$ of order n and $\{\hat{A},\hat{B},\hat{C},\hat{E}\}$ is a realization of $T(s)$ of the same order, prove that $\{\hat{A},\hat{B},\hat{C},\hat{E}\}$ is also a minimal realization of $T(s)$.

4-20 By applying the polynomial division algorithm (Section 2.5) to each entry, $t_{ij}(s) = \frac{r_{ij}(s)}{p_{ij}(s)}$, of a rational transfer matrix $T(s)$; i.e. by (uniquely) expressing each $\frac{r_{ij}(s)}{p_{ij}(s)}$ as $e_{ij}(s) + \frac{\hat{r}_{ij}(s)}{p_{ij}(s)}$ where $\partial[\hat{r}_{ij}(s)] < \partial[p_{ij}(s)]$, show that $T(s)$ can be (uniquely) represented as the sum of its <u>strictly proper part</u>, namely $\hat{T}(s) = \frac{\hat{r}_{ij}(s)}{p_{ij}(s)}$, and its polynomial matrix <u>quotient</u>, $E(s) = [e_{ij}(s)]$; i.e. $T(s) = \hat{T}(s) + E(s)$ where $E(s)$ represents that (unique) portion of $T(s)$ which does not vanish in the limit as $s \to \infty$. Verify that if $T(s) = \begin{bmatrix} \frac{s^2+2}{s}, & 1 \\ \frac{3s}{s^2+2}, & \frac{2s^2+2s-1}{s+1} \end{bmatrix}$ then $\hat{T}(s) = \begin{bmatrix} \frac{2}{s}, & 0 \\ \frac{3s}{s^2+2}, & \frac{-1}{s+1} \end{bmatrix}$ and $E(s) = \begin{bmatrix} s & 1 \\ 0 & 2s \end{bmatrix}$.

CHAPTER 5
DIFFERENTIAL OPERATOR REPRESENTATIONS

5.1 INTRODUCTION

The primary goal of this chapter is to present an in-depth study of the third common method for representing the dynamical behavior of linear systems, namely the differential operator representation. We begin by considering the various transfer relations between any two of the three primary methods used to describe dynamical system behavior, noting that certain of the (six) relations have already been given in the earlier chapters. In particular, in Section 5.2, we mathematically define a notion of "equivalence" between state-space systems and the more general class of differential operator systems. We note that this notion of equivalence implies a number of desirable necessary conditions for equivalence. For example, system order and mode or pole locations are preserved as well as the ability to set initial conditions on either equivalent system. We further note that our definition of equivalence reduces to the standard definition 3.4.4 when both systems are in state-space form. We then present an algorithm for obtaining equivalent state-space representations of any systems whose dynamics are expressed in the more general differential operator form. Paramount to the development are the notions of row proper and column proper polynomial matrices which were introduced in Section 2.5, and the structure theorem which was introduced in Section 4.3.

In Section 5.3, we extend the concepts of controllable and observable state-space systems, in a rather natural way, to include systems whose dynamics are expressed in the more general differential operator form. The notion of a greatest common right (and left) divisor of two polynomial matrices is then employed to reduce a given differential operator representation to minimal form.

The results given in Section 5.3 are utilized in Section 5.4 in order to outline a new technique for obtaining minimal realizations of proper transfer matrices based on an initial frequency domain reduction. The notions of "right and left inverse systems" are then introduced in Section 5.5, and related to the intuitive, analogous concepts of "output function controllability" and "input function observability". Certain alternative procedures for constructing inverse systems are outlined and discussed from a practical point of view. Some pertinent background material is then presented in the final section along with a number of concluding remarks.

5.2 TRANSFER AND EQUIVALENCE RELATIONS

As stated in the introduction to Chapter 3, the most general class of dynamical systems which will be considered in this text are those whose dynamical behavior can be represented by a set of linear, time-invariant, ordinary differential equations in the differential operator form 3.1.1; i.e.

$$P(D)z(t) = Q(D)u(t); \qquad 5.2.1a\ (3.1.1a)$$

$$y(t) = R(D)z(t) + W(D)u(t), \qquad 5.2.1b\ (3.1.1b)$$

where $u(t)$ and $y(t)$ are the m-dimensional input and p-dimensional output, respectively, $z(t)$ is the q-dimensional partial state of the system, and $P(D)$, $Q(D)$, $R(D)$, and $W(D)$ are $q \times q$, $q \times m$, $p \times q$, and $p \times m$ polynomial matrices, respectively, in the differential operator $D = \frac{d}{dt}$, with $P(D)$ assumed nonsingular. One of our first goals in this chapter will be to develop the various transfer relationships between this differential operator representation for modelling the dynamical behavior of physical systems and the other two methods (the state representation and the transfer matrix) for representing linear systems. This goal is in accord with our previously stated objective of displaying the similarities between various system representations in order to provide the control engineer with a choice of analysis and design procedures.

We note that we have already presented the (two) transfer relations between the state representation and the transfer matrix in the previous chapter. We further recall (from Section 3.2) that the state representation 3.2.1, i.e. $\dot{x}(t) = Ax(t) + Bu(t)$; $y(t) = Cx(t) + Eu(t)$, merely defines a subclass of the more general differential operator representation 5.2.1 with $\{DI-A,B,C,E\} = \{P(D),Q(D),R(D),W(D)\}$, a fact which immediately resolves the transfer from the state-space to the differential operator representation.

We can also readily resolve the transfer from the differential operator representation 5.2.1 to the transfer matrix $T(s)$ of the system. In particular, if we assume zero initial conditions of the partial state $z(t)$ and all its derivatives, and then take the Laplace transform of 5.2.1, we immediately obtain:

$$P(s)z(s) = Q(s)u(s); \qquad 5.2.2a$$

$$y(s) = R(s)z(s) + W(s)u(s) \qquad 5.2.2b$$

Since $|P(s)| \neq 0$, we further note that

$$z(s) = P^{-1}(s)Q(s)u(s) \qquad 5.2.3$$

Substituting 5.2.3 for $z(s)$ in 5.2.2b, we immediately obtain an expression for the transfer matrix $T(s)$ of the system in terms of the differential operator representation $\{P(D),Q(D),R(D),W(D)\}$ (actually $\{P(s),Q(s),R(s),W(s)\}$); i.e.

$$y(s) = [R(s)P^{-1}(s)Q(s) + W(s)]u(s) = T(s)u(s), \qquad 5.2.4$$

where

$$T(s) = R(s)P^{-1}(s)Q(s) + W(s) \qquad 5.2.5$$

The relation 5.2.5 represents the desired transfer relation from the differential operator representation to the transfer matrix. In light of the above, it is clear that the differential operator D and the Laplace operator s can and will be interchanged freely whenever initial conditions are irrelevant to the analysis. It might be noted that the above expression (5.2.5) for $T(s)$ does not necessarily imply a proper transfer matrix as in the case of the state representation 3.2.1, where $T(s) = C(sI-A)^{-1}B + E$ is always proper.

In view of the preceding, we need only establish the (two) transfer relations from the differential operator representation to the state representation, and from the transfer matrix to the differential operator representation in order to complete the set of (six) transfer relations.

We therefore begin by considering the first of these unresolved transfer questions; i.e. given any differential operator representation 5.2.1, find an "equivalent" representation, sometimes called a <u>normal form</u> representation, of the state form:

$$\dot{x}(t) = Ax(t) + Bu(t); \qquad 5.2.6a$$

$$y(t) = Cx(t) + E(D)u(t) \qquad 5.2.6b$$

It might be noted that 5.2.6 is slightly more general than 3.2.1 in that $E(D)$ is no longer restricted to be constant, but is allowed to be a $(p \times m)$ polynomial matrix in the differential operator D. We further note that the transfer matrix $T(s)$ associated with 5.2.6 is given by

$$T(s) = C(sI-A)^{-1}B + E(s), \qquad 5.2.7$$

where $C(sI-A)^{-1}B$ is the strictly proper part of $T(s)$ and $E(s)$ is its quotient (see Problem 4-20). It might be recalled that if $T(s)$ is proper,

$$E(s) = E = \lim_{s \to \infty} T(s) \qquad 5.2.8$$

and, in general, that $E(s)$ represents that (unique) portion of $T(s)$ which does not vanish in the lim as $s \to \infty$. We will soon show that allowing $E(D)$ in 5.2.6b to be a polynomial matrix permits us to equate systems with rational transfer matrices which are not necessarily proper--such as those which can arise (via 5.2.5) from the differential operator representation 5.2.1, to the more general state representation 5.2.6. In order to establish any result with respect to this question, however, we must first define "equivalence" of systems whose dynamics can be represented by either 5.2.1 or 5.2.6.

DEFINITION 5.2.9: We will say that systems 5.2.1 and 5.2.6 are <u>equivalent</u> if and only if the following two conditions hold:

(i) Let $u(t)$ be any known measurable input defined on $[t_o,\infty)$, and $z(t)$ be the solution of 5.2.1a corresponding to this input for known initial conditions on $z(t)$ and its derivatives. Then $z(t)$ can also be expressed as:

$$z(t) = C_o x(t) + H(D)u(t), \qquad 5.2.10$$

for some $q \times n$ real constant matrix, C_o, and some $q \times m$ polynomial matrix $H(D)$, where $z(t)$ is the solution of 5.2.6a and 5.2.10 corresponding to the same input, $u(t)$, and <u>unique</u> initial state, $x(t_o)$. Conversely, if $z(t)$ is the solution of 5.2.6a and 5.2.10 corresponding to some known $x(t_o)$ and $u(t)$ for $t \varepsilon [t_o,\infty)$, then $z(t)$ also satisfies 5.2.1a for the same input, $u(t)$, and some appropriate <u>unique</u> set of initial conditions.

(ii) $y(t)$ in 5.2.6b "satisfies" 5.2.1b, 5.2.6a, and 5.2.10; i.e. if condition (i) holds, then substituting 5.2.10 for $z(t)$ in 5.2.1b yields:

$$y(t) = R(D)[C_o x(t) + H(D)u(t)] + W(D)u(t), \qquad 5.2.11$$

and evaluating 5.2.11 in light of 5.2.6a, we obtain:

$$y(t) = \bar{C}x(t) + \bar{E}(D)u(t) \qquad 5.2.12$$

for some $p \times n$ real constant matrix, $\bar{C}$, and some $p \times m$ polynomial matrix $\bar{E}(D)$. For the two systems to be equivalent, we will require that $\bar{C} = C$ for some $\bar{C}$ obtained via 5.2.11, and that $\bar{E}(D) = E(D)$, where C and $E(D)$ are given in 5.2.6b. It is important to note that since the degree of the determinant of $P(D)$ represents the number of independent initial conditions that can be set on $z(t)$ and its derivatives, $\partial[|P(D)|]$ must equal n, the order of any equivalent state-space system of the form 5.2.6.

The above definition of equivalence directly implies a number of desirable necessary conditions for equivalence, which will now be enumerated. In particular,

5.2.13a: <u>Partial-state/input transfer matrix equivalence</u>; i.e. the fact that $C_o(sI-A)^{-1}B + H(s) = P(s)^{-1}Q(s)$, where $|sI-A| = \alpha|P(s)|$ for some nonzero scalar α, follows directly from the fact that $\partial[|P(D)|] = n$ and condition (i) of the equivalence definition by simply ignoring all initial conditions while equating the Laplace transform of $z(t)$ in terms of $u(s)$ in 5.2.6a and 5.2.10 to the equivalent Laplace transformed expression in 5.2.1a. <u>Input-output transfer matrix equivalence</u>; i.e. the fact that $C(sI-A)^{-1}B + E(s) = R(s)P(s)^{-1}Q(s) + W(s)$ then follows directly from equivalence condition (ii) by equating the same output to input Laplace transformed expressions. Since $|sI-A| = \alpha|P(s)|$, it is seen that <u>all (n) modes or poles of the differential operator representation 5.2.1 are preserved by the equivalent state-space system 5.2.6</u>.

5.2.13b: <u>Complete observability of the system defined by 5.2.6a and 5.2.10</u> follows directly from the uniqueness of $x(t_o)$. More specifically, if the pair $\{C_o, A\}$ was not completely observable, it would be possible to obtain the same time function, $z(t)$, as the solution to 5.2.6a and 5.2.10, for more than one (unique) value of the initial state, $x(t_o)$; i.e. the completely unobservable modes of 5.2.10 would not affect $z(t)$. This cannot be the case if equivalence condition (i) is to hold, since $x(t_o)$ is assumed to be unique.

5.2.13c: <u>$x(t) = L(D)z(t) + M(D)u(t)$</u> for some pair, $\{L(D), M(D)\}$, of polynomial matrices. This relationship follows directly from (5.2.13b), the observability of the pair $\{C_o, A\}$, since by repeated differentiation of 5.2.10, it is clear that the entire state vector, $x(t)$, can be reconstructed from $z(t)$, $u(t)$ and their derivatives.

5.2.13d: If 5.2.1a is in state-space form, i.e. if $P(D) = DI-\hat{A}$, and $Q(D) = \hat{B}$, then $q = n$, C_o is nonsingular, and $H(D) = 0$. In particular, if we substitute 5.2.13c for $x(t)$ in 5.2.10, we find that $z(t) = C_oL(D)z(t) + [C_oM(D) + H(D)]u(t)$, which directly implies that $L(D) = C_o^{-1}$ and $C_oM(D) = -H(D)$. Furthermore, since both systems are in state form, it also follows that $H(D) = M(D) = 0$ (see Problem 5-16). We therefore conclude that <u>this definition of equivalence (5.2.9) "reduces" to the standard equivalence definition (3.4.3) when both systems are in state form</u>.

In view of these relationships, and 5.2.13d in particular, we can now define equivalence of any two systems in differential operator form; i.e. two systems of the form 5.2.1 are <u>equivalent</u> if and only if they are both equivalent to the same state-space system.

Now that equivalence has been formally defined, the question of transfer from the differential operator representation 5.2.1 to an equivalent state representation of the form 5.2.6 can be constructively resolved.

THEOREM 5.2.14: <u>Any differential operator system of the form 5.2.1</u>

<u>has an equivalent state-space representation of the form 5.2.6.</u>

<u>Proof</u>: The proof is constructive and will be presented in the form of a five step algorithm. Briefly, we will reduce the given system to a special equivalent differential operator form (Steps 1 through 3) which will then enable us to apply the observable version of the structure theorem to directly obtain an equivalent, observable state-space representation of the form 5.2.6 (Step 4). Equivalence condition (ii) will then be satisfied by employing 5.2.11 to determine an appropriate C and $E(D)$ (Step 5).

<u>The Algorithm</u>

<u>Step 1</u>: If $P(D)$ is row proper (see Section 2.5), this step can be omitted. If $P(D)$ is not row proper, we premultiply 5.2.1a by any unimodular matrix, $U_L(D)$, which reduces $P(D)$ to row proper form. An appropriate algorithm is given in Section 2.5. It follows directly from the uniqueness of solutions to linear differential equations that premultiplication of 5.2.1a by a unimodular matrix has no effect whatsoever on the solution, $z(t)$, provided the same (n) initial conditions are set on $z(t)$ and its derivatives, and the same input, $u(t)$, is applied. Consequently, this step does not affect equivalence in any way; i.e. the system

$$U_L(D)P(D)z(t) = U_L(D)Q(D)u(t); \tag{5.2.15a}$$

$$y(t) = R(D)z(t) + W(D)u(t), \tag{5.2.15b}$$

is equivalent to 5.2.1.

<u>Step 2</u>: Let

$$z_o(t) = \Gamma_r z(t) \tag{5.2.16}$$

where Γ_r is the $q \times q$ nonsingular constant matrix consisting of the coefficients of the highest degree D terms in each row of $U_L(D)P(D)$, as defined in Section 2.5. If $\Gamma_r = I$ this step can be omitted. If not, we substitute $\Gamma_r^{-1} z_o(t)$ for $z(t)$ in 5.2.15 to obtain:

$$P_o(D)z_o(t) = Q_o(D)u(t); \qquad 5.2.17a$$

$$y(t) = R_o(D)z_o(t) + W(D)u(t), \qquad 5.2.17b$$

where $P_o(D) = U_L(D)P(D)\Gamma_r^{-1}$, $Q_o(D) = U_L(D)Q(D)$, and $R_o(D) = R(D)\Gamma_r^{-1}$. It is clear that this step, which represents a coordinate transformation, also does not affect equivalence and, therefore, that systems 5.2.1, 5.2.15, and 5.2.17 are equivalent to one another. We now note that the matrix $P_o(D)$ is in a particularly useful form; i.e.

$$P_o(D) = U_L(D)P(D)\Gamma_r^{-1} = \begin{bmatrix} D^{\bar{d}_1}+\ldots, & \ldots, & \ldots, & \ldots \\ \ldots, & D^{\bar{d}_2}+\ldots, & \ldots, & \ldots \\ \vdots & \vdots & \ddots & \vdots \\ \ldots, & \ldots, & \ldots, & D^{\bar{d}_q}+\ldots \end{bmatrix} \qquad 5.2.18$$

where the ... denotes polynomials of lower degree than $\bar{d}_k$ in each (k-th) row of $P_o(D)$. We assume at this point that $\bar{d}_k > 0$ for all $k = 1,2,\ldots,q$.[†]

Step 3: We now consider the partial-state/input transfer matrix of 5.2.17a; i.e.

$$z_o(s) = P_o^{-1}(s)Q_o(s)u(s) \qquad 5.2.19$$

If $P_o^{-1}(s)Q_o(s)$ is strictly proper (if $\partial_r[Q_o(s)] < \partial_r[P_o(s)]$), this step can be omitted. If not, we write 5.2.19 as:

[†]This is almost always the case in practice. If, however, $d_k = 0$ for any k, then the corresponding partial state element $z_{ok}(t)$, of $z_o(t)$ represents what we will term a non-essential state; i.e. $z_{ok}(t)$ will be a function of only $u(t)$ and its derivatives and can easily be "eliminated" from the dynamical equations of the system. Example 5.2.32 and Problem 5-7 serve to illustrate this point.

$$z_o(s) = [P_o^{-1}(s)\bar{Q}_o(s) + H_o(s)]u(s), \qquad 5.2.20$$

where $P_o^{-1}(s)\bar{Q}_o$ represents the strictly proper part of $P_o^{-1}(s)Q_o(s)$ and $H_o(s)$ its quotient (see Problem 4-20). In view of 5.2.20 it now follows that $\bar{Q}_o(s) = Q_o(s) - P_o(s)H_o(s)$ and if we define

$$\bar{z}_o(t) = z_o(t) - H_o(D)u(t), \qquad 5.2.21$$

we can write 5.2.17 as

$$P_o(D)\bar{z}_o(t) = \bar{Q}_o(D)u(t); \qquad 5.2.22a$$

$$\begin{aligned} y(t) &= R_o(D)[\bar{z}_o(t) + H_o(D)u(t)] + W(D)u(t) \\ &= R_o(D)\bar{z}_o(t) + \bar{W}(D)u(t), \end{aligned} \qquad 5.2.22b$$

where $\bar{W}(D) = R_o(D)H_o(D) + W(D)$. We further note that this "new" system representation is equivalent to 5.2.17 (see Problem 5-9) and, therefore, that systems 5.2.1, 5.2.15, 5.2.17, and 5.2.22 are equivalent to one another.

Step 4: Once $P_o(D)$ and $\bar{Q}_o(D)$ have been determined, we can directly obtain an observable realization $\{A_o, B_o, \bar{C}_o\}^\dagger$ of the strictly proper transfer matrix, $P_o^{-1}(s)\bar{Q}_o(s)$, of the system 5.5.22a via the observable version of the structure theorem (4.3.17). Furthermore, when compared to the differential operator representation 5.2.22a, the observable realization will satisfy condition (i) of equivalence definition 5.2.9. In particular, we write the $q \times q$ matrix $P_o(s)$ as:

† It should be noted that whenever we consider a strictly proper transfer matrix T(s), the matrix E can be omitted from any state-space realization {A,B,C,E} of T(s) for convenience, and we can meaningfully speak of the triple {A,B,C} which realizes T(s).

$$P_o(s) = \begin{bmatrix} s^{\bar{d}_1} & & & 0 \\ & s^{\bar{d}_2} & & \\ & & \ddots & \\ 0 & & & s^{\bar{d}_q} \end{bmatrix} - \bar{S}(s)\hat{A}_q \qquad 5.2.23$$

where $\bar{d}_k \geq 1$ for all $k = 1,2,\ldots q$, $\hat{A}_q$ is a $(n \times q)$ constant real matrix, and $\bar{S}(s)$ is the following $(q \times n)$ matrix of single term monic polynomials in s; i.e.

$$\bar{S}(s) = \begin{bmatrix} 1 & s \ldots s^{\bar{d}_1-1} & 0 & 0 \ldots 0 & 0 \ldots & & 0 \\ 0 & 0 \ldots 0 & 1 & s \ldots s^{\bar{d}_2-1} & 0 \ldots & & 0 \\ \vdots & \vdots \quad \vdots & \vdots & \vdots & \vdots & & \vdots \\ 0 & 0 \ldots 0 & 0 & \ldots 0 & 0 \ldots 1 & s \ldots & s^{\bar{d}_q-1} \end{bmatrix} \qquad 5.2.24$$

Since $P_o^{-1}(s)\bar{Q}_o(s)$ is a strictly proper transfer matrix, it follows by 4.3.24 (with strict inequality) that $\bar{Q}_o(s)$ can be written as

$$\bar{Q}_o(s) = \bar{S}(s)\hat{B} \qquad 5.2.25$$

for some $(n \times m)$ constant real matrix $\hat{B}$. In view of the observable version of the structure theorem, and equations 4.3.16 through 4.3.19 in particular, we now note that $P_o(s)$, $\bar{Q}_o(s)$, and I_q can be directly equated to $\bar{\delta}(s)$, $\bar{S}(s)\hat{B}$, and $\hat{C}_p$ respectively. More specifically an n-th order observable realization, $\{A_o, B_o, \bar{C}_o\}$, of $P_o^{-1}(s)\bar{Q}_o(s)$ can be obtained as follows: Define $\bar{\sigma}_k = \sum_{i=1}^{k} \bar{d}_i$ for $k = 1,2,\ldots q$. Replace the $(q)\bar{\sigma}_k$ columns of the $(n \times n)$ matrix $\begin{bmatrix} 0 \ldots 0 \\ I_{n-1} \quad \vdots \\ \quad 0 \end{bmatrix}$ by the (q) ordered columns of $\hat{A}_q$. The resulting matrix is an appropriate A_o.

Let $B_o = \hat{B}$ as given by 5.2.25. An appropriate $\overline{C}_o$ corresponding to the A_o and B_o thus selected is the matrix obtained by replacing the $(q)\overline{\sigma}_k$ columns of the $(q \times n)$ null matrix by the (q) ordered columns of I_q. We thus obtain an observable realization $\{A_o, B_o, \overline{C}_o\}$ of $P_o^{-1}(s)\overline{Q}_o(s)$ which, when compared to the differential operator representation 5.2.22a, clearly satisfies condition (i) of equivalence definition 5.2.9. It therefore follows that the following state representations:

$$\dot{x}_o(t) = A_o x_o(t) + B_o u(t); \tag{5.2.26}$$

with

$$\overline{z}_o(t) = \overline{C}_o x_o(t), \tag{5.2.27a}$$

or in view of 5.2.21, with

$$z_o(t) = \overline{C}_o x_o(t) + H_o(D)u(t), \tag{5.2.27b}$$

or by 5.2.16, with

$$z(t) = \Gamma_r^{-1} z_o(t) = C_o x_o(t) + H(D)u(t), \tag{5.2.27c}$$

where $C_o = \Gamma_r^{-1}\overline{C}_o$ and $H(D) = \Gamma_r^{-1}H_o(D)$ are equivalent to the differential operator systems 5.2.22a, 5.2.17a, and 5.2.1a respectively. It is also clear that

$$C_o(sI-A_o)^{-1}B_o + H(s) = P^{-1}(s)Q(s), \tag{5.2.28}$$

where $|sI-A_o| = |P_o(s)| = |U_L(s)| \times |\Gamma_r^{-1}| \times |P(s)| = \alpha|P(s)|$, and that 5.2.26 and 5.2.27c thus represent a state-space system of the form 5.2.6a and 5.2.10 which is equivalent to the given differential operator representation 5.2.1a.

<u>Step 5</u>: Equation 5.2.11 can now be used to find an appropriate C and $E(D)$ and thus properly define $y(t)$ in accordance with condition (ii) of the equivalence definition. In particular, if we substitute $z(t)$, as given by 5.2.27c, in 5.2.1b we obtain the relation:

$$y(t) = R(D)C_o x_o(t) + [R(D)H(D) + W(D)]u(t) \tag{5.2.29}$$

By now employing 5.2.26 to eliminate any derivatives of $x_o(t)$, it follows that

$$y(t) = Cx_o(t) + E(D)u(t) \tag{5.2.30}$$

for some C and $E(D)$--the remaining two members of the equivalent quadruple $\{A_o, B_o, C, E(D)\}$. The subscript o, which implies observability of the equivalent system given by 5.2.26 and 5.2.30, should now be dropped since there is no guarantee that this equivalent system is an observable one.

Before presenting two examples which illustrate this algorithm, two important observations will be made. In particular, we first note that the dual of this algorithm can now be developed rather easily (Problem 5-4) and also employed to obtain equivalent state representations. Furthermore, it should be emphasized that if, in the differential operator representation 5.2.1, $P(D)$ is column (row) proper, $W(D) = 0$, and $Q(D) = I_q$ $(R(D) = I_q)$ with $\partial_c[R(D)] < \partial_c[(P(D)]$ $(\partial_r[Q(D)] < \partial_r[P(D)])$, a condition which often occurs in practice, then the derivation of an equivalent state representation via this algorithm (or its dual) can be accomplished virtually by inspection since Steps 1, 3, and 5 can be omitted (see Problem 5-10).

To illustrate the algorithm, we now consider the following two examples:

EXAMPLE 5.2.31: Consider a differential operator system with $R(D) = I$;

$$P(D) = \begin{bmatrix} D^2+3D+1, & 2D+3 \\ D^3+3D^2+D, & 3D^2+3D+6 \end{bmatrix}; \; Q(D) = \begin{bmatrix} 1, & 0 \\ D+1, & D+3 \end{bmatrix}; \text{ and } \; W(D) = 0.$$

Step 1: Since $P(D)$ is not row proper, we first premultiply both $P(D)$ and $Q(D)$, i.e. 5.2.1a, by any unimodular matrix which "reduces" $P(D)$ to row proper form. $U_L(D) = \begin{bmatrix} -D & 1 \\ 1 & 0 \end{bmatrix}$ is such a matrix and can be found using the results given in Section 2.5. We thus obtain:

$$U_L(D)P(D) = \begin{bmatrix} 0 , & D^2+6 \\ D^2+3D+1, & 2D+3 \end{bmatrix} \text{ and } U_L(D)Q(D) = \begin{bmatrix} 1, & D+3 \\ 1, & 0 \end{bmatrix}.$$

Step 2: By inspection of $U_L(D)P(D)$, it is clear that $\Gamma_r = \begin{bmatrix} 0 & 1 \\ 1 & 0 \end{bmatrix}$ and, therefore, that $P_o(D) = U_L(D)P(D)\Gamma_r^{-1} = \begin{bmatrix} D^2+6, & 0 \\ 2D+3, & D^2+3D+1 \end{bmatrix}$. Since $R(D) = I$, $R_o(D) = \Gamma_r^{-1} = \Gamma_r$, and $Q_o(D) = U_L(D)Q(D)$, as given in Step 1.

Step 3: This step can be omitted; i.e. it is clear (since $\partial_r[Q_o(s)] < \partial_r[P_o(s)]$) that $P_o^{-1}(s)Q_o$ is a strictly proper transfer matrix. Therefore, $H_o(s) = 0$, and $\bar{Q}_o(s) = Q_o(s)$.

Step 4: By inspection of $P_o(s)$, it is clear that $\bar{d}_1 = \bar{d}_2 = 2$, and using 5.2.23 and 5.2.25:

$$P_o(s) = \begin{bmatrix} s^2 & 0 \\ 0 & s^2 \end{bmatrix} - \underbrace{\begin{bmatrix} 1 & s & 0 & 0 \\ 0 & 0 & 1 & s \end{bmatrix}}_{\bar{S}(s)} \underbrace{\begin{bmatrix} -6 & 0 \\ 0 & 0 \\ -3 & -1 \\ -2 & -3 \end{bmatrix}}_{\hat{A}_q} ; \quad \bar{Q}_o(s) = Q_o(s) = \underbrace{\begin{bmatrix} 1 & s & 0 & 0 \\ 0 & 0 & 1 & s \end{bmatrix}}_{\bar{S}(s)} \underbrace{\begin{bmatrix} 1 & 3 \\ 0 & 1 \\ 1 & 0 \\ 0 & 0 \end{bmatrix}}_{\hat{B}}$$

Clearly, $\bar{\sigma}_1 = \bar{d}_1 = 2$ and $\bar{\sigma}_2 = \bar{d}_1 + \bar{d}_2 = n = 4$, and completeing Step 4 we see that: $H(D) = \Gamma_r^{-1} H_o(D) = 0$,

$$A_o = \begin{bmatrix} 0 & -6 & 0 & 0 \\ 1 & 0 & 0 & 0 \\ 0 & -3 & 0 & -1 \\ 0 & -2 & 1 & -3 \end{bmatrix}, \quad B_o = \begin{bmatrix} 1 & 3 \\ 0 & 1 \\ 1 & 0 \\ 0 & 0 \end{bmatrix}, \quad \bar{C}_o = \begin{bmatrix} 0 & 1 & 0 & 0 \\ 0 & 0 & 0 & 1 \end{bmatrix}, \text{ and } C_o =$$

$$\Gamma_r^{-1}\bar{C}_o = \begin{bmatrix} 0 & 0 & 0 & 1 \\ 0 & 1 & 0 & 0 \end{bmatrix}.$$ When employed in 5.2.26 and 5.2.27c, this quadruple, $\{A_o, B_o, C_o, H(D)\}$, represents an observable state-space system

which is equivalent to the given differential operator system.

Step 5: Note that 5.2.11 need not be employed to determine C and $E(D)$ in 5.2.6b (or 5.2.30); i.e. since $y(t) = z(t)$ in this example, $C = C_o$ and $E(D) = H(D) = 0$.

Since the pair $\{C_o, A_o\}$ is completely observable, we can, by repeated differentiation of 5.2.10, employing 5.2.6a to eliminate explicit reference to any derivatives of $x(t)$, obtain the expression, 5.2.13c, for $x(t)$ in terms of $z(t)$, $u(t)$, and their derivatives. For this example, we readily determine that

$$x_o(t) = x(t) = \underbrace{\begin{bmatrix} 0, & D \\ 0, & 1 \\ D+3, & 2 \\ 1, & 0 \end{bmatrix}}_{L(D)} z(t) + \underbrace{\begin{bmatrix} 0 & -1 \\ 0 & 0 \\ 0 & 0 \\ 0 & 0 \end{bmatrix}}_{M(D)} u(t)$$

The next example is more complex than the first, and also illustrates the notion of non-essential states.

EXAMPLE 5.2.32: Consider the differential operator representation:

$$P(D) = \begin{bmatrix} D^3-1, & -D^3+1, & 0 \\ D^2+D, & -D^2+1, & D-1 \\ D^2+D, & -D^2-1, & D+1 \end{bmatrix}, \quad Q(D) = \begin{bmatrix} D^3-2D, & D^2+3D \\ \frac{1}{2}D - \frac{3}{2}, & D+2 \\ -\frac{1}{2}D - \frac{5}{2}, & D+4 \end{bmatrix},$$

$$R(D) = \begin{bmatrix} D-1, & D+2, & -2D-3 \\ D, & -D, & 0 \\ D+1, & -D+1, & 2 \end{bmatrix} \quad \text{and} \quad W(D) = 0.$$

Step 1: Since $P(D)$ is not row proper, we first premultiply both $P(D)$ and $Q(D)$ (5.2.1a) by any unimodular matrix which "reduces" $P(D)$ to row proper form.

$$U_L(D) = \begin{bmatrix} 1, & -D, & 0 \\ 0, & 1, & -1 \\ 0, & 0, & 1 \end{bmatrix} \quad \text{is one such matrix; i.e.}$$

$$U_L(D)P(D) = \begin{bmatrix} -D^2-1, & -D+1, & -D^2+D \\ 0, & 2, & -2 \\ D^2+D, & -D^2-1, & D+1 \end{bmatrix} \quad \text{and} \quad U_L(D)Q(D) = \begin{bmatrix} D^3-\frac{1}{2}D^2-\frac{1}{2}D, & D \\ D+1, & -2 \\ -\frac{1}{2}D-\frac{5}{2}, & D+4 \end{bmatrix}.$$

Step 2: $\Gamma_r[U_L(D)P(D)] = \begin{bmatrix} -1 & 0 & -1 \\ 0 & 2 & -2 \\ 1 & -1 & 0 \end{bmatrix}$, $\Gamma_r^{-1} = \dfrac{\begin{bmatrix} -2 & 1 & 2 \\ -2 & 1 & -2 \\ -2 & -1 & -2 \end{bmatrix}}{4}$, and

$$U_L(D)P(D)\Gamma_r^{-1} = \begin{bmatrix} D^2, & -\frac{1}{2}D, & -1 \\ 0, & 1, & 0 \\ -D, & -\frac{1}{2}, & D^2 \end{bmatrix}$$ which is in the desired form 5.2.18.

In this example, however, we note that $\bar{d}_2 = 0$, contrary to the assumption that all $\bar{d}_k > 0$. To deal with this complication, we now employ the notion of non-essential states. In particular, we first zero all off-diagonal, second column elements of $U_L(D)P(D)\Gamma_r^{-1}$ by elementary row operations; i.e. by premultiplying the matrix differential equation $U_L(D)P(D)\Gamma_r^{-1}z(t) = U_L(D)Q(D)u(t)$ by the (second) unimodular matrix $U_{L2}(D) = \begin{bmatrix} 1, & \frac{1}{2}D, & 0 \\ 0, & 1, & 0 \\ 0, & \frac{1}{2}, & 1 \end{bmatrix}$, which yields the equivalent differential operator representation:

$$\underbrace{\begin{bmatrix} D^2, & 0, & -1 \\ 0, & 1, & 0 \\ -D, & 0, & D^2 \end{bmatrix}}_{P_o(D) = U_{L2}(D)U_L(D)P(D)\Gamma_r^{-1}} z_o(t) = \underbrace{\begin{bmatrix} D^3, & 0 \\ D+1, & -2 \\ -2, & D+3 \end{bmatrix}}_{Q_o(D) = U_{L2}(D)U_L(D)Q(D)} u(t);$$

$$y(t) = \underbrace{\begin{bmatrix} 1, & D+1, & D \\ 0, & 0, & D \\ -2, & 0, & D-1 \end{bmatrix}}_{R_o(D) = R(D)\Gamma_r^{-1}} z_o(t) + \underbrace{\begin{bmatrix} 0 & 0 \\ 0 & 0 \\ 0 & 0 \end{bmatrix}}_{W(D)} u(t)$$

The partial state element $z_{02}(t)$ is, by definition, a non-essential state, so defined for rather obvious reasons. In particular, we now note that in view of this latter differential operator representation, the non-essential state $z_{02}(t)$ does not satisfy any differential equation and can be equated to a linear function of the input $u(t)$ and its derivatives (in this case, $z_{02}(t) = u_1(t) + \dot{u}_1(t) - 2\, u_2(t)$). Therefore, the output $y(t)$ need no longer display any explicit reference to $z_{02}(t)$. In particular, if we substitute $[D+1, \ -2]u(t)$ for $z_{02}(t)$, $y(t) = R_o(D)z_o(t) + W(D)u(t)$ can be rewritten as:

$$y(t) = \underbrace{\begin{bmatrix} 1, & 0, & D \\ 0, & 0, & D \\ -2, & 0, & D-1 \end{bmatrix}}_{\hat{R}_o(D)} z_o(t) + \underbrace{\begin{bmatrix} D^2+2D+1, & -2D-2 \\ 0 \quad , & 0 \\ 0 \quad , & 0 \end{bmatrix}}_{\hat{W}(D)} u(t)$$

<u>Step 3</u>: We now compute the quotient, $H_o(s)$, associated with

$$P_o^{-1}(s)Q_o(s) = \frac{\begin{bmatrix} s^5-2 \quad , & s+3 \\ s^5+s^4-s^2-s, & -2s^4+2s \\ s^4-2s^2 \quad , & s^3+3s^2 \end{bmatrix}}{s^4-s}$$

by applying the polynomial division algorithm to each entry of this rational transfer matrix. We thus obtain $H_o(s) = \begin{bmatrix} s & 0 \\ s+1 & -2 \\ 1 & 0 \end{bmatrix}$. It therefore follows that

$$\overline{Q}_o(D) = Q_o(D) - P_o(D)H_o(D) = \begin{bmatrix} 1 & 0 \\ 0 & 0 \\ -2 & D+3 \end{bmatrix}.$$

<u>Step 4</u>: By inspection of $P_o(D)$ and $\overline{Q}_o(D)$, we now directly determine that the state-space system: $\dot{x}_o(t) = A_o x_o(t) + B_o u(t)$; $\overline{z}_o(t) = \overline{C}_o x_o(t)$, with

$$A_o = \begin{bmatrix} 0 & 0 & 0 & 1 \\ 1 & 0 & 0 & 0 \\ 0 & 0 & 0 & 0 \\ 0 & 1 & 1 & 0 \end{bmatrix}, \quad B_o = \begin{bmatrix} 1 & 0 \\ 0 & 0 \\ -2 & 3 \\ 0 & 1 \end{bmatrix}, \quad \text{and} \quad \overline{C}_o = \begin{bmatrix} 0 & 1 & 0 & 0 \\ 0 & 0 & 0 & 0 \\ 0 & 0 & 0 & 1 \end{bmatrix}$$

is equivalent to the differential operator representation: $P_o(D)\overline{z}_o(t) = \overline{Q}_o(D)u(t)$. By 5.2.27b, we further note that $z_o(t) = \overline{C}_o x_o(t) + H_o(D)u(t)$ and, therefore, that $z(t) = \Gamma_r^{-1} z_o(t) = C_o x_o(t) + H(D)u(t)$, where

$$C_o = \Gamma_r^{-1}\overline{C}_o = \frac{\begin{bmatrix} 0 & -2 & 0 & 2 \\ 0 & -2 & 0 & -2 \\ 0 & -2 & 0 & -2 \end{bmatrix}}{4} \quad \text{and} \quad H(D) = \Gamma_r^{-1} H_o(D) = \frac{\begin{bmatrix} -D+3, & -2 \\ -D-1, & -2 \\ -3D-3, & 2 \end{bmatrix}}{4}.$$

The state-space system, $\{A_o, B_o, C_o, H(D)\}$, thus obtained, is equivalent to the given differential system: $P(D)z(t) = Q(D)u(t)$.

Step 5: We finally employ the relation 5.2.1b along with the state equations 5.2.26 and 5.2.27c to determine C and $E(D)$.

In particular,

$$y(t) = \underbrace{\begin{bmatrix} 0 & 1 & 0 & D \\ 0 & 0 & 0 & D \\ 0 & -2 & 0 & D-1 \end{bmatrix}}_{R(D)C_o} x_o(t) + \underbrace{\begin{bmatrix} D^2+4D+1, & -2D-2 \\ D \quad , & 0 \\ -D-1 \quad , & 0 \end{bmatrix}}_{R(D)H(D)+W(D)} u(t).$$

or since $\dot{x}_{04}(t) = x_{02}(t) + x_{03}(t) + u_2(t)$,

$$y(t) = \underbrace{\begin{bmatrix} 0 & 2 & 1 & 0 \\ 0 & 1 & 1 & 0 \\ 0 & -1 & 1 & -1 \end{bmatrix}}_{C} x_o(t) + \underbrace{\begin{bmatrix} D^2+4D+1, & -2D-1 \\ D \quad , & 1 \\ -D-1 \quad , & 1 \end{bmatrix}}_{E(D)} u(t).$$

The state-space system, $\{A_o, B_o, C, E(D)\}$, thus obtained, is equivalent to the given differential operator system: $P(D)z(t) = Q(D)u(t)$; $y(t) = R(D)z(t)$.

The final transfer relation which remains as yet unresolved

involves the transition from any transfer matrix $T(s)$ to an equivalent <u>differential operator realization</u> of $T(s)$; i.e. a quadruple, $\{P(D),Q(D),R(D),W(D)\}$, in the form 5.2.1, which satisfies 5.2.5. It is clear that if $T(s)$ is proper, the realization algorithm of Section 4.4 can be used to find a state-space realization $\{A,B,C,E\}$ of $T(s)$, and therefore a differential operator quadruple $\{P(D),Q(D),R(D),W(D)\} = \{DI-A,B,C,E\}$ which also realizes $T(s)$. Therefore, the only question which remains unresolved is what to do when $T(s)$ is not proper. In these cases, however, we recall that we can always subtract the quotient, $E(s)$, from $T(s)$, and thus obtain the (unique) strictly proper transfer matrix $\hat{T}(s)$; i.e.

$$\hat{T}(s) = T(s) - E(s) \qquad 5.2.33$$

If we then obtain a state-space realization $\{\hat{A},\hat{B},\hat{C}\}$ of $\hat{T}(s)$ via the algorithm given in Section 4.4, it is clear that the state-space quadruple $\{\hat{A},\hat{B},\hat{C},E(D)\}$ realizes $T(s)$ and, therefore, that the differential operator representation $\{P(D),Q(D),R(D),W(D)\} = \{DI-\hat{A},\hat{B},\hat{C},E(D)\}$ also realizes $T(s)$. This procedure for resolving the question of transition from the transfer matrix to a differential operator realization is not, generally speaking, the easiest or most efficient technique to use, and other procedures for resolving this question will be presented in Section 5.4.

We have now developed techniques for transferring from any dynamical system representation (the transfer matrix, the state form, or the differential operator form) to any other representation. We recall, from the various transfer procedures which have been developed, that the notion of equivalence is important whenever a transfer from one time domain representation to the other is made, while the concept of minimality is important when we transfer from a frequency domain transfer matrix to a time domain state-space realization. The notion of minimality can also arise when a transfer from a frequency domain transfer matrix to a time domain differential operator realization is made as we will later show (Section 5.4). Before we can do so, how-ever, we must first extend the notions of controllability and observability, thus far defined for only state-space systems (Section 3.5), to include the more general class of differential operator systems. We will do this in the next section.

5.3 DIFFERENTIAL OPERATOR CONTROLLABILITY AND OBSERVABILITY

Now that we have established an equivalence relationship between systems whose dynamics are represented in the general differential operator form 5.2.1 and those whose dynamics are expressed in the more restrictive state form 5.2.6, it becomes possible to extend the notions of controllability and observability to the more general class of systems. In particular, by employing certain of the polynomial matrix definitions and results given in Section 2.5, we can now state and establish:

THEOREM 5.3.1: _Consider any system of the form_ 5.2.1. _Any equivalent state-space representation of the form_ 5.2.6 _is_ (a) _completely controllable_, (b) _completely observable_, (c) _minimal if and only if_ (a) _any g.c.l.d._ $G_L(D)$ _of_ $P(D)$ _and_ $Q(D)$ _is unimodular_, (b) _any g.c.r.d._ $G_R(D)$ _of_ $R(D)$ _and_ $P(D)$ _is unimodular_, (c) _both_ (a) _and_ (b) _hold. Furthermore, the uncontrollable (unobservable) modes of 5.2.6 are the zeros of_ $|G_L(s)|$ $(|G_R(s)|)$.

Proof: To establish part (a) of this theorem, we need only consider system 5.2.1a since the output, y(t), plays no role insofar as system controllability is concerned. We now employ a "negative proof" in the following sense: We will first show that if any g.c.l.d. of $P(D)$ and $Q(D)$ is _not_ unimodular, then any equivalent state-space system of the form 5.2.6 is not completely controllable. Conversely, we will also establish that if any equivalent state-space system of the form 5.2.6 is not completely controllable, then we can find a g.c.l.d. of $P(D)$ and $Q(D)$ which is not unimodular. In particular, assume first that $G_L(D)$, a greatest common left divisor of $P(D)$ and $Q(D)$ is not a unimodular matrix or that

$$P^{-1}(s)Q(s) = \tilde{P}^{-1}(s)G_L^{-1}(s)G_L(s)\tilde{Q}(s) = \tilde{P}^{-1}(s)\tilde{Q}(s) \qquad 5.3.2$$

where $|\tilde{P}(s)|$ divides but is not divided by $|P(s)|$; i.e.

$$|G_L(s)| \times |\tilde{P}(s)| = |P(s)|, \qquad 5.3.3$$

with $\partial[|G_L(s)|] \geq 1$. By using the algorithm given in the previous section, we can now find equivalent, completely observable, state-space representations, $\{A_o,B_o,C_o,H(D)\}$, and $\{\tilde{A}_o,\tilde{B}_o,\tilde{C}_o,H(D)\}$, for the two differential operator systems defined by $P(D)z(t) = Q(D)u(t)$ and $\tilde{P}(D)z(t) = \tilde{Q}(D)u(t)$ respectively. Note that, by 5.3.2, both of these systems will have the same transfer matrix (after all possible pole-zero "cancellations" have been made); i.e.

$$\begin{aligned} P^{-1}(s)Q(s) = C_o(sI-A_o)^{-1}B_o + H(s) &= \tilde{C}_o(sI-\tilde{A}_o)^{-1}\tilde{B}_o + H(s) \\ &= \tilde{P}^{-1}(s)\tilde{Q}(s) \end{aligned} \qquad 5.3.4$$

Note further that both state-space systems will be completely observable, but the former system, $\{A_o,B_o,C_o,H(D)\}$, will be of greater order than the latter system, $\{\tilde{A}_o,\tilde{B}_o,\tilde{C}_o,H(D)\}$; i.e. $|sI-A_o| = |P(s)|$ and $|sI-\tilde{A}_o| = |\tilde{P}(s)|$, with $\partial|P(s)| > \partial|\tilde{P}(s)|$. We now recall (see Section 4.4) that any rational transfer matrix, $T(s)$, has a state-space realization of least order, $\{\tilde{A},\tilde{B},\tilde{C},E(s)\}$, called a minimal realization; i.e.

$$\tilde{C}(sI-\tilde{A})^{-1}\tilde{B} + E(s) = T(s) \qquad 5.3.5$$

with $|sI-\tilde{A}|$ of least possible degree. Furthermore, by Theorem 4.4.15, any such minimal realization is both controllable and observable and conversely, any controllable and observable realization of $T(s)$ is a minimal one. Since both state-space representations, $\{A_o,B_o,C_o,H(D)\}$ and $\{\tilde{A}_o,\tilde{B}_o,\tilde{C}_o,H(D)\}$, of the same transfer matrix, $P^{-1}(s)Q(s) = \tilde{P}^{-1}(s)\tilde{Q}(s)$, are completely observable it follows that the former system is uncontrollable; i.e. the ability to "factor" the non-unimodular g.c.l.d., $G_L(D)$, from both $P(D)$ and $Q(D)$, with a resultant decrease in system order, represents a lack of complete controllability of the higher order system $\{A_o,B_o,C_o,H(D)\}$. In view of 5.3.3, it is also clear that the zeros of $|G_L(s)|$ represent the uncontrollable modes of the higher order system.

We next assume that any completely observable system, $\{A_o,B_o,C_o,H(D)\}$, of the form 5.2.6, which is equivalent to 5.2.1a, is not completely controllable. By 5.2.13a, $C_o(sI-A_o)^{-1}B_o + H(s) = P^{-1}(s)Q(s)$, where $|sI-A_o| = \alpha|P(s)|$ for some nonzero scalar α. By applying the structure theorem to only the controllable and observable "portion" of $\{A_o,B_o,C_o,H(D)\}$; i.e. to the quadruple $\{A_{co},B_{co},C_{co},H(D)\}$, of reduced (state) dimension $\hat{n}$ in observable companion form, $C_o(sI-A_o)^{-1}B_o + H(s) = C_{co}(sI-A_{co})^{-1}B_{co} + H(s)$ can be expressed as $P_{co}^{-1}(s)Q_{co}(s)$, where I_q is a g.c.l.d. of $P_{co}(s)$ and $Q_{co}(s)$. Furthermore, $|sI-A_{co}|$ and $|P_{co}(s)|$ are polynomials of degree $\hat{n}$, which differ by only a nonzero scalar. Therefore, $P_{co}^{-1}(s)Q_{co}(s) = P^{-1}(s)Q(s)$, or $Q(s) = P(s)P_{co}^{-1}(s)Q_{co}(s)$, where the polynomial matrix $P(s)P_{co}^{-1}(s) = G_L(s)$ (see Problem 5-13) is a non-unimodular g.c.l.d. of $P(s)$ and $Q(s)$. Finally, the determinant of $G_L(s)$ is equal to $|P(s)| \div |P_{co}(s)| = |P_{co}(s)||P_{\overline{c}o}(s)| \div |P_{co}(s)| = |P_{\overline{c}o}(s)|$, a polynomial whose zeros represent the uncontrollable but observable modes of the equivalent system $\{A_o,B_o,C_o,H(D)\}$. Part (a) of Theorem 5.3.1

is thus established. By duality (Section 3.5) an analogous relationship can also be established between any g.c.r.d. of $R(D)$ and $P(D)$ and the observability of an equivalent state-space system, thus establishing part (b) of the theorem. Part (c) then follows from parts (a) and (b) by Theorem 4.4.15, thus establishing the complete theorem.

COROLLARY 5.3.6: *Any state-space system of the form* 5.2.6 is (a) *completely controllable*, (b) *completely observable*, (c) *minimal if and only if* (a) *any g.c.l.d. of* $(sI-A)$ *and* B *is unimodular*, (b) *any g.c.r.d. of* $(sI-A)$ *and* C *is unimodular*, (c) *both* (a) *and* (b) *hold*.

Proof: The proof of this corollary is a direct consequence of Theorem 5.3.1 and the fact that the state-space representation 5.2.6 is simply a special case of 5.2.1, where $P(D) = (DI-A)$, $Q(D) = B$, $R(D) = C$, and $W(D) = E(D)$.

In view of the above, it now becomes possible to rather naturally extend the notions of controllability and observability, which have thus far been confined solely to state-space systems, to include systems whose dynamics are expressed in the more general differential operator form. In particular, we now have:

DEFINITION 5.3.7: Consider the system 5.2.1. The quadruple, $\{P(D),Q(D),R(D),W(D)\}$, will be called (a) an *observable differential operator representation*, (b) a *controllable differential operator representation*, (c) a *minimal differential operator representation* if and only if (a) any g.c.r.d. of $P(D)$ and $R(D)$ is unimodular, (b) any g.c.l.d. of $P(D)$ and $Q(D)$ is unimodular, (c) both (a) and (b) hold.

It might be noted that $W(D)$ plays no role in this definition, just as $E(D)$ does not affect controllability and observability of state-space systems of the form 5.2.6. Since we impose no requirements on the relative degrees of the numerator and denominator elements of the transfer matrix of 5.2.1, namely $R(s)P^{-1}(s)Q(s) + W(s)$, it also follows that the qualitative properties of controllability and observability are not confined solely to time-invariant systems whose transfer matrices are proper. For example, we can now speak of the controllability of (scalar) systems such as: $p(D)y(t) = q(D)u(t)$, where the degree of $p(D)$ can be less than the degree of $q(D)$; i.e. the representation, $\{p(D), q(D), 1, 0\}$, is a controllable one if and only if $p(D)$ and $q(D)$ are relatively prime, regardless of their respective degrees. Indeed, it can now be shown that when $p(D)$ and $q(D)$ are relatively prime, there does exist a scalar control, $u(t)$, which transfers $y(t)$ and its first $(n-1)$ derivatives (where n equals the degree of $|p(D)|$) from any initial "state" at time t_o to any desired final "state" at any time $t_1 > t_o$. To illustrate this fact, consider the following elementary example:

EXAMPLE 5.3.8: Consider a scalar system of the form 5.2.1a, where $p(D) = D + 1$ and $q(D) = D^2$. Since $p(D)$ and $q(D)$ are relatively prime, it should be possible to drive $z(t_o)$ to any desired output, $z(t_1)$, via $u(t)$ defined over any finite time interval $[t_o, t_1]$. This is indeed the case; i.e. if we apply the algorithmic proof of Theorem 5.2.14, we obtain the equivalent, completely controllable state-space system: $\dot{x}(t) = -x(t) + u(t)$; $z(t) = x(t) + \dot{u}(t) - u(t)$. Clearly, the (controllable) state, $x(t)$, can be driven from any $x(t_o)$ to any desired $x(t_1)$, where $t_1 > t_o$ via $u(t)$. Furthermore, if $u(t)$ is then suddenly brought to zero and held there for all $t \geq t_1^+$, it is clear that $z(t_1^+) = x(t_1)$, since the output $(x(t))$ of an integrator cannot change instantaneously if only bounded inputs are applied. If $p(D)$ and $q(D)$ were not relatively prime, this result would not hold, regardless of their respective degrees, since any equivalent state representation would not be controllable.

An analogous statement can also be made in the more general

matrix case regarding the (n) components of $L(D)z(t) = x(t) - M(D)u(t)$ (see 5.2.13c) whenever any g.c.l.d. of $P(D)$ and $Q(D)$ is unimodular; i.e. we can transfer the equivalent controllable state, $x(t)$, in 5.2.6a from any arbitrary initial state, x_o, at time t_o to any desired, arbitrary final state, x_1, at any time $t_1 > t_o$ via an appropriate control, $u(t)$, defined over the time interval $[t_o, t_1]$. If we then instantaneously bring $u(t)$ to zero at $t = t_1^+$ and hold it there, it follows from the fact that integrator outputs cannot change instantaneously, that

$$L(D)z(t)\Big|_{t=t_1^+} = x(t)\Big|_{t=t_1^+} = x_1, \qquad 5.3.9$$

the desired final state. To formalize the above, we now conclude that <u>if any g.c.l.d. of $P(D)$ and $Q(D)$ is unimodular, an input $u(t)$ can be found which transfers $L(D)z(t)$ at time t_o to any desired "state" at any time t_1^+, where $t_1 > t_o$</u>. We therefore see that controllability of systems whose dynamics are expressed in the differential operator form 5.2.1a cannot only be defined, but can also be given a useful physical interpretation. A dual result involving observability also holds; i.e. <u>if any g.c.r.d. of $P(D)$ and $R(D)$ is unimodular, then knowledge of the input $u(t)$ and output $y(t)$ of 5.2.1 over any finite time interval $[t_o, t_1]$ is sufficient to completely determine all (n) components of $L(D)z(t)$ at $t = t_o$</u>.

5.4 REALIZATION THEORY (FREQUENCY DOMAIN REDUCTION)

Thus far, most of our attention has focused on various questions related to state-space systems of the form 5.2.6 which are equivalent to differential operator systems whose dynamical behavior is given by 5.2.1. The results which have been presented, however, can also be applied to the question of obtaining realizations, and in particular, minimal state-space realizations as well as minimal differential operator realizations of known transfer matrices, a question which, in part, has already been considered in Section 4.4 and partially resolved via the structure theorem and time domain reduction.

In this section, we will employ a new approach based on first representing the given transfer matrix, $T(s)$, as the sum of its quotient, $E(s)$, and its strictly proper part, $\hat{T}(s)$ (see Problem 4-20). $\hat{T}(s)$ will then be represented as $R^*(s)P_m^{-1}(s)I_m$ $(I_pP_p^{-1}(s)Q^*(s))$ and then reduced to minimal form by "factoring out" any non-unimodular g.c.r.d. (g.c.l.d.) of $R^*(s)$ and $P_m(s)$ ($P_p(s)$ and $Q^*(s)$), after which the algorithm given in Section 5.2 can be employed. More specifically, we first establish:

THEOREM 5.4.1: <u>Any</u> $(p \times m)$ <u>rational matrix</u>, $T(s)$, <u>can be represented as</u> $R(s)P_c^{-1}(s)I_m + E(s)$ $(I_pP_o^{-1}(s)Q(s) + E(s))$, <u>where any g.c.r.d. of</u> $R(s)$ <u>and</u> $P_c(s)$ (<u>any g.c.l.d. of</u> $P_o(s)$ <u>and</u> $Q(s)$) <u>is unimodular</u>.

<u>Proof</u>: By duality, we need only establish that $\hat{T}(s) = T(s) - E(s)$ can be represented as $R(s)P_c^{-1}(s)I_m$, where $R(s)$ and $P_c(s)$ are relatively right prime. Consequently, consider any $(p \times m)$ strictly proper transfer matrix $\hat{T}(s) = [\hat{r}_{ij}(s)/\hat{p}_{ij}(s)]$, where $\hat{r}_{ij}(s)$ and $\hat{p}_{ij}(s)$ are relatively prime polynomials with the degree of $\hat{r}_{ij}(s)$ strictly less than that of $\hat{p}_{ij}(s)$ for all i and j. Repeating the procedure employed in Section 4.4, let $g_j(s)$, for $j = 1,2,\ldots,m$, denote the least common (monic) multiple of the (p) denominator polynomials, $\hat{p}_{ij}(s), \hat{p}_{2j}(s),\ldots,\hat{p}_{pj}(s)$, which appear in each (j-<u>th</u>) column of $\hat{T}(s)$. $\hat{T}(s)$ can then be written as (see 4.4.4 also):

$$\hat{T}(s) = R^*(s)P_m^{-1}(s), \tag{5.4.2}$$

where $P_m(s) = \text{diag}[g_j(s)]$, while each element of $R^*(s)$, namely $r^*_{ij}(s) = \hat{r}_{ij}(s)g_j(s)/\hat{p}_{ij}(s)$, where $\hat{p}_{ij}(s)$ divides $g_j(s)$. If $G_R(s)$ is any non-unimodular g.c.r.d. of $R^*(s)$ and $P_m(s)$, $\hat{T}(s)$ can be "reduced" by factoring $G_R(s)$ from both; i.e.

$$\hat{T}(s) = R^*(s)P_m^{-1}(s) = R(s)G_R(s)G_R^{-1}(s)P_c^{-1}(s) = R(s)P_c^{-1}(s)I_m, \qquad 5.4.3$$

and, consequently

$$T(s) = R(s)P_c^{-1}(s)I_m + E(s), \qquad 5.4.4$$

where $R(s)G_R(s) = R^*(s)$ and $P_c(s)G_R(s) = P_m(s)$. Since $G_R(s)$ is a <u>greatest</u> common right divisor of $R^*(s)$ and $P_m(s)$, any g.c.r.d. of $P_c(s)$ and $R(s)$ must be unimodular, thus establishing the theorem.

It is now clear, in view of Theorem 5.4.1, that <u>any rational transfer matrix $T(s)$ has a minimal differential operator realization</u> $\{P_c(D), I_m, R(D), E(D)\}$ (or $\{P_o(D), Q(D), I_p, E(D)\}$) which can be found by using the procedure just outlined. This fact clarifies the final point made in Section 5.2 regarding the notion of minimality whenever a transfer from some rational $T(s)$ to a differential operator realiza- is made.

The algorithm outlined in Section 5.2 (or its dual--see Example 5.4.6) can now be employed to directly yield a minimal state-space realization of $T(s)$ from the minimal differential operator realization $\{P_c(D), I_m, R(D), E(D)\}$ (or $\{P_o(D), Q(D), I_p, E(D)\}$); i.e. a quadruple $\{\tilde{A}, \tilde{B}, \tilde{C}, E(D)\}$ such that

$$\tilde{C}(sI-\tilde{A})^{-1}\tilde{B} + E(s) = R(s)P_c^{-1}(s)I_m + E(s) = I_pP_o^{-1}(s)Q(s) + E(s) = T(s), \qquad 5.4.5$$

where $|sI-\tilde{A}| = \alpha|P_c(s)| = \beta|P_o(s)|$, and all three polynomials have degree $\tilde{n}$ corresponding to the minimal (state) dimension.

EXAMPLE 5.4.6: To illustrate this technique for determining both a minimal differential operator realization and a minimal state-space realization of any rational transfer matrix, we again consider the same $T(s) = \begin{bmatrix} \frac{1}{s+1} & \frac{s}{s-2} \\ 2 & 0 \\ \frac{2}{s-2} & 1 \end{bmatrix}$ used in both examples 4.4.11 and 4.4.14.

In particular, we first note that the quotient of $T(s)$, namely

$$E = \lim_{s \to \infty} T(s) = \begin{bmatrix} 0 & 1 \\ 2 & 0 \\ 0 & 1 \end{bmatrix}$$ and, therefore, that its strictly proper part

$$\hat{T}(s) = T(s) - E = \begin{bmatrix} \frac{1}{s+1} & \frac{2}{s-2} \\ 0 & 0 \\ \frac{2}{s-2} & 0 \end{bmatrix}.$$ We now determine that $g_1(s) =$ $(s+1)(s-2) = s^2-s-2$ and $g_2(s) = s-2$. Therefore, we can factor $\hat{T}(s)$ as the product of $R^*(s) = \begin{bmatrix} s-2 & 2 \\ 0 & 0 \\ 2s+2 & 0 \end{bmatrix}$ and $P_m^{-1}(s) = \begin{bmatrix} s^2-s-2, & 0 \\ 0\ , & s-2 \end{bmatrix}^{-1}$.

We next ascertain (using the algorithm given in Section 2.5) that $G_R(s) = I_2$ is a g.c.r.d. of $R^*(s)$ and $P_m(s)$; i.e. that $R^*(s)$ and $P_m(s)$ are relatively right prime. It is thus clear that $R(s) = R^*(s)$ and $P_c(s) = P_m(s)$, and together with I_2 and E, they constitute a minimal differential operator realization of $T(s)$.

To find a minimal state-space realization of this $T(s)$, we now apply the dual of the algorithm given in Section 5.2; i.e. by directly equating $P_c(s)$, $R(s)$, I_2, and E to $\delta(s)$, $\hat{C}S(s)$, $\hat{B}_m$, and E (of Section 4.3) respectively, we immediately determine that the following quadruple represents a minimal state-space realization of $T(s)$:

$$A = \begin{bmatrix} 0 & 1 & 0 \\ 2 & 1 & 0 \\ 0 & 0 & 2 \end{bmatrix}, \quad B = \begin{bmatrix} 0 & 0 \\ 1 & 0 \\ 0 & 1 \end{bmatrix}, \quad C = \begin{bmatrix} -2 & 1 & 2 \\ 0 & 0 & 0 \\ 2 & 2 & 0 \end{bmatrix}, \text{ and } E = \begin{bmatrix} 0 & 1 \\ 2 & 0 \\ 0 & 1 \end{bmatrix}.$$

5.5 SYSTEM INVERTIBILITY AND FUNCTIONAL REPRODUCIBILITY

We introduced the notion of a system which is output controllable, or one whose output is functionally reproducible in Section 3.5, stating at that time that if a system is to possess this property there must exist an input which transfers the output of the system along any continuous precribed path. Closely coupled to this notion of output function controllability is the concept of "input function observability"; i.e. a system will be called <u>input function observable</u>, or the input to the system will be called <u>functionally reproducible</u> if knowledge of the dynamical equations of the system, all initial conditions, and the output time function, $y(t)$, completely specify the (unique) input which produced the particular output trajectory. Both of these definitions are somewhat heuristic and will require additional clarification which we will provide in this section.

We begin the study of these two notions by formally introducing the notions of a "left (or right) inverse" of a system characterized by a rational transfer matrix. In particular, if we now consider the $(p \times m)$ rational transfer matrix, $T(s)$, of some linear dynamical system, it is clear that if $T(s)$ has full rank (over $\mathscr{P}$), then $\rho[T(s)] = \min(p,m)$, and if $T(s)$ does not have full rank, $\rho[T(s)] < \min(p,m)$.

DEFINITION 5.5.1: Any dynamical system with an $(m \times p)$ (a $(p \times m)$) rational transfer matrix, $T_{LI}(s)$ $(T_{RI}(s))$, is said to be a <u>left (right) inverse</u> of a system with the $(p \times m)$ rational transfer matrix $T(s)$ if and only if

$$T_{LI}(s)T(s) = I_m \quad (T(s)T_{RI}(s) = I_p). \qquad 5.5.2$$

For convenience, we will call a transfer matrix, $T_{LI}(s)$ $(T_{RI}(s))$, which satisfies 5.5.2 a <u>left (right) inverse of $T(s)$</u>.

In view of the above, we can now state and establish a fundamental result pertaining to the existence of inverse systems.

THEOREM 5.5.3: <u>A (system with a) $p \times m$ rational transfer matrix, $T(s)$, has a left (right) inverse if and only if</u>

$$\rho[T(s)] = m (= p). \qquad 5.5.3$$

Proof: We will formally establish only that part of the theorem statement which is not in parenthesis since the other half follows directly by duality. Necessity is readily established by noting that if $\rho[T(s)] < m$ but $\hat{T}(s)T(s) = I_m$ for some rational matrix, $\hat{T}(s)$, then by Sylvester's inequality (see Problem 2-16) we have a contradiction, since $m = \rho[\hat{T}(s)T(s)] \leq \min(\rho[\hat{T}(s)],\rho[T(s)]) < m$.

Sufficiency will now be constructively established. In particular, if $\rho[T(s)] = m$, then there exists an $(m \times p)$ constant matrix G such that $GT(s)$ is nonsingular; e.g. one such G is the matrix which annihilates all but the first m linearly independent rows of $T(s)$. It then follows that

$$T_{LI}(s) = [GT(s)]^{-1}G \qquad 5.5.4$$

is a left inverse of $T(s)$ as the reader can readily verify. Theorem 5.5.3 is thus established. It should now be clear that whenever $\rho[T(s)] = m = p$,

$$T_{LI}(s) = T_{RI}(s) = T^{-1}(s), \qquad 5.5.5$$

which we simply call the inverse of $T(s)$, the transfer matrix of an invertible system.

In view of the above, we can now formalize the notions of input function observability and output function controllability in order to establish a fundamental relationship between these notions and inverse systems. To begin, we first define the class of dynamical systems which will be considered in our treatment as those whose dynamical behavior can be represented by the controllable[†] differential operator representation:

$$P(D)z(t) = u(t); \qquad 5.5.6a$$

[†]The case of uncontrollable (partially controllable) systems will be considered in the problem section of this chapter--see Problem 5-21 in particular.

$$y(t) = R(D)z(t), \qquad 5.5.6b$$

with $P(D)$ column proper and $\partial_c[R(D)] \leq \partial_c[P(D)]$, a condition which is satisfied by most physical systems. It might be noted that this representation can be obtained from any controllable state space quadruple, $\{A,B,C,E\}$, via the results presented in Section 4.3 and, as we will show, represents a convenient means of developing a number of relationships between functional reproducibility and system invertibility.

We now formally state the question of input function observability; i.e. given (i) the differential operator representation 5.5.6, (ii) the n initial conditions on $z(t)$ and its derivatives implicit in 5.5.6a (see Section 5.2), and (iii) $y(t)$, the output of the given system defined for all $t \geq t_o$ and subject to the n known initial conditions and some unknown input $u(t)$, can this specific $u(t)$ be determined for all $t \geq t_o$? Closely coupled to the question of input function observability is the question of output function controllability which we will also formally state at this time; i.e. given (i) the differential operator representation 5.5.6 and (ii) any desired but arbitrary p-dimensional output function, $\hat{y}(t)$, defined and "sufficiently differentiable" for all $t \geq t_o$, can this specific $\hat{y}(t)$ be obtained as the output of the given system subject to some bounded input $u(t)$ and an appropriate set of (n) initial conditions on $z(t)$ and its derivatives?

A fundamental result which constructively resolves these two questions will now be formally established.

THEOREM 5.5.7: The question of input function observability (output function controllability) has an affirmative resolution if and only if the given system 5.5.6 has a left (right) inverse; i.e. in view of Theorem 5.5.3, if and only if

$$\rho[T(s)] = \rho[R(s)P^{-1}(s)] = m(= p). \qquad 5.5.7$$

Proof: Unlike Theorem 5.5.3, where only the first half of the theorem had to be explicitly established due to duality, it should be noted that duality cannot be employed here since input function observability and output function controllability are not dual notions. Therefore, we must formally establish both parts of Theorem 5.5.7.

We begin by establishing necessity of that portion of the theorem statement which is not in parenthesis. In particular, suppose that $\rho[T(s)] = \hat{p} < m$ and $y(s)$ is the Laplace transform of $y(t)$, the output of the given system subject to some unknown input, $u(t)$, and zero initial conditions on $z(t)$ and all of its derivatives; i.e.

$$y(s) = T(s)u(s) \tag{5.5.8}$$

If $\hat{p} = p$, we can extend the row dimension and rank of $T(s)$ to m by appending to $T(s)$, $m-p$ additional, linearly independent rows. If we then call the resulting transfer matrix $T_e(s)$ and define $y_e(s)$ as $T_e(s)u(s)$, it follows that

$$u(s) = T_e^{-1}(s)y_e(s), = T_e^{-1}(s)\begin{bmatrix} y(s) \\ \hline \tilde{y}(s) \end{bmatrix}, \tag{5.5.9}$$

where $\tilde{y}(s)$ is an $(m-p)$-vector. If $y_{e1}(s) = \begin{bmatrix} y(s) \\ \hline \tilde{y}_1(s) \end{bmatrix}$ and $y_{e2} = \begin{bmatrix} y(s) \\ \hline \tilde{y}_2(s) \end{bmatrix}$ represent known m-vectors with $\tilde{y}_1 \neq \tilde{y}_2(s)$, it then follows in view of 5.5.8 and 5.5.9 that $u_1(s) = T_e^{-1}(s)y_{e1}(s) \neq u_2(s) = T_e^{-1}(s)y_{e2}(s)$ although both $u_1(s)$ and $u_2(s)$ will produce $y(s)$ as the output of the given system with zero initial conditions on $z(t)$ and its derivatives. Note that if $\hat{p} \neq p$, we can eliminate all but the first $\hat{p}$ linearly independent rows of $T(s)$, which represent the maximum number of independent outputs, and then repeat the above procedure to again establish the non-uniqueness of $u(t)$ whenever $\rho[T(s)] \neq m$. Necessity of the first part of the theorem is thus established.

To establish necessity of the part in parenthesis, we note that if $\rho[T(s)] < p$, then the defined system outputs, $y_i(s)$, must be linearly dependent over $\mathscr{P}$; i.e.

$$\sum_1^p f_i(s)y_i(s) = 0 \qquad 5.5.10$$

for one or more nonzero rational functions, $f_i(s)$. It is therefore clear that we cannot "drive" each of the (p) system outputs over mutually independent, arbitrary paths and still satisfy 5.5.10. Necessity is therefore completely established.

The proof of sufficiency is constructive and will also be presented in two parts, first for the case of input function observability via a left inverse system. In particular, we first note that the system defined by 5.5.6 has a left inverse if and only if $\rho[T(s) = R(s)P^{-1}(s)] = m$ or, since $\rho[P(s)] = m$, if and only if $\rho[R(s)] = m$ (see Problem 5-18). If this latter condition holds, we can define $\hat{R}(D)$ as any $(m \times m)$ nonsingular matrix obtained by premultiplying $R(D)$ by an appropriate $(m \times p)$ constant matrix G; i.e.

$$\hat{R}(D) = GR(D). \qquad 5.5.11$$

We will now verify that the differential operator system:

$$\hat{R}(D)\hat{z}(t) = Gy(t); \qquad 5.5.12a$$

$$\hat{u}(t) = P(D)\hat{z}(t), \qquad 5.5.12b$$

which is "driven" by the output of the given system, as depicted in Figure 5.5.13, represents a left inverse of the given system whose output $\hat{u}(t) \equiv u(t)$ for all $t \geq t_0$ provided the appropriate initial conditions are set on $\hat{z}(t)$ and its derivatives. In particular, in light of 5.5.11 and Figure 5.5.13, it can readily be established that

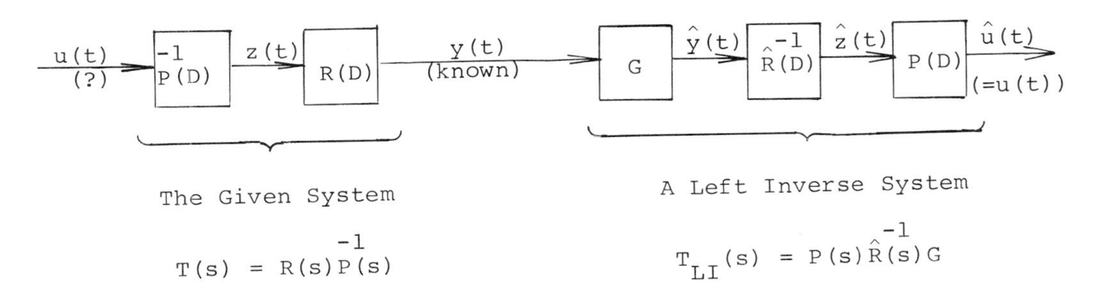

FIGURE 5.5.13

INPUT FUNCTION OBSERVABILITY VIA A LEFT INVERSE SYSTEM

the transfer matrix of this composite, series connected system is I_m; i.e.

$$\hat{u}(s) = T_{LI}(s)T(s)u(s) = P(s)\hat{R}^{-1}(s)GR(s)P^{-1}(s)u(s) = I_m u(s). \qquad 5.5.14$$

It is therefore clear that the system defined by 5.5.12 is a left inverse of the given system. Furthermore, it should be noted that if a "separation" is now made in the composite system depicted in Figure 5.5.13 at the input to $\hat{R}^{-1}(D)$, which we have denoted as $\hat{y}(t)$, then the systems defined to either side of this "separation" will satisfy identical matrix differential equations; i.e. $P(D)z(t) = u(t)$; $\hat{y}(t) = GR(D)z(t) = \hat{R}(D)z(t)$ and $\hat{R}(D)\hat{z}(t) = \hat{y}(t)$; $\hat{u}(t) = P(D)\hat{z}(t)$ respectively. Therefore, if all of the initial conditions associated with 5.5.12a are set equal to those corresponding initial conditions on $z(t)$ and its derivatives, then $\hat{u}(t) = u(t)$ for all $t \geq t_o$, a fact which clearly follows from the uniqueness of solutions to linear differential equations of this type [C2]. Sufficiency of the first half of Theorem 5.5.7 is therefore constructively established.

In order to establish sufficiency of the second half of the theorem we assume that $\rho[T(s) = R(s)P^{-1}(s)] = p$, a condition which holds if and only if $\rho[R(s)] = p$ (see Problem 5-18). Therefore, if this condition does hold, we can define $\hat{R}(D)$ as any $(m \times m)$ nonsingular polynomial matrix obtained by appending to $R(D)$ $(m-p)$ linearly independent rows of column degree less than or equal to the column degree of $P(D)$. We might note that $R(D)$ can be recovered from $\hat{R}(D)$ by premultiplication (of $\hat{R}(D)$) by the constant $(p \times m)$ matrix G which is identically zero except for its first p columns which equal I_p; i.e.

$$R(D) = G\hat{R}(D) \tag{5.5.15}$$

We will now verify that the system "driven" by the desired output, $y(t)$, of the given system and defined by the differential operator representation:

$$\hat{R}(D)\hat{z}(t) = G^T\hat{y}(t); \tag{5.5.16a}$$

$$u(t) = P(D)\hat{z}(t), \tag{5.5.16b}$$

as depicted in Figure 5.5.17, represents a right inverse of the given system 5.5.6. In particular, in light of 5.5.15 and Figure 5.5.17,

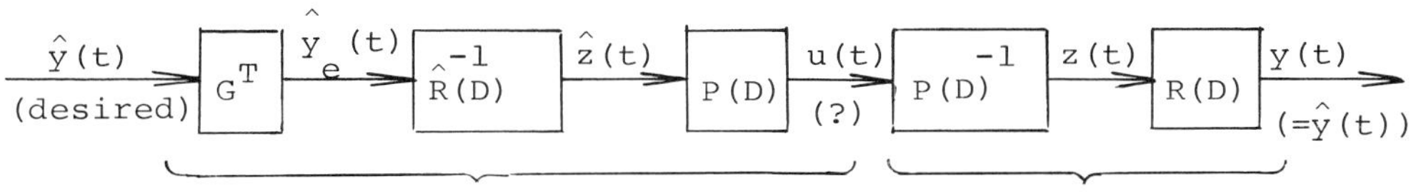

A Right Inverse System: $T_{RI}(s) = P(s)\hat{R}(s)^{-1}G^T$

The Given System: $T(s) = R(s)P(s)^{-1}$

FIGURE 5.5.17

OUTPUT FUNCTION CONTROLLABILITY VIA A RIGHT INVERSE SYSTEM

it follows that the transfer matrix of this composite series connected system is I_p; i.e. since $R(s)\hat{R}(s)^{-1} = G$ and $GG^T = I_p$,

$$y(s) = T(s)T_{RI}(s)\hat{y}(s) = R(s)P(s)^{-1}P(s)\hat{R}(s)^{-1}G^T\hat{y}(s) = I_p\hat{y}(s) \tag{5.5.18}$$

The system defined by 5.5.16 is therefore a right inverse of the given system. It should be noted that by appending $(m-p)$ additional rows to $R(D)$ in the construction of this right inverse system, we have, in a sense, enlarged the dimension of the desired output from p to m; i.e. the final $(m-p)$ rows of $\hat{R}(D)\hat{z}(t) = \hat{y}_e(t)$ will be identically zero for all time which implies, in turn, that we now require that the final $(m-p)$ rows of $y_e(t) \triangleq \hat{R}(D)z(t) \equiv 0$ for all $t \geq t_o$. As in the case of the left inverse, we can readily show that both systems satisfy identical matrix differential equations, namely $\hat{R}(D)\hat{z}(t) = \hat{y}_e(t)$ (or $R(D)\hat{z}(t) = \hat{y}(t)$); $u(t) = P(D)\hat{z}(t)$ and $P(D)z(t) = u(t)$; $y_e(t) = \hat{R}(D)z(t)$ (or $y(t) = R(D)z(t)$) respectively. The problem of establishing this final portion of the theorem thus reduces to the determination of an appropriate set of initial conditions for $z(t)$, $\hat{z}(t)$, and their derivatives so that $y_e(t) \equiv \hat{y}_e(t)$ for all $t \geq t_o$.

To resolve the question of initial conditions, we now employ a result which will be formally established, under less restrictive assumptions, in Section 7.4 (see Theorem 7.4.3), namely that $P(D)$ can be expressed as

$$P(D) = N(D) + M(D)\hat{R}(D), \tag{5.5.19}$$

and

$$\partial_c[N(D)] < \partial_c[\hat{R}(D)]^{\dagger} \tag{5.5.20}$$

In view of this result, the equations which define the right inverse system (5.5.16) can now be written as:

$$\hat{R}(D)\hat{z}(t) = G^T\hat{y}(t) = \hat{y}_e(t); \tag{5.5.21a}$$

$$u(t) = N(D)\hat{z}(t) + M(D)\hat{y}_e(t), \tag{5.5.21b}$$

---

† This result actually follows quite readily here if we express the transfer matrix, $P(s)\hat{R}^{-1}(s)$, as the sum of its quotient, $M(s)$, and its strictly proper part, $N(s)\hat{R}^{-1}(s)$ (see Problems 4-20 and 5-8), and then postmultiply the resulting expression by $\hat{R}(s)$. Example 5.5.24 illustrates this procedure.

and represented pictorially in Figure 5.5.22.

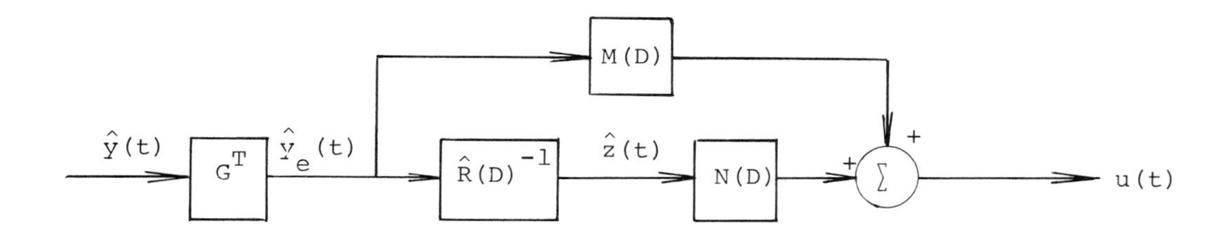

FIGURE 5.5.22

A RIGHT INVERSE SYSTEM

In view of 5.5.20 and the fact that $\partial_c[\hat{R}(D)] \leq \partial_c[P(D)]$, it now follows that $\partial_c[N(D)] < \partial_c[P(D)]$ and, therefore, that the determinants of both $P(D)$ and the product $M(D)\hat{R}(D) = P(D) - N(D)$ have degree n; i.e. $n = \partial[|P(D)|] = \partial[|M(D)\hat{R}(D)|]$ and, furthermore, that

$$n = \partial[|M(D)\hat{R}(D)|] = \partial[|M(D)|] + \partial[|\hat{R}(D)|] = d_m + d_{\hat{r}}, \qquad 5.5.23$$

where $d_m = \partial[|M(D)|]$ and $d_{\hat{r}} = \partial[|\hat{R}(D)|]$. In view of the results given in Section 5.2, it is clear that $d_{\hat{r}} = \partial[|\hat{R}(D)|]$ represents the number of integrators required to realize this right inverse system. Furthermore, $d_m = \partial[|M(D)|]$ represents the number of differentiators required, a fact which can readily be established once we show that $M(s)^{-1}$ is a proper transfer matrix (see Problem 5-22).

It is now important to note that since $\hat{y}_e(t)$ is not specified until a sufficient number of derivatives of $\hat{z}(t)$ to define $\hat{R}(D)\hat{z}(t)$ has been obtained, the $d_{\hat{r}}$ initial conditions (on the lower order derivatives of $z(t)$) which are implicit in Equation 5.5.21a can be arbitrarily set. However, $\hat{y}_e(t)$ must equal $\hat{R}(D)\hat{z}(t)$ for all time, including the initial time, t_o. Furthermore, $\hat{y}_e(t)$ must be sufficiently differentiable to permit a solution (for the higher order derivatives of $\hat{z}(t_o)$) to the following sequence of differential equations which is obtained by repeated differentiation of $\hat{y}_e(t)$: $D\hat{y}_e(t_o) = D\hat{R}(D)\hat{z}(t_o)$, $D^2\hat{y}_e(t_o) = D^2\hat{R}(D)z(t_o)$, etc., terminating with the equation, $M(D)\hat{y}_e(t_o) = -N(D)\hat{z}(t_o) + u(t_o)$. Since $\hat{y}_e(t)$ is completely specified for all $t \geq t_o$ and $\partial[|M(D)|] = d_m$, it follows that these latter d_m initial conditions on the higher order derivatives of $\hat{z}(t)$ are uniquely determined via this procedure. These d_m "constrained" initial conditions in combination with the $d_{\hat{r}}$ "arbitrary" initial conditions implicit in 5.5.21a thus constitute a complete set of $n = d_{\hat{r}} + d_m$ initial conditions which can now be placed on both $z(t)$ and $\hat{z}(t)$ and their derivatives in both systems depicted in Figure 5.5.17 in order to obtain $y_e(t) \equiv \hat{y}_e(t)$ as the desired system output. Example 5.5.24, at the conclusion of this section, illustrates the construction of a right inverse system as well as the details associated with the derivation and placement of an appropriate set of initial conditions on both systems. Before presenting the example, however, a number of remarks are in order.

R 1: It should be noted that, as in the case of the right inverse system defined by 5.5.21, the output, $\hat{u}(t)$, of the left inverse system defined by 5.5.12 can also be expressed as the sum of $N(D)\hat{z}(t)$ and $M(D)\hat{y}(t)$ with $\partial_c[N(D)] < \partial_c[\hat{R}(D)]$; i.e. in view of 5.5.19 and 5.5.20 we can construct the left inverse system using $d_{\hat{r}}(= \partial[|\hat{R}(D)|])$ integrators and $d_m(= \partial[|M(D)|])$ differentiators in virtually the same manner employed in the construction of the right inverse system depicted in Figure 5.5.22.

R 2: With respect to the controllability properties of either inverse system, we leave, as an exercise for the reader, the formal verification of the fact that any left inverse, 5.5.12, of the given system 5.5.6 will be controllable. However, a right inverse, 5.5.16, of 5.5.6 need not be controllable, although any uncontrollable modes associated with a right inverse system can be arbitrarily assigned (see Problem 5-27).

Regarding the observability of either inverse system, the reader can readily verify that if the given system 5.5.6 is observable, then any right inverse system will also be observable, although a left inverse system need not be observable. On the other hand, if the given system is unobservable, then any left inverse will also be unobservable, although a right inverse may be observable (see Problem 5-24). In any case, it is worth noting that a lack of complete observability in either inverse system does not present any difficulties from the point of view of either input function observability or output function controllability, since unobservable inverse system modes have no effect whatsoever on the output, $u(t)$, of the inverse system. We might further note here that any unobservable modes of either inverse system simply imply either a lack of uniqueness of the initial conditions on the unobservable modes or a reduced order inverse (see Problem 5-25).

R 3: We next note that the zeros of $|\hat{R}(s)|$ represent the poles of either inverse system, and, furthermore, that the integer $d_m = n - \partial[|\hat{R}(s)|]$ represents the number of differentiators required to construct either inverse system. In view of these observations and the obvious desirability that either inverse system not only be stable but also involve as few differentiators as possible, we would like to select $\hat{R}(D)$ so that $|\hat{R}(s)|$ is a Hurwitz polynomial of highest possible degree, whenever such a choice is possible. Therefore, the selection of $(m-p)$ independent rows to add to $R(D)$ to define $\hat{R}(D)$ in the case of a right inverse or the choice of a premultiplier, G, to reduce $R(D)$ to $\hat{R}(D)$ in the case of a left inverse should be made so that, whenever possible, $|\hat{R}(s)|$ is a Hurwitz polynomial of highest degree (see Problems 5-19 and 5-20). To determine whether or not $|\hat{R}(s)|$ can be chosen to be a Hurwitz polynomial, i.e. whether or not a stable inverse can be achieved, we will now completely identify the "invariant poles" of either inverse. In particular, in the case of a right inverse system we will now show that if $R(D)$ and $P(D)$ are relatively right prime[†] and we reduce $R(D)$ to lower left triangular form via some unimodular matrix, $U_R(D)$, as in the constructive proof of Theorem 2.5.11, then the zeros of the determinant of the lower left triangular matrix, $R_{Rp}(s)$, formed by the first p columns of $R(s)U_R(s) \triangleq R_R(s)$ will represent the <u>invariant poles</u> or the <u>invariant modes</u> of any right inverse of the given system 5.5.6; i.e. those poles which are intrinsic to any $T_{RI}(s)$ and, therefore, cannot be cancelled or altered regardless of our choice for $\hat{R}(s)$. To show this, recall that if $T_{RI}(s)$ is any right inverse of $T(s) = R(s)P^{-1}(s)$, then in view of 5.5.2, $R(s)P^{-1}(s)T_{RI}(s) = R_R(s)P_R^{-1}(s)T_{RI}(s) = I_p$, where $P_R(s) = P(s)U_R(s)$. If we now append to $R_R(s)$, the $(m-p)$ ordered constant, standard basis row vectors (see Section 2.2): $e_{p+1}, e_{p+2}, \ldots, e_m$, and denote the resulting nonsingular, lower left triangular matrix as $R_{Re}(s)$, it follows that $R_{Re}(s)P_R^{-1}(s)T_{RI}(s) = Q_R(s)$, with the first p rows of $Q_R(s)$ equal to I_p. Since $R_R(s)$ and $P_R(s)$ are relatively right prime (by assumption), $R_{Re}(s)$ and $P_R(s)$ are also relatively right prime. Furthermore, $R_{Re}(s)$ and $Q_R(s)$ are relatively left prime since the first p rows of $Q_R(s)$ and the final $m-p$ rows of

---

[†] If $P(D)$ and $R(D)$ are not relatively right prime, i.e. if the given differential operator system is not observable, the unobservable modes can be "eliminated" by first defining $z_o(t) = G_R(D)z(t)$, where $G_R(D)$ is any g.c.r.d. of $P(D)$ and $R(D)$, and then considering the reduced order observable system: $P_o(D)z_o(t) = u(t)$; $y(t) = R_o(D)z(t)$, with $P_o(D) = P(D)G_R^{-1}(D)$ and $R_o(D) = R(D)G_R^{-1}(D)$--see Problem 5-25 also.

$R_{Re}(s)$ constitute the identity matrix I_m. In view of the fact that $T_{RI}(s) = P_R(s)R_{Re}^{-1}(s)Q_R(s)$, it is thus clear that $\{R_{Re}(D), Q_R(D), P_R(D), 0\}$ represents a minimal differential operator realization of $T_{RI}(s)$ of least possible order, and, therefore, that the zeros of $|R_{Re}(s)|$, which include the zeros of $|R_{Rp}(s)|$, represent poles of any right inverse system. Since the differential operator representation $\{R_{Re}(D), Q_R(D), P_R(D), 0\}$ does represent a right inverse of the given system, it follows that the zeros of $|R_{Rp}(s)|$ represent the (only) invariant poles of a right inverse system.

A similar result can also be established in the case of a left invertible system; i.e. if $T(s) = R(s)P^{-1}(s)$ is the $(p \times m)$ proper transfer matrix of a left invertible system, with $R(s)$ and $P(s)$ relatively right prime, then the zeros of the determinant of the upper right triangular matrix, $R_{Lm}(s)$, formed by the first m rows of $R_L(s) = U_L(s)R(s)$ (where $U_L(s)$ is any unimodular matrix which reduces $R(s)$ to upper right triangular form) will represent the <u>invariant poles</u> or the <u>invariant modes</u> of any left inverse system, a fact which we leave as an exercise for the reader to formally verify (Problem 5-28).

<u>R 4</u>: Finally, it should be noted that our development of both time domain inverse systems assumes a given system with a proper transfer matrix since this is usually the case in practice and this assumption allows us to draw certain conclusions regarding the number of integrators and differentiators required to construct either inverse system (see Equation 5.5.23). However, the results given in this section can be rather easily extended to include the more general case when $T(s) = R(s)P^{-1}(s)$ need not be proper (see Problem 5-26).

EXAMPLE 5.5.24: To illustrate the development and employment of left and right inverse systems to input and output functional reproducibility let us consider a particular state-space system of the form 3.2.1 with

$$A = \begin{bmatrix} -4 & -2 & -2 \\ 0 & 0 & 1 \\ 0 & -4 & 1 \end{bmatrix}, \quad B = \begin{bmatrix} 1 & 0 \\ 0 & 0 \\ 0 & 1 \end{bmatrix}, \quad C = \begin{bmatrix} -10 & -3 & -3 \\ 4 & 0 & -1 \end{bmatrix} \quad \text{and} \quad E = \begin{bmatrix} 2 & 0 \\ 0 & 0 \end{bmatrix}.$$

It is clear that this system is already in controllable companion form and, therefore, that the results given in Section 4.3 can be directly employed to obtain an equivalent differential operator representation of the form 5.5.6. In particular, from 4.3.9 we determine that $P(D) = \begin{bmatrix} D+4, & 2D+2 \\ 0, & D^2-D+4 \end{bmatrix}$ and from 4.3.8 that $R(D) = \begin{bmatrix} 2D-2, & D+1 \\ 4, & -D \end{bmatrix}$. It also follows from the results given in Section 5.2 that $z_1(t) = x_1(t)$, $z_2(t) = x_2(t)$, and $\dot{z}_2(t) = x_3(t)$.

Since $\rho[T(s) = R(s)P(s)^{-1}] = \rho[R(s)] = m = p = 2$ in this example, we will obtain the same equations for either inverse system; i.e. the given system has both a left and a right inverse, or simply an inverse, and the dynamical equations which define its inverse are given by 5.5.22 with $\hat{R}(D) = R(D)$ and $G = I$. To determine $M(D)$ and $N(D)$ we first obtain the quotient of the transfer matrix

$$P(s)R(s)^{-1} = \frac{\begin{bmatrix} s^2+12s+8, & -3s^2+5s+8 \\ 4s^2-4s+16, & -2s^3+4s^2-10s+8 \end{bmatrix}}{2s^2+2s+4}$$

by applying the scalar division algorithm (Section 2.6) to each rational entry of $P(s)R(s)^{-1}$.

In this way, we readily find that the quotient $M(s) = \begin{bmatrix} \frac{1}{2}, & -\frac{3}{2} \\ 2, & -s+3 \end{bmatrix}$

and that the strictly proper part,

$$N(s)R(s)^{-1} = P(s)R(s)^{-1} - M(s) = \frac{\begin{bmatrix} 11s+6, & 8s+14 \\ -8s+8, & -12s-4 \end{bmatrix}}{2s^2+2s+4}.$$

By postmultiplying

this latter expression by $R(s)$, we obtain $N(s) = \begin{bmatrix} 11, & \frac{3}{2} \\ -8, & 2 \end{bmatrix}$. Once $\hat{R}(D) = R(D)$, $M(D)$, $N(D)$, and G have been determined they can be employed as indicated in Figure 5.5.22 to realize either a left or right inverse of the given system. In the former (left inverse) case, $y(t)$, the output of the given system would be known for all $t \geq t_o$ along with $z_1(t_o)$, $z_2(t_o)$, and $\dot{z}_2(t_o)$. Therefore, since $\hat{R}(D) = R(D)$ and $|R(D)| = -2D^2-2D-4$, a polynomial of second ($= d_{\hat{r}}$) degree, we must set two ($= d_{\hat{r}}$) initial conditions on the integrators which are used to realize the left inverse system. Since $R(D)\hat{z}(t)$ in this example involves $\hat{z}_1(t)$, $\hat{z}_2(t)$, and their first derivatives, it follows that we would set $\hat{z}_1(t_o) = z_1(t_o) = x_1(t_o)$ and $\hat{z}_2(t_o) = z_2(t_o) = x_2(t_o)$ as the initial conditions on the two integrators of the inverse system (see Section 5.2). Note that the left inverse system would also involve one ($= d_m$) differentiation since $|M(D)| = -\frac{1}{2}D + \frac{9}{2}$.

Insofar as a right inverse system is concerned, we now assume knowledge of some "sufficiently differentiable" desired output, $\hat{y}(t)$. For example, suppose we were asked to find an input, $u(t)$, as well as a set of initial conditions on $z_1(t)$, $z_2(t)$, and $\dot{z}_2(t)$ of the given system in order to achieve $\hat{y}(t) = \begin{bmatrix} e^{-t} + e^{-2t} \\ \cos 4t \end{bmatrix}$ as the output of the given system. Since the $(d_{\hat{r}})$ initial conditions associated with the integrators of the inverse system can be arbitrarily set, we will choose these initial conditions to be: $\hat{z}_1(t_o) = \hat{z}_2(t_o) = 0$. Therefore, $z_1(t_o) = z_2(t_o) = x_1(t_o) = x_2(t_o) = 0$ in the given system as well. To determine $\dot{z}_2(t_o)$, we now employ 5.5.22a, i.e. $\hat{y}_e(t) = \hat{y}(t) = R(D)\hat{z}(t)$ (or $y(t) = R(D)z(t)$). We thus obtain

$$\begin{bmatrix} e^{-t} + e^{-2t} \\ \cos 4t \end{bmatrix} = \begin{bmatrix} 2D-2, & D+1 \\ 4, & -D \end{bmatrix} \begin{bmatrix} z_1(t) \\ z_2(t) \end{bmatrix} \quad \text{or at} \quad t = t_o,$$

$$\begin{bmatrix} e^{-t_o} + e^{-2t_o} \\ \cos 4t_o \end{bmatrix} = \begin{bmatrix} 2 & 1 \\ 0 & -1 \end{bmatrix} \begin{bmatrix} \dot{z}_1(t_o) \\ \dot{z}_2(t_o) \end{bmatrix},$$ which clearly implies that $\dot{z}_2(t_o) = -\cos 4t_o$. If these three initial conditions are now placed on the given system, and it is "driven by the right inverse system with zero initial conditions and input $\hat{y}(t)$, then $y(t) \equiv \hat{y}(t)$ for all $t \geq t_o$ and the output, $u(t)$, of the right inverse system will represent an appropriate input to the given system.

5.6. CONCLUDING REMARKS AND REFERENCES

We have now presented all (six) possible transfer relations between any two of the three methods commonly employed for representing the dynamical behavior of linear, time-invariant, dynamical systems, namely the (frequency domain) transfer matrix and the (time domain) state and differential operator representations. Table 5.6.1 summarizes these various transfer relations, indicating where appropriate transfer algorithms can be found. As noted, the notion of equivalence plays an important role whenever a transfer from one time domain representation to another is made, while the concept of minimality becomes important whenever a transfer from the frequency domain to the time domain is made.

A considerable amount of effort was directed at defining and determining state representations which are equivalent to the more general class of differential operator representations. The definition of equivalence which we employed was motivated, in large part, by a similar but less inclusive one due to Polak [P2], and was shown to directly imply a number of desirable necessary conditions for equivalence such as transfer matrix equivalence, preservation of system order and all (n) modes, the ability to equate the state of the equivalent system to the partial state, input, and their derivatives of the differential operator system, and the "reduction" to the conventional equivalence definition whenever both systems are in state form.

The notions of (complete state) controllability and observability were then extended, via the equivalence definition and Theorem 5.2.14, to include the more general class of differential operator systems. The fact that such an extension is possible was first noted, but not clearly established, by Rosenbrock [R2][R3]. In particular, Rosenbrock [R2] was first to state and prove the equivalent of Corollary 5.3.6 in terms of the "Smith canonical form" [G1] of the composite matrix $\begin{bmatrix} sI-A \\ \hdashline C \end{bmatrix}$ rather than a g.c.r.d. of (sI-A) and C. He also noted that the matrices P(s) and Q(s) (P(s) and R(s)) are related to the controllability (observability) of differential operator representations, although he was rather elusive in defining controllability (observability) of the more general class of systems--his "second form". We have attempted to clarify this point here by our definition of equivalence, Theorem 5.3.1, and the notions of controllable, observable, and minimal differential operator representations. Rosenbrock [R3] also appears to have been first to establish the fact that a state form equivalent system is minimal if and only if the "Smith canonical form" [G1] of the composite matrices [P(s),Q(s)] and $\begin{bmatrix} P(s) \\ \hdashline R(s) \end{bmatrix}$ are [I, 0] and $\begin{bmatrix} I \\ \hdashline 0 \end{bmatrix}$ respectively, although Popov [P5] has also obtained analogous results. We have established essentially the same fact here by employing the notions of greatest common (left or right) divisors of polynomial matrices rather than the "Smith canonical form". This significantly facilitates the computations required to determine minimality.

The extension of the notions of controllability and observability to include differential operator representations naturally led to a new technique for obtaining realizations and, in particular, minimal realizations of rational transfer matrices, the subject of a number of recent investigations (e.g. [W4][K5][Y1][H2][M2]). The approach taken here in Section 5.4 was based on an initial frequency domain reduction, rather than a final time domain reduction as in Chapter 4, although both approaches involve approximately the same amount of numerical computation. It should perhaps be noted that the results given in Sections 5.2 through 5.4 were, for the most part, first presented in [W5].

The question considered in Section 5.5--namely, inverting dynamical systems, has been of interest to control engineers for over two decades, and the applicability of a system inverse either implicitly or explicitly to a multitude of problems in control and network synthesis has been demonstrated repeatedly [B2][B3][A2][F1][S1][S3]. The results on inverse systems presented in Section 5.5 represent a compilation of a number of individual contributions, including the author's. It appears that Brockett and Mesarovic [B2] were first to define and present a necessary and sufficient matrix rank condition for output function controllability (functional reproducibility) in terms of the state-space quadruple $\{A,B,C,E\}$. Brockett [B3] later related his results on functional reproducibility to the inverse system in those cases when $p = m$. Dorato [D2] later simplified the matrix rank condition of Brockett and Mesarovic somewhat, again employing a state-space description of the given system while considering only the case when $m = p$. Sain and Massey [S3] also simplified the original criterion given in [B2] and, furthermore, extended the relationship between invertibility and functional reproducibility to include the case when $m \neq p$. They were first to speak of "functional input observability", or what we call input function observability, as representing the "dual" of "functional output controllability", or what we call output function controllability, but restricted their discussion of inverse systems almost exclusively to what we call the left inverse system here; i.e. they do not distinguish between the two, and speak of "functional output controllability" and system invertibility as "dual" notions, which is not completely consistent with the approach employed here. They also outline a procedure for the construction of "L-integral inverse systems", which can be used to produce a proper form of inverse system.

Silverman [S1] also restricts his results to the case when $m = p$, but considers time-varying systems as well as the time-invariant case. He also outlines a procedure for the construction of the inverse system when $m = p$ and appears to have been first to discover that the inverse of a square system can be implemented via a total of n integrators and differentiators. Other investigations involving the inverse system have focused either on specific applications of the inverse system or on improving the computational procedures involved in the construction of the inverse (e.g. [A2][F1][H3][M3][S4]).

It might finally be noted that conditions for determining the existence of a right or left inverse system of a state-space system in terms of the quadruple $\{A,B,C,E\}$ do exist [B2][S1][S3][O1], as well as constructive procedures for implementing either inverse system when $p = m$ [S1], or an "L-integral inverse" when $m < p$ [S3]. However, these existence tests are rather unwieldy when compared to the computation of the rank of $T(s)$ (Theorem 5.5.3).

TO DETERMINE

GIVEN	A STATE REPRESENTATION $\{A,B,C,E(D)\}$	A DIFFERENTIAL OPERATOR REPRESENTATION $\{P(D),Q(D),R(D),W(D)\}$	THE TRANSFER MATRIX $T(s)$
$\{A,B,C,E(D)\}$	----	$P(D) = DI-A;\ Q(D) = B$ $R(D) = C;\ W(D) = E(D)$ (SECTION 3.2)	$T(s) = C(sI-A)^{-1}B + E(s)$ (SECTION 4.2)
$\{P(D),Q(D),R(D),W(D)\}$	USE ALGORITHM GIVEN IN SECTION 5.2	----	$T(s) = R(s)P^{-1}(s)Q(s)+W(s)$ (SECTION 5.2)
$T(s)$	USE TIME DOMAIN REDUCTION ALGORITHM (SECTION 4.4) USE FREQUENCY DOMAIN REDUCTION ALGORITHM (SECTION 5.4)	EXPRESS $T(s)$ AS $R(s)P_c^{-1}(s) + E(s)$ (SECTION 5.4)	----

TABLE 5.6.1

DYNAMICAL SYSTEM TRANSFER RELATIONS

PROBLEMS - CHAPTER 5

5-1 Draw a block diagram of the system described by 5.2.1.

5-2 Determine the quotients and the strictly proper parts of the following transfer matrices:

(a) $T(s) = \begin{bmatrix} \frac{3s-1}{s+2} \\ 4s \end{bmatrix}$ (b) $T(s) = \begin{bmatrix} \frac{s^2-2}{2s^3+3}, & \frac{s^3}{3s+1} \\ 2s-2, & 0 \end{bmatrix}$

(c) $T(s) = P^{-1}(s)Q(s)$, where $P(s) = \begin{bmatrix} 2s^2-2, & 2s^2+1 \\ -s, & -s-3 \end{bmatrix}$ and $Q(s) = \begin{bmatrix} s^3-s^2+s-1 \\ s+1 \end{bmatrix}$.

5-3 Show that equivalence, in the sense of Definition 5.2.9, implies both zero-state and zero-input equivalence (defined in Section 3.4).

5-4 Outline the "dual" of the algorithm employed to establish Theorem 5.2.14 when $Q(D)$ is unimodular.

5-5 Verify that the transfer function associated with the differential operator system: $P(D) = \begin{bmatrix} 2D^2-2, & 2D^2+1 \\ -D, & D+3 \end{bmatrix}$, $Q(D) = \begin{bmatrix} D-1 \\ D^2+2 \end{bmatrix}$, $R(D) = [D, -D]$, and $W(D) = D^2 - 3/2D + 17/4$ is strictly proper.

5-6 Find a state-space system which is equivalent to the following differential operator system: $P(D) = \begin{bmatrix} D^3-D, & D^2-1 \\ -D-2, & 0 \end{bmatrix}$,

$Q(D) = \begin{bmatrix} D-1, & -2D+2 \\ 1, & 3D \end{bmatrix}$, $R(D) = \begin{bmatrix} 2D^2+D+2, & 2D \\ -D-2, & 0 \end{bmatrix}$, and

$W(D) = \begin{bmatrix} -1, & 3D+4 \\ -1, & -3D \end{bmatrix}$. Is the equivalent state-space system controllable? observable? Verify that the transfer matrices of both systems are equal.

5-7 Consider the differential operator system: $P(D) = \begin{bmatrix} D & 0 & -1 \\ 0 & 1 & 0 \\ -D & 0 & D^2 \end{bmatrix}$, $Q(D) = \begin{bmatrix} 1, & -D+1 \\ D+1, & -1 \\ 1, & 0 \end{bmatrix}$, $R(D) = [2D-1, \ -2, \ D+3]$, and $W(D) = [D-1, \ -3]$. Show that the second column of $R(D)$ can be zeroed by "appropriately modifying" $W(D)$. Find an equivalent state-space system.

5-8 Show that if $R(s)P(s)^{-1}$ $(P(s)^{-1}Q(s))$ is a rational proper transfer matrix, then $\partial_c[R(s)] \leq \partial_c[P(s)]$ $(\partial_r[Q(s)] \leq \partial_r[P(s)])$ whether or not $P(s)$ is column (row) proper. Also show that if $P(s)$ is column (row) proper and $\partial_c[R(s)] \leq \partial_c[P(s)]$ $(\partial_r[Q(s)] \leq \partial_r[P(s)])$ then $R(s)P(s)^{-1}$ $(P(s)^{-1}Q(s))$ will be proper, and verify, by example, that this result does not necessarily hold if $P(s)$ is not column (row) proper. Show that strict inequality holds whenever $R(s)P(s)^{-1}$ is strictly proper.

5-9 Show that Step 3 in the constructive proof of Theorem 5.2.14 does not affect system equivalence.

5-10 Determine (by inspection) an observable state-space system which is equivalent to the differential operator system:

$P(D) = \begin{bmatrix} D^2+D-2, & 3D-1 \\ 2D^2+2, & D^3-2D \end{bmatrix}$, $Q(D) = \begin{bmatrix} 1, & 3D-2 \\ -D^2, & 2D \end{bmatrix}$, $R(D) = I_2$, and $W(D) = 0$. In view of Problem 5-4, find a (controllable) state-space system which is equivalent to the differential operator system:

$$P(D) = \begin{bmatrix} D+1, & 2D-3 \\ 1, & D^2 \end{bmatrix}, \quad Q(D) = I_2, \quad R(D) = \begin{bmatrix} 2, & D-3 \\ 1, & 2D \end{bmatrix}, \text{ and } W(D) = 0.$$

5-11 Show that the C and $E(D)$ obtained in the final step of Example 5.2.32 can also be obtained from the relations: $y(t) = \hat{R}_o(D)z_o(t) + \hat{W}(D)u(t)$, $\dot{x}_o(t) = A_o x_o(t) + B_o u(t)$, and $z_o(t) = \bar{C}_o x_o(t) + H_o(D)u(t)$.

5-12 By employing Theorem 5.3.1, find any and all uncontrollable and/or unobservable modes associated with the differential operator system defined in Problem 5-6. Compare your findings to the results obtained in Problem 5-6.

5-13 Show that $P(s)P_{co}^{-1}(s)$ (employed in the proof of Theorem 5.3.1) is a polynomial matrix, $G_L(s)$, which represents a g.c.l.d. of $P(s)$ and $Q(s)$. (Hint: Use the dual of the result established in Problem 2-25 to show that $Q_{co}(s)M(s) + P_{co}(s)N(s) = I_q$, and by then premultiplying by $P(s)P_{co}^{+}(s)$, show that $|P_{co}(s)|$ must divide $P(s)P_{co}^{+}(s)$).

5-14 Consider the differential equation:

$$\ddot{y}(t) + \dot{y}(t) - 2\,y(t) = \dddot{u}(t) - \dot{u}(t) + u(t)$$

Find an equivalent state-space representation and verify that it is both controllable and observable. Express the equivalent state, $x(t)$, as a linear function of $y(t)$, $u(t)$, and their derivatives (see 5.2.13c). Repeat the above for the differential operator system:

$$\begin{bmatrix} D, & 1 \\ -D, & D+3 \end{bmatrix}\begin{bmatrix} z_1(t) \\ z_2(t) \end{bmatrix} = \begin{bmatrix} u_1(t) \\ u_2(t) \end{bmatrix}; \quad \begin{bmatrix} y_1(t) \\ y_2(t) \end{bmatrix} = \begin{bmatrix} D^2, & 2 \\ 1, & 0 \end{bmatrix}\begin{bmatrix} z_1(t) \\ z_2(t) \end{bmatrix}.$$

5-15 Express the transfer matrix $T(s) = \begin{bmatrix} \frac{s^2-2}{s^2-1}, & \frac{3s}{s+1} \\ 1, & \frac{2}{s-1} \end{bmatrix}$ as the product $R(s)P_c^{-1}(s) + E(s)$, where $E(s)$ is the quotient of $T(s)$ and $R(s)$ and $P_c(s)$ are relatively right prime. Express $T(s)$ as the product $P_o^{-1}(s)Q(s) + E(s)$, where $P_o(s)$ and $Q(s)$ are relatively left prime. Find a minimal realization of $T(s)$.

Repeat the above for $T(s) = \begin{bmatrix} \frac{2}{s+1} & \frac{1}{s+1} \\ \frac{4}{s+1} & \frac{2}{s+1} \end{bmatrix}$.

5-16 Show that if system 5.2.1a with $P(D) = DI-\hat{A}$ and $Q(D) = \hat{B}$ is equivalent to system 5.2.6a, in the sense of Definition 5.2.9(i), then $H(D)$ in 5.2.10 will be identically zero.

5-17 Find the invariant poles of the inverse system developed in Example 5.5.24 and verify that the inverse system is asymptotically stable.

5-18 Using Sylvester's Inequality (Problem 2-16), verify that $\rho[T(s)] = \rho[R(s)P^{-1}(s)] = \rho[R(s)]$ in 5.5.7.

5-19 Consider a dynamical system characterized by the differential operator representation 5.5.6 with $P(D) = \begin{bmatrix} D-2, & 0 \\ 0, & D^2 \end{bmatrix}$ and $R(D)$ given below. For each of these cases determine if the given system is input function observable, output function controllable, or both and whenever an affirmative answer is obtained, determine the invariant poles of the inverse system and find an appropriate inverse system.

$$R(D) = [D+1, \quad D-1] \qquad\qquad R(D) = \begin{bmatrix} D, & -1 \\ 2, & 1 \end{bmatrix}$$

(a) (b)

$$R(D) = \begin{bmatrix} D-2, & 3D-6 \\ 1, & 3 \end{bmatrix} \qquad\qquad R(D) = \begin{bmatrix} D, & 1 \\ 2, & D \\ D+1, & 0 \end{bmatrix}$$

(c) (d)

5-20 Show that any arbitrary set of (2) poles can be obtained for a controllable and observable left inverse of the system defined in case (d) of the preceding problem. Show that any set of (3) poles can be obtained for a controllable and observable right inverse of the system defined in case (a) of Problem 5-19. Illustrate, by example, that a right inverse of order zero ($= \partial[|\hat{R}(D)|]$) can also be defined in case (a), although such an inverse would require three ($= m - \partial[|\hat{R}(D)|]$) differentiators.

5-21 Suppose that the dynamical behavior of an m-input, p-output uncontrollable multivariable system is represented via the differential operator representation:

$$\underbrace{\begin{bmatrix} P_c(D) & P_{c\bar{c}}(D) \\ 0 & P_{\bar{c}}(D) \end{bmatrix}}_{P(D)} \begin{bmatrix} z_c(t) \\ z_{\bar{c}}(t) \end{bmatrix} = \underbrace{\begin{bmatrix} I_m \\ 0 \end{bmatrix}}_{Q(D)=Q} u(t); \quad y(t) = \underbrace{[R_c(D) \mid R_{\bar{c}}(D)]}_{R(D)} \begin{bmatrix} z_c(t) \\ z_{\bar{c}}(t) \end{bmatrix}$$

where $P(D)$ is $r \times r$ and nonsingular and the subscript c is used to denote the controllable "portion" of the system. Show that a left (right) inverse of this system can be developed using essentially the same procedure employed in Section 5.5 provided $\rho[R(s)P^{-1}(s)] = r(\rho[R_c(s)P_c^{-1}(s) = T(s)] = p)$ and then employed to resolve the question of input function observability (output function controllability).

5-22 Prove that if $M(D)$ is obtained via 5.5.19, then $M^{-1}(s)$ will be a proper transfer matrix which, consequently, can be realized through the employment of $d_m = \partial[|M(s)|]_{-1}$ integrators. In view of this observation, show that if $M(s)$ is proper then $M(s)$ can be realized by d_m differentiators by simply replacing all $(d_m)_{-1}$ integrators of any realization of $M(s)$ by differentiators while simultaneously reversing the directions of all signal flows, taking care to "appropriately alter" the summation junction signs in order to maintain the same functional relationships. Show that this procedure for realizing $M(s)$ is unnecessary if $M(s)$ is column proper.

5-23 In view of Problem 5-22, illustrate (by example) that a differential operator, $M(D)$, cannot always be realized by $d_m = \partial[|M(D)|]$ differentiators.

5-24 Show that if a system defined by 5.5.6 is unobservable then any left inverse of the form 5.5.12 will also be unobservable although a right inverse, as given by 5.5.16, may not be unobservable.

5-25 Show that the zeros of the determinant of any g.c.r.d., $G_R(D)$, of $\hat{R}(D)$ and $N(D)$ represent the unobservable modes of either inverse system and, furthermore, that the initial conditions associated with any unobservable modes of both the given system and either inverse can be arbitrarily altered without affecting the overall input/output behavior of the composite series connection. Show that this latter observation implies the ability to eliminate any and all unobservable modes in the construction of either inverse system, thereby obtaining an inverse system of reduced order ($< n$).

5-26 Show that the condition: $\partial_c[R(D)] \leq \partial_c[P(D)]$, which was imposed early in the development of inverse systems can actually be omitted and, consequently, that systems which do not have proper transfer matrices can also have well-defined inverse systems provided their transfer matrix is of full rank. In view of this observation, show that a left (right) inverse of a right (left) inverse of a given system is the given system itself.

5-27 Show that any left inverse of the controllable system 5.5.6 must also be controllable, although a right inverse need not be. Also show that, other than the invariant modes, any other modes of a right inverse of 5.5.6 must necessarily arise from the addition, to $R(D)$, of the $(m-p)$ independent but arbitrary row vectors used to define $\hat{R}(D)$ and, therefore, that all of these non-invariant modes will represent uncontrollable modes of the right inverse system which can be arbitrarily assigned.

5-28 Verify the final statement made in R3 of Section 5.5 regarding the invariant poles of any left inverse system and show that all other poles of a left inverse system can be arbitrarily assigned.

5-29 Define the dual of the differential operator representation $\{P(D),Q(D),R(D),W(D)\}$ and show that the dual of a left (right) invertible system is a right (left) invertible system.

5-30 Verify, by example, that the invariant poles of either inverse system need not appear explicitly as zeros of the transfer function entries which comprise the given system.

5-31 Show that any rational $(p \times m)$ transfer matrix, $T(s)$, can be uniquely factored as the product: $R(s)P(s)^{-1}$, where $R(s)$ and $P(s)$ are relatively right prime and $P(s)$ is in lower left triangular form (as in Theorem 2.5.11) with $\Gamma_r[P(s)] = I_m$. State the dual of this fact.

5-32 Prove that any two minimal (controllable and observable) differential operator systems are equivalent if and only if they have the same transfer matrix. (Hint: First show that two controllable and observable state space systems are equivalent if and only if they have the same transfer matrix, and then employ Theorem 5.2.14.)

5-33 If the zeros of a system characterized by a full rank, $p \times m$ rational transfer matrix $T(s) = R(s)P(s)^{-1}$, with $R(s)$ and $P(s)$ relatively right prime, are defined as the invariant poles of any right (left) inverse of $T(s)$, show that the zeros of the system are also given by those zeros which are common to all of the nonzero, $p \times p$ $(m \times m)$ minors of $R(s)$. In view of this definition, show that the zeros of $T(s)$ are also equal to those s_i in $\mathscr{C}$, the complex field, for which $\rho[R(s_i)] < p(m)$ over $\mathscr{C}$.

5-34 In view of the previous problem, verify that the system with the transfer matrix $T(s) = \begin{bmatrix} \frac{s+1}{s^2} \\ \frac{s}{s^2-1} \end{bmatrix}$ has no zeros, while the system whose transfer matrix $T(s) = \dfrac{\begin{bmatrix} s^2+s & -s-1 \\ s^2+2s & -s-2 \\ s+3 & s^2+3s \end{bmatrix}}{s^2+1}$ has a single zero at $s = -3$.

CHAPTER 6

LINEAR STATE VARIABLE FEEDBACK

6.1 INTRODUCTION

With this chapter, we begin the study of the synthesis or design of compensation schemes for linear dynamical systems. Thus far, our attention has focused almost exclusively on a variety of procedures used in the analysis of these systems as well as the interrelations between various system representations. These initial investigations and the results which were obtained will now prove to be instrumental in the development of various alternative control and compensation techniques. The remainder of this text will therefore be devoted to the detailed development of a number of general control schemes including various types of feedforward and feedback compensators, as well as certain combinations of the two. We will employ both time domain and frequency domain methods for compensation, displaying any analogies between the two approaches when and where appropriate.

The need for the compensation or control of physical systems is self-evident, and examples of control systems can be found in virtually every walk of life; e.g. biological systems, transportation systems, industrial processes, aerospace vehicles, as well as economic and social systems are all subject to some form of regulatory behavior (control). For our point of view, we will be concerned here with controlling or improving the performance of systems whose dynamical behavior is represented via any one of the three methods which have been discussed in the previous three chapters, and although we will not consider any specific physical systems here, it should be noted that the dynamical behavior of a large number of practical systems is often approximated by one or more of the three mathematical representations which we have discussed.

In this chapter, we will be primarily concerned with a very common state-space technique for compensating linear dynamical systems, namely linear state variable feedback. In Section 6.2 we motivate this notion via linear quadratic optimal control and, in particular, by formulating and displaying the solution to the "output regulator problem". We then discuss the implications of linear state variable feedback from the point of view of its influence on the controllable companion form representation which was introduced in Section 3.6, and also outline a procedure for completely and arbitrarily assigning the closed loop poles of controllable systems via this compensation technique. Certain questions concerning the practical employment of a linear state variable feedback compensation scheme when the entire state of the system is not directly measurable are then considered in Section 6.4, where we introduce and discuss the notion of a Luenberger observer as an alternative to linear state variable feedback. We then conclude with some background material and references in Section 6.5.

6.2 QUADRATIC OPTIMIZATION--THE INITIAL MOTIVATION

Although virtually all forms of linear compensation can be represented, in part, by an appropriate linear state variable feedback control law, as we will later illustrate, it appears that the first "explicit" instance of the use of linear state variable feedback (l.s.v.f.) for the compensation of state-space systems was provided by the solutions to "quadratic optimization" problems, such as the "output regulator problem". We will therefore discuss this problem in detail in this section in order to motivate the more general employment of this very common synthesis procedure. In particular, we first consider dynamical systems in the state form 3.2.1, with $E = 0$, or

$$\dot{x}(t) = Ax(t) + Bu(t); \; y(t) = Cx(t), \qquad 6.2.1$$

which is repeated here for convenience. We further assume that this system, or the pair $\{A,B\}$, is (completely state) controllable, the pair $\{A,C\}$ is (completely state) observable, and that B is of full rank $m \leq n$.

If, for some reason, the unforced ($u(t) \equiv 0$) dynamical response of this system, due to nonzero initial conditions, was not satisfactory (e.g. it might be underdamped or even unstable), then some form of compensation would be in order. Generally speaking, it seems reasonable to assume that a control engineer might be interested in designing a control system which (i) causes the entire state, $x(t)$, of the compensated system to rapidly seek the origin of the state space in response to any nonzero initial conditions with relatively little oscillatory behavior, and (ii) does not involve an excessive amount of control action which might cause saturation (nonlinear behavior) or involve an excessive utilization of fuel. Furthermore, it might be desirable to assign different weights to certain states and controls relative to the others, thus penalizing the control system design for excessive variations of certain selected states and controls (e.g. it is important in manned vehicle maneuvers to avoid excessive accelerations). The goal or design objective, thus stated, is a heuristic one which will now be given a mathematical interpretation; i.e. the goal of the <u>output regulator problem</u> is to find a control $u^*(t)$, if one exists, which minimizes the <u>quadratic performance index</u>, J, when

$$J = \int_{t_o}^{\infty} [x^T(t)C^TCx(t) + u^T(t)Ru(t)]dt, \qquad 6.2.2$$

subject to the constraint that $x(t)$ in 6.2.2 be the (unique) solution to 6.2.1 with any arbitrary initial condition, $x(t_o)$.

In 6.2.2, we will require that R be a real, symmetric, <u>positive definite</u> matrix; i.e. that $R = M^TM$ for some nonsingular $(m \times m)$ matrix M with elements in $\mathcal{R}$. This assumption implies that

$u^T(t)Ru(t) > 0$ for all $u(t) \neq 0$, and that $u^T(t)Ru(t) = 0$ only when $u(t) \equiv 0$, and, together with the assumption of system controllability, is necessary to insure the existence as well as the uniqueness of the optimal control, $u^*(t)$. The observability assumption is necessary to insure that all (n) states of the system contribute to the final value of J. In many cases, either R is chosen to be (or C^TC happens to be) a diagonal matrix; i.e. $R = \text{diag}[r_i]$ for $i = 1,2,\ldots,m$ (or $C^TC = \text{diag}[c_j]$ for $j = 1,2,\ldots,p$). It might be noted that in these cases, a relatively large value of r_k (or c_k) for some k clearly implies a desire to avoid excessive excursions of the state $x_k(t)$ from the origin in the state space (or excessive utilization of the control $u_k(t)$).

The solution to the output regulator problem, thus formulated, is well known [K1][A1] and can now be formally stated:

THEOREM 6.2.3: Consider the performance index J, given by 6.2.2, where $x(t)$ is the solution to 6.2.1 with arbitrary initial state, $x(t_o)$, under the control $u(t)$. Under the assumptions noted, the (optimal) control, $u^*(t)$, which minimizes J exists, is unique, and can be expressed as a linear function of the state, $x(t)$; i.e.

$$u^*(t) = -R^{-1}B^TKx(t), \tag{6.2.4}$$

where K is the unique, positive definite solution to the matrix Riccati equation,

$$KA + A^TK - KBR^{-1}B^TK = -C^TC \tag{6.2.5}$$

Furthermore, the closed loop poles of the (optimal) system are the zeros of $|sI-A + BR^{-1}B^TK|$, and lie in the half-plane, $\text{Re}(s) < 0$, and the (minimum) value of the performance index J is equal to $x^T(t_o)Kx(t_o)$.

Proof: We merely remark here that this theorem represents a special case of a more general result, and its proof can be obtained in more than one way, but goes beyond the intended scope of this text. The primary reason for presenting the output regulator problem here has been to introduce and motivate the use of l.s.v.f. as a general feedback design procedure. We note that the control law 6.2.4 does indeed represent a closed loop or feedback control law, irrespective of the particular initial conditions on the state of the system, and can be represented pictorially as we have done in Figure 6.2.5, the dotted lines denoting the original, open loop or uncompensated system.

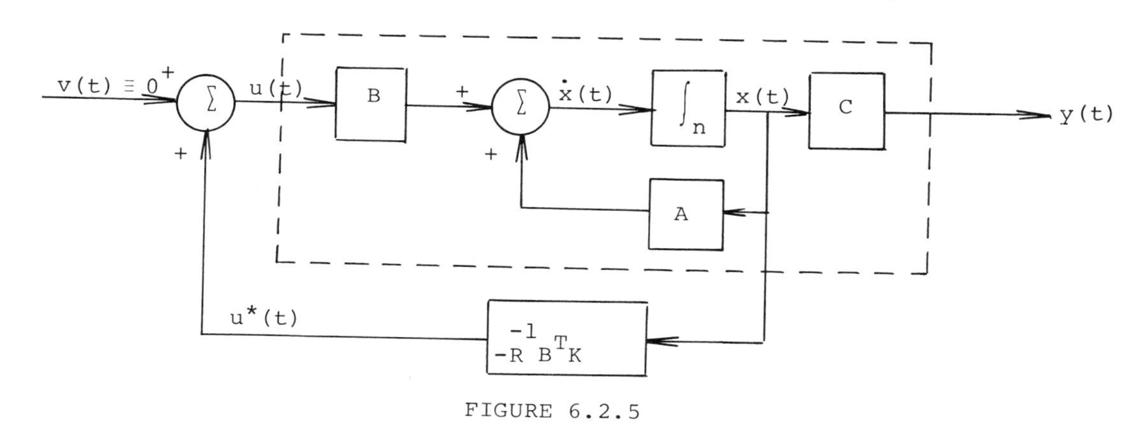

FIGURE 6.2.5

THE OPTIMAL OUTPUT REGULATOR SYSTEM

It is clear that the closed loop dynamical behavior of the optimal output regulator system can be found by substituting $u^*(t)$, as given by 6.2.4, for $u(t)$ in 6.2.1, and is thus given by:

$$\dot{x}(t) = [A - BR^{-1}B^TK]x(t); \; y(t) = Cx(t) \qquad 6.2.6$$

We further note that the optimal control law 6.2.4 does not involve any external feedforward control; i.e. $v(t)$ in Figure 6.2.5 is identically zero and actually need not be present. However, in order to speak of the closed loop transfer matrix associated with the optimal output regulator system or, for that matter, the closed loop transfer matrix of any system compensated via l.s.v.f., we must assume, an external input, $v(t)$.

DEFINITION 6.2.7: In view of this observation, we will now formally define <u>linear state variable feedback</u>, as the control law,

$$u(t) = Fx(t) + Gv(t), \qquad 6.2.7$$

where F and G are $m \times n$ and $m \times q$ matrices, respectively, with elements in $\mathscr{R}$.

EXAMPLE 6.2.8: In order to illustrate the preceding results, we now consider the controllable and observable state form system: $\dot{x}(t) = Ax(t) + Bu(t)$; $y(t) = Cx(t)$, where $A = \begin{bmatrix} 0 & 1 \\ 0 & 0 \end{bmatrix}$, $B = \begin{bmatrix} 0 \\ 1 \end{bmatrix}$, and $C = \begin{bmatrix} 1 & 0 \\ 0 & 2 \end{bmatrix}$. Suppose we wish to find the optimal control $u^*(t)$ which minimizes the performance index $J = \int_0^\infty [x^T(t)C^TCx(t) + u^Tu]dt = \int_0^\infty [y_1^2(t) + y_2^2(t) + u^2(t)]dt$, with $R = 1$, a scalar in this example. The control $u^*(t)$ is given by 6.2.4, where everything but K, the positive definite solution of 6.2.5, is known. We must therefore solve 6.2.5 for K in order to find $u^*(t)$. In general, this would involve solving a suitable nonlinear matrix differential equation, usually with the aid of a digital computer [Al]. For this relatively simple example, however, we can simply set $K = \begin{bmatrix} k_1 & k_2 \\ k_2 & k_3 \end{bmatrix}$, with $k_1 > 0$ and $k_1k_3 > k_2^2$, the two conditions which must be satisfied if K is to be positive definite [H1], and solve 6.2.5 for K by direct substitution in light of these relations; i.e.

$$\underbrace{\begin{bmatrix} k_1 & k_2 \\ k_2 & k_3 \end{bmatrix}}_{K}\underbrace{\begin{bmatrix} 0 & 1 \\ 0 & 0 \end{bmatrix}}_{A} + \underbrace{\begin{bmatrix} 0 & 0 \\ 1 & 0 \end{bmatrix}}_{A^T}\underbrace{\begin{bmatrix} k_1 & k_2 \\ k_2 & k_3 \end{bmatrix}}_{K} - \underbrace{\begin{bmatrix} k_1 & k_2 \\ k_2 & k_3 \end{bmatrix}}_{K}\underbrace{\begin{bmatrix} 0 \\ 1 \end{bmatrix}}_{B}\underbrace{[0 \quad 1]}_{B^T}\underbrace{\begin{bmatrix} k_1 & k_2 \\ k_2 & k_3 \end{bmatrix}}_{K}$$

$= -\underbrace{\begin{bmatrix} 1 & 0 \\ 0 & 2 \end{bmatrix}}_{C}\underbrace{\begin{bmatrix} 1 & 0 \\ 0 & 2 \end{bmatrix}}_{C}$. Performing the above indicated operations, we readily arrive at the following nonlinear matrix equation:

$\begin{bmatrix} -k_2^2, & k_1-k_2k_3 \\ k_1-k_2k_3, & 2k_2-k_3^2 \end{bmatrix} = \begin{bmatrix} -1 & 0 \\ 0 & -4 \end{bmatrix}$, which immediately implies the nonlinear set of equations: $k_2^2 = 1$, $k_1 = k_2k_3$, and $k_3^2 - 2k_2 = 4$. If we (arbitrarily) choose $k_2 = +1$, it follows that k_3 is equal to either plus or minus $\sqrt{6}$, and that $k_1 = k_3 = \pm\sqrt{6}$. Since k_1 must be greater than zero, $k_1 = k_3 = +\sqrt{6}$ if $k_2 = +1$, and we can now verify that this triple satisfies the required relations noted above for K to be positive definite. Therefore $K = \begin{bmatrix} \sqrt{6} & 1 \\ 1 & \sqrt{6} \end{bmatrix}$, and the optimal control $u^*(t) = -R^{-1}B^TKx(t) = [-1, \; -\sqrt{6}]x(t) = -x_1(t) - \sqrt{6}\,x_2(t)$. We might also note that the poles of this system are moved from their open loop locations at $s_1 = 0$, $s_2 = 0$ to the "optimal locations" $s_1 = \frac{-\sqrt{3}+1}{\sqrt{2}}$, $s_2 = \frac{-\sqrt{3}-1}{\sqrt{2}}$ in the (stable) half-plane $Re(s) < 0$, which is consistent with the final statement of Theorem 6.2.3.

6.3 POLE ASSIGNMENT VIA THE CONTROLLABLE COMPANION FORM

We will now consider the general employment of linear state variable feedback (l.s.v.f.) for the compensation of dynamical systems, limiting our discussion in this chapter to the case where the system dynamics are expressed in the state form:

$$\dot{x}(t) = Ax(t) + Bu(t); \qquad 6.3.1a$$

$$y(t) = Cx(t) + Eu(t) \qquad 6.3.1b$$

In particular, if the l.s.v.f. control law

$$u(t) = Fx(t) + Gv(t), \qquad 6.3.2$$

is employed to "improve the performance" of the open loop system, 6.3.1, we can readily obtain a state-space representation for the dynamical behavior of the compensated system by simply substituting 6.3.2 for $u(t)$ in 6.3.1; i.e.

$$\dot{x}(t) = (A + BF)x(t) + BGv(t); \qquad 6.3.3a$$

$$y(t) = (C + EF)x(t) + EGv(t)$$

This closed loop system is represented pictorially in Figure 6.3.4,

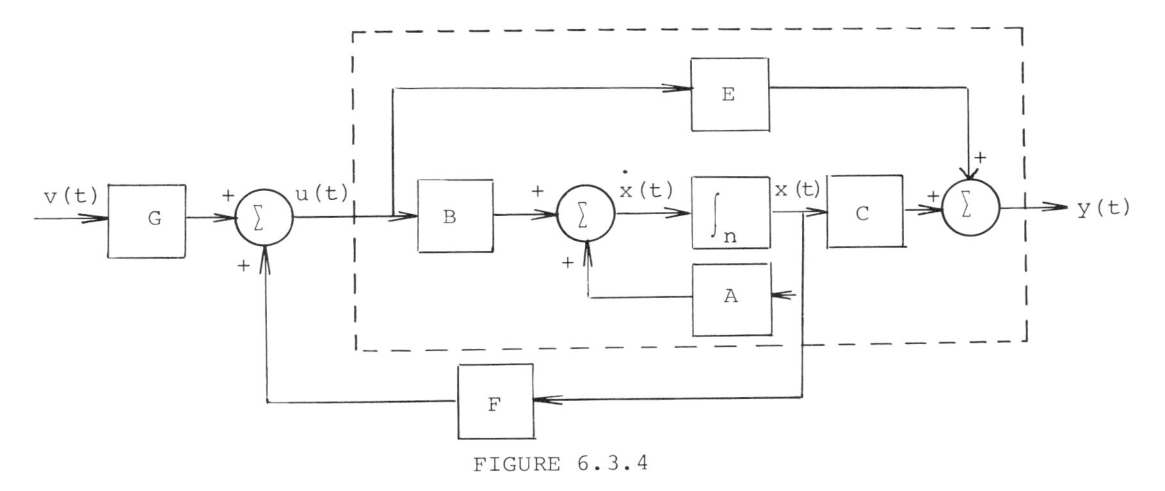

FIGURE 6.3.4

THE CLOSED LOOP SYSTEM

with dotted lines enclosing the original (uncompensated) portion of the system. It might be noted that general l.s.v.f., as defined by 6.3.2, involves both feedback (through F) and feedforward (through G) compensation, although the term "feedforward" does not appear in our definition of l.s.v.f. The reason for this apparent misnomer is that the feedback gain matrix F plays the dominant role in this compensation scheme, as we will soon see.

In general, it is not at all clear what effect the control law 6.3.2 has on the state-space system 6.3.1, since the closed loop quadruple $\{A+BF, BG, C+EF, EG\}$ appears to have been completely and perhaps indiscernibly altered via l.s.v.f. This not only appears to be the case, but actually is, if we consider any arbitrary, "unstructured" open loop representation, $\{A, B, C, E\}$. However, if the open loop system, 6.3.1, is in controllable companion form (Section 3.6), the effect of l.s.v.f. on the closed loop system can be clarified considerably as we will now illustrate.

In particular, if we first reduce a given system of the form 6.3.1 to controllable companion form, assuming for the moment that the pair $\{A, B\}$ is controllable and that $\rho[B] = m$; i.e. if we let

$$\hat{x}(t) = Qx(t) \quad \text{or} \quad x(t) = Q^{-1}\hat{x}(t), \tag{6.3.5}$$

for an appropriate, nonsingular matrix transformation Q, the equivalent system

$$\dot{\hat{x}}(t) = \hat{A}\hat{x}(t) + \hat{B}u(t); \tag{6.3.6a}$$

$$y(t) = \hat{C}\hat{x}(t) + Eu(t) \tag{6.3.6b}$$

(actually the pair $\{\hat{A}, \hat{B}\}$) assumes the structure represented by 3.6.9, where all information regarding the state matrix $\hat{A}$ is contained in the (m) ordered controllability indices d_i, and the (m) ordered σ_k rows of $\hat{A}$, which we have defined as the $(m \times n)$ matrix $\hat{A}_m$ (see Section 4.3). Furthermore, only these same (m) ordered σ_k rows of $\hat{B}$, which we have defined as $\hat{B}_m$, are nonzero. In terms of this equivalent new state, $\hat{x}(t)$, the l.s.v.f. control law 6.3.2 becomes

$$u(t) = \hat{F}\hat{x}(t) + Gv(t), \tag{6.3.7}$$

where $\hat{F} = FQ^{-1}$, and the dynamical behavior of the equivalent closed loop system can therefore be represented in state form as:

$$\dot{\hat{x}}(t) = (\hat{A} + \hat{B}\hat{F})\hat{x}(t) + \hat{B}Gv(t); \qquad 6.3.8a$$

$$y(t) = (\hat{C} + E\hat{F})\hat{x}(t) + EGv(t). \qquad 6.3.8b$$

If we now define $\bar{F} = \hat{B}_m\hat{F}$, where $\bar{F} = [\bar{f}_{ij}]$, it follows from the particular structure (3.6.9) for the controllable companion form pair $\{\hat{A},\hat{B}\}$, that each of the (mn) nontrivial entries, $\hat{a}_{mij}$, of $\hat{A}_m$ is replaced by $\hat{a}_{mij} + \bar{f}_{ij}$ under the l.s.v.f. control law 6.3.7; i.e.

$$\hat{A} + \hat{B}\hat{F} = \hat{A} + \hat{B}\hat{B}_m^{-1}\bar{F}, \qquad 6.3.9$$

with $\hat{A} + \hat{B}\hat{F}$ retaining the exact same structure (3.6.9a) as $\hat{A}$. It is now important to note that in view of this observation, <u>every one of the (mn) nontrivial elements of $\hat{A}$, denoted by an "x" in 3.6.9, can be arbitrarily altered via</u> $\hat{F}(= \hat{B}_m^{-1}\bar{F})$. If we further define $\bar{G} = \hat{B}_mG$, then $\hat{B}G = \hat{B}\hat{B}_m^{-1}\bar{G}$, and in view of the particular structure (3.6.9b) for $\hat{B}$, it follows that each of the (m) ordered σ_k rows of $\hat{B}$ is replaced by the corresponding row of $\bar{G}$ under the control law 6.3.7. It is thus clear that <u>the m ordered σ_k rows of $\hat{B}$ can be arbitrarily altered via</u> $G(= \hat{B}_m^{-1}\bar{G})$. The controllable companion form, 3.6.9, thus enables us to relate the (mn) individual entries of $\hat{F}$ (actually $\bar{F} = \hat{B}_m\hat{F}$) to those corresponding, nontrivial entries of $\hat{A} + \hat{B}\hat{F}$, and the (mq) individual entries of G (actually $\bar{G} = \hat{B}_mG$) to those corresponding nontrivial entries of $\hat{B}G$, while preserving the general structure (3.6.9) of both, a most important and useful observation. We further note that if the pair $\{A,B\}$ is not controllable we can still transform the system, via Q, to an equivalent one in which the controllable and completely uncontrollable portions of the system are separated (see 3.6.13). The l.s.v.f. control law 6.3.7 then affects the $\bar{n}$-dimensional controllable part, $\{\hat{A}_c,\hat{B}_c\}$, of the system in exactly the same way noted above in the controllable case, but has no effect whatsoever on $\hat{A}_{\bar{c}}$, the completely uncontrollable portion of the state matrix. These points will be further clarified when we establish Theorem 6.3.10.

It might be noted at this time that there are several important implications associated with the preceding observations which cannot be fully discussed until after we have considered a frequency domain interpretation of the above via the structure theorem, which we will do in the next chapter. For the purposes of this chapter and, more specifically, this section, we can and will now present one important result which can readily be established in view of the results obtained thus far.

THEOREM 6.3.10: Consider the system 6.3.1 and the l.s.v.f. control law 6.3.2. All $(\bar{n})$ controllable poles of the closed loop system 6.3.3 can be completely and arbitrarily assigned via l.s.v.f., while the $(n-\bar{n})$ uncontrollable poles of the system are unaffected by l.s.v.f.

Proof (Constructive): Assume that we have already transformed (via Q) the given system, 6.3.1, to the form 3.6.13; i.e. $\hat{A} = QAQ^{-1} = \left[\begin{array}{c|c} \hat{A}_c & \hat{A}_{c\bar{c}} \\ \hline 0 & \hat{A}_{\bar{c}} \end{array}\right]$ and $\hat{B} = QB = \left[\begin{array}{c} \hat{B}_c \\ \hline 0 \end{array}\right]$, where the pair $\{\hat{A}_c, \hat{B}_c\}$ is an $\bar{n}$-dimensional controllable companion form as represented by 3.6.9, while $\hat{A}_{\bar{c}}$ represents the completely uncontrollable portion of the state matrix. As we have previously noted, all (m) σ_k rows of $\hat{A}_c + \hat{B}_c\hat{F}$ can be completely and arbitrarily altered via $\hat{F}$. In view of the structured representation 3.6.9, therefore, we can choose the first $\bar{n}$ columns, $\hat{F}_c$, of $\hat{F}$ such that

$$\hat{A}_c + \hat{B}_c\hat{F}_c = \begin{bmatrix} 0 & 1 & 0 & \cdots & 0 \\ 0 & 0 & 1 & 0\cdots & 0 \\ & & \cdot & & \\ & & & \cdot & \\ & & & & 1 \\ -a_o & -a_1 & \cdots & & -a_{\bar{n}-1} \end{bmatrix}, \qquad 6.3.11$$

an $\bar{n}$-dimensional companion matrix, where the scalars $a_o, a_1, \ldots, a_{\bar{n}-1}$ represent the coefficients of the desired closed loop denominator polynomial; i.e. the $(\bar{n})$ zeros of the polynomial $|\lambda I-\hat{A}_c-\hat{B}_c\hat{F}_c| = \lambda^{\bar{n}}+a_{\bar{n}-1}\lambda^{\bar{n}-1}+\ldots+a_1\lambda+a_o$ are equivalent to the (controllable) closed loop poles of the system (see Section 4.2). Since the remaining $(n-\bar{n})$ columns of $\hat{F}$ affect only $\hat{A}_{c\bar{c}}$, the final $n-\bar{n}$ rows of $\hat{A}$ are completely unaffected by $\hat{F}$, which implies that the $(n-\bar{n})$ eigenvalues of $\hat{A}_{\bar{c}}$, or, equivalently, <u>the uncontrollable poles of the system, remain unaltered by l.s.v.f.</u> This follows formally from the fact that all (n) poles of the closed loop system are equivalent to the zeros of

$$|\lambda I-A-BF| = |\lambda I-\hat{A}-\hat{B}\hat{F}| = |\lambda I-\hat{A}_c-\hat{B}_c\hat{F}_c| \times |\lambda I-\hat{A}_{\bar{c}}|, \qquad 6.3.12$$

as noted earlier (see 3.6.14).

In order to explicitly determine an $\hat{F}$ (or $F = \hat{F}Q$) which yields the controllable part of the closed loop system matrix as represented by 6.3.11, we let $\hat{A}_m^*$ denote the (m) ordered σ_k rows of $\hat{A}_c + \hat{B}_c\hat{F}_c$, as given by 6.3.11, and define $\hat{A}_{cm}$ and $\hat{B}_{cm}$ as these same ordered σ_k rows of $\hat{A}_c$ and $\hat{B}_c$, respectively. It therefore follows that

$$\hat{A}_{cm} + \hat{B}_{cm}\hat{F}_c = \hat{A}_m^*, \qquad 6.3.13$$

or that the control law 6.3.7, with the first $\bar{n}$ columns of $\hat{F}$ given by

$$\hat{F}_c = \hat{B}_{cm}^{-1}[\hat{A}_m^* - \hat{A}_{cm}] \qquad 6.3.14$$

yields the desired $\bar{n}$-dimensional closed loop system submatrix 6.3.11.

The final $(n-\bar{n})$ columns of $\hat{F}$ play no part in closed loop pole assignment, since they affect only $\hat{A}_{c\bar{c}}$ which, in turn, has no effect on the eigenvalues of the closed loop system matrix. We can therefore set the final $(n-\bar{n})$ columns of $\hat{F}$ equal to zero in order to complete our assignment of all (mn) entries of an appropriate $\hat{F}$. In view of 6.3.5, the state feedback gain matrix F associated with the original system, 6.3.1, and the feedback control law, 6.3.2, is given by

$$u(t) = \hat{F}\hat{x}(t) + Gv(t) = \hat{F}Qx(t) + Gv(t) = Fx(t) + Gv(t), \quad 6.3.15$$

with

$$F = \hat{F}Q. \quad 6.3.16$$

Theorem 6.3.10 is thus constructively established.

It should be noted that neither y(t) nor G play any role here with respect to the question of arbitrary pole placement; i.e. all results concerning this question have involved only the triple {A,B,F} (or $\{\hat{A},\hat{B},\hat{F}\}$).

DEFINITION 6.3.17: In view of Theorem 6.3.10, we will say that the system 6.3.1, or the pair {A,B}, is asymptotically stabilizable via l.s.v.f. if and only if the $(n-\bar{n})$ uncontrollable poles of the system (the eigenvalues of $\hat{A}_{\bar{c}}$) lie in the half plane Re(s) (or Re(λ)) < 0. We finally remark that the ability to simply stabilize a system, by either l.s.v.f. or any other scheme, regardless of whether or not certain poles can be arbitrarily assigned, is usually of paramount initial concern in any system design.

EXAMPLE 6.3.18: To illustrate the above constructive procedure for finding a state feedback gain matrix, F, which yields any arbitrary set of $(\bar{n})$ closed loop poles, consider the following particular system in the state form 6.3.1; i.e.

$$A = \begin{bmatrix} -1 & 0 & 0 & -6 & 3 & -1 \\ 1 & -2 & 1 & 0 & -1 & -1 \\ 1 & 1 & 0 & 6 & -2 & 1 \\ 1 & 0 & 0 & 0 & 0 & 0 \\ -1 & 2 & -1 & 0 & 2 & 1 \\ -2 & 0 & 0 & -2 & 0 & -1 \end{bmatrix}, \quad B = \begin{bmatrix} 0 & 1 \\ -1 & -2 \\ 0 & -1 \\ 0 & 0 \\ 1 & 2 \\ 0 & 0 \end{bmatrix}, \quad C = \begin{bmatrix} 0 & 0 & 0 & 0 & 1 & 0 \\ 1 & 0 & 0 & 0 & 0 & 0 \end{bmatrix},$$

and $E = 0$. We first transform this system to an equivalent one of the form 3.6.13 via Q, noting that the controllability matrix $\mathscr{C}$ for this system has rank $5 < n = 6$, and therefore that the system is not completely controllable. Note that we restrict our attention here to the pair $\{A,B\}$ exclusively, since $y(t)$ plays no role in the question of pole assignment via l.s.v.f. Therefore, in view of the results given in Section 3.6, if

$$Q = \begin{bmatrix} 1 & 0 & 1 & 0 & 0 & 0 \\ 0 & 1 & 0 & 0 & 1 & 0 \\ 0 & 0 & 0 & 0 & 1 & 0 \\ 0 & 0 & 0 & 1 & 0 & 0 \\ 1 & 0 & 0 & 0 & 0 & 0 \\ 0 & 0 & 0 & 2 & 0 & 1 \end{bmatrix}, \text{ then } \hat{A} = QAQ^{-1} = \left[\begin{array}{ccccc|c} 0 & 1 & 0 & 0 & 0 & 0 \\ 0 & 0 & 1 & 0 & 0 & 0 \\ -1 & 2 & 0 & -2 & 0 & 1 \\ 0 & 0 & 0 & 0 & 1 & 0 \\ 0 & 0 & 3 & -4 & -1 & -1 \\ \hline 0 & 0 & 0 & 0 & 0 & -1 \end{array}\right] =$$

$$\left[\begin{array}{c|c} \hat{A}_c & \hat{A}_{c\bar{c}} \\ \hline 0 & \hat{A}_{\bar{c}} \end{array}\right] \text{ and } \hat{B} = QB = \left[\begin{array}{cc} 0 & 0 \\ 0 & 0 \\ 1 & 2 \\ 0 & 0 \\ 0 & 1 \\ \hline 0 & 0 \end{array}\right] = \left[\begin{array}{c} \hat{B}_c \\ \hline 0 \end{array}\right].$$

Clearly the pair $\{A_c, B_c\}$ is in controllable companion form, with $d_1 = 3$ and $d_2 = 2$. Therefore, $\sigma_1 = 3$ and $\sigma_2 = 5$. We further note that $\hat{A}_{\bar{c}} = [-1]$, or that the $(n-\bar{n}) = (6-5) = 1$ uncontrollable pole at $s = -1$ is an asymptotically stable one. This system is therefore asymptotically stabilizable via l.s.v.f. If we now require that the five controllable closed loop poles of the system be given by: $s_1 = -0.1$, $s_2 = -0.2$ $s_{3,4} = -1 \pm j$, and $s_5 = -2$, it follows that we will require an $\hat{F}_c$

such that $|\lambda I-\hat{A}_c-\hat{B}_c\hat{F}_c| = (\lambda+0.1)(\lambda+0.2)(\lambda^2+2\lambda+2)(\lambda+2) = \lambda^5 + 4.3\lambda^4 + 7.22\lambda^3 + 5.88\lambda^2 + 1.32\lambda + 0.08\lambda$ or, in light of our procedure, that

$$\hat{A}_c + \hat{B}_c\hat{F}_c = \begin{bmatrix} 0 & 1 & 0 & 0 & 0 \\ 0 & 0 & 1 & 0 & 0 \\ 0 & 0 & 0 & 1 & 0 \\ 0 & 0 & 0 & 0 & 1 \\ -.08 & -1.32 & -5.88 & -7.22 & -4.3 \end{bmatrix}, \text{ which implies that}$$

$\hat{A}_m^* = \begin{bmatrix} 0 & 0 & 0 & 1 & 0 \\ -.08 & -1.32 & -5.88 & -7.22 & -4.3 \end{bmatrix}$. Since $\hat{A}_{cm} = \begin{bmatrix} -1 & 2 & 0 & -2 & 0 \\ 0 & 0 & 3 & -4 & -1 \end{bmatrix}$

and $\hat{B}_{cm} = \begin{bmatrix} 1 & 2 \\ 0 & 1 \end{bmatrix}$, which are evident by inspection of $\hat{A}$ and $\hat{B}$, respectively, it follows from 6.3.14 that

$$\hat{F}_c = \underbrace{\begin{bmatrix} 1 & -2 \\ 0 & 1 \end{bmatrix}}_{\hat{B}_{cm}^{-1}} \underbrace{\begin{bmatrix} 1 & -2 & 0 & 3 & 0 \\ -.08 & -1.32 & -8.88 & -3.22 & -3.3 \end{bmatrix}}_{\hat{A}_m^* - \hat{A}_{cm}} =$$

$\begin{bmatrix} 1.16 & .64 & 17.76 & 9.44 & 6.6 \\ -.08 & -1.32 & -8.88 & -3.22 & -3.3 \end{bmatrix}$. An appropriate F can now be found by adding a zero sixth column to $\hat{F}_c$, and postmultiplying the resulting matrix, $\hat{F}$, by Q as indicated by 6.3.16; i.e. if

$$F = \underbrace{\begin{bmatrix} 1.16 & .64 & 17.76 & 9.44 & 6.6 & 0 \\ -.08 & -1.32 & -8.88 & -3.22 & -3.3 & 0 \end{bmatrix}}_{\hat{F}} \underbrace{\begin{bmatrix} 1 & 0 & 1 & 0 & 0 & 0 \\ 0 & 1 & 0 & 0 & 1 & 0 \\ 0 & 0 & 0 & 0 & 1 & 0 \\ 0 & 0 & 0 & 1 & 0 & 0 \\ 1 & 0 & 0 & 0 & 0 & 0 \\ 0 & 0 & 0 & 2 & 0 & 1 \end{bmatrix}}_{Q},$$

which is equal to $\begin{bmatrix} 7.76 & .64 & 1.16 & 9.44 & 18.4 & 0 \\ -3.38 & -1.32 & -.08 & -3.22 & 10.2 & 0 \end{bmatrix}$, then the zeros of $|\lambda I-\hat{A}-\hat{B}\hat{F}| = |\lambda I-A-BF|$, which are equivalent to the poles of the closed loop system, are equal to the $(\bar{n} = 5)$ desired ones, namely

-0.1, -0.2, $-1\pm j$, and -2, and the uncontrollable, but asymptotically stable one at $s = -1$.

6.4 ASYMPTOTIC STATE ESTIMATION

As noted in the previous two sections, linear state variable feedback (l.s.v.f.) is an important compensation technique in the synthesis of linear dynamical systems, a point which will be illustrated still further in the remainder of this text. The astute reader, however, may have noted one important factor concerning l.s.v.f. which could, in many cases, prevent its direct employment for closed loop compensation. In particular, on closer inspection of Figure 6.3.4, it is apparent that the feedback path from the state $x(t)$ through the gain matrix F crosses the boundary which encloses the original system. This clearly implies the ability to directly measure the entire internal n-dimensional state vector $x(t)$. In general, however, only the external m-dimensional input, $u(t)$, and the p-dimensional output, $y(t)$, are directly measurable, so that the l.s.v.f. compensation scheme depicted in Figure 6.3.4 is not directly realizable. Unfortunately, this factor limits the practicality of various l.s.v.f. compensation schemes, although it should be noted that additional sensors (e.g. rate gyros, accelerometers, pressure transducers, flow meters, etc.) can often be employed to measure additional components of the state if they are not directly available at the output. From a practical point of view, however, these additional sensors mean increased costs which are to be avoided if alternative control schemes can be employed.

In 1964, D. G. Luenberger [L4] described a technique for implementing the "equivalent of" a l.s.v.f. control law, which employs only the available directly measurable input and output signals. Not surprisingly, his scheme which is commonly referred to as

"Luenberger state estimation (or observation)", relies on the (complete state) observability of a system and is based on the dual of Theorem 6.3.10, which we will now merely state without formal proof.

THEOREM 6.4.1: Consider the system 6.3.1. All (n) eigenvalues of $A + KC$ can be completely and arbitrarily assigned via K if and only if the pair $\{A,C\}$ is observable, any unassignable eigenvalues corresponding to the unobservable modes of the system.

Assuming that the system 6.3.1 is observable, we can therefore dualize the results given in the previous section to find a K[†] which arbitrarily positions all (n) eigenvalues of $A + KC$ in the half plane $Re(\lambda) < 0$. If we then construct the following n-dimensional dynamical system which is driven by the input, $u(t)$, and the output, $y(t)$, of the given system 6.3.1; i.e.

$$\dot{\tilde{x}}(t) = (A+KC)\tilde{x}(t) + (B + KE)u(t) - Ky(t), \quad 6.4.2$$

we find, by subtracting 6.4.2 from 6.3.1a, that

$$\dot{x}(t) - \dot{\tilde{x}}(t) = Ax(t) + Bu(t) - A\tilde{x}(t) - KC\tilde{x}(t) - Bu(t) - KEu(t) + Ky(t), \quad 6.4.3$$

or, by combining terms in view of 6.3.1b, that

$$\dot{x}(t) - \dot{\tilde{x}}(t) = (A + KC)(x(t) - \tilde{x}(t)) \quad 6.4.4$$

In view of the results presented in Section 3.2 it is thus clear that

$$x(t) - \tilde{x}(t) = e^{[A+KC](t-t_o)}[x(t_o)-\tilde{x}(t_o)] \quad 6.4.5$$

Since K was chosen such that all (n) eigenvalues of $A + KC$ lie in the half-plane $Re(\lambda) < 0$, it follows that $x(t)-\tilde{x}(t)$ will exponen-

[†]This K is not related to and therefore should not be confused with the solution to the matrix Riccati equation, 6.2.5.

tially approach zero with increasing time or, equivalently, that $\tilde{x}(t)$ will approach $x(t)$ exponentially with time, regardless of the input, $u(t)$, the output, $y(t)$, or the initial conditions on either state vector. In view of the above, we therefore call the system 6.4.2 an exponential estimator or a Luenberger observer of (the entire state of) the given system 6.3.1. A pictorial representation of this open loop Luenberger observer is given in Figure 6.4.6. It might be noted that

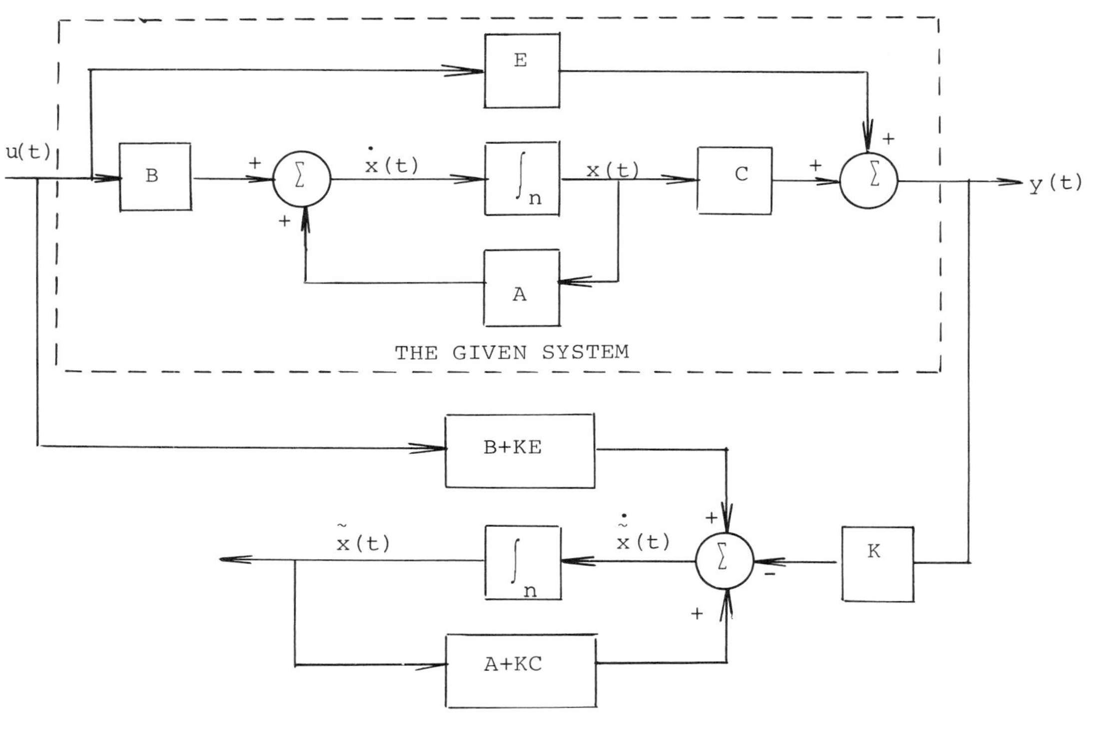

FIGURE 6.4.6

OPEN LOOP EXPONENTIAL ESTIMATION

(LUENBERGER OBSERVATION)

the term "open loop" is used to signify that the estimate, $\tilde{x}(t)$, of the true state of the given system is not (as yet) employed in any feedback control configuration and can be considered to represent a part of the output of the composite 2n-dimensional system depicted in Figure 6.4.6 whose complete dynamical behavior is given by:

$$\begin{bmatrix} \dot{x}(t) \\ \dot{\tilde{x}}(t) \end{bmatrix} = \begin{bmatrix} A & 0 \\ -KC & A+KC \end{bmatrix} \begin{bmatrix} x(t) \\ \tilde{x}(t) \end{bmatrix} + \begin{bmatrix} B \\ B \end{bmatrix} u(t); \qquad 6.4.7a$$

$$\begin{bmatrix} y(t) \\ \tilde{y}(t) \end{bmatrix} = \begin{bmatrix} C & 0 \\ 0 & I \end{bmatrix} \begin{bmatrix} x(t) \\ \tilde{x}(t) \end{bmatrix} + \begin{bmatrix} E \\ 0 \end{bmatrix} u(t) \qquad 6.4.7b$$

It should perhaps be noted that controllability plays no role in the design of an open loop state estimator and, in fact, we do not even require that the system have a defined input in order to construct a state estimator. The (open loop) state estimator has been pictured in Figure 6.4.6 for illustrative purposes only, and serves no useful control function; i.e. the dynamical behavior of the given system is unaffected by the presence of the open loop estimator. The important practical utility of a Luenberger observer is to approximate a closed loop l.s.v.f. control law via $\tilde{x}(t)$ instead of $x(t)$, as we will now show.

In particular, suppose that we now use the control law

$$u(t) = F\tilde{x}(t) + Gv(t) \qquad 6.4.8$$

instead of the actual l.s.v.f. control law 6.2.7 to compensate the given system (e.g. to attempt to arbitrarily assign all of the controllable eigenvalues of the closed loop system). What can be said regarding the closed loop behavior of 2n-dimensional system, when compared to the employment of an actual state feedback control law? To answer this question, we first note that the dynamical behavior of the given system compensated by the exponential estimator as depicted in Figure 6.4.9, can be determined by substituting 6.4.8 for $u(t)$ in 6.4.7; i.e.

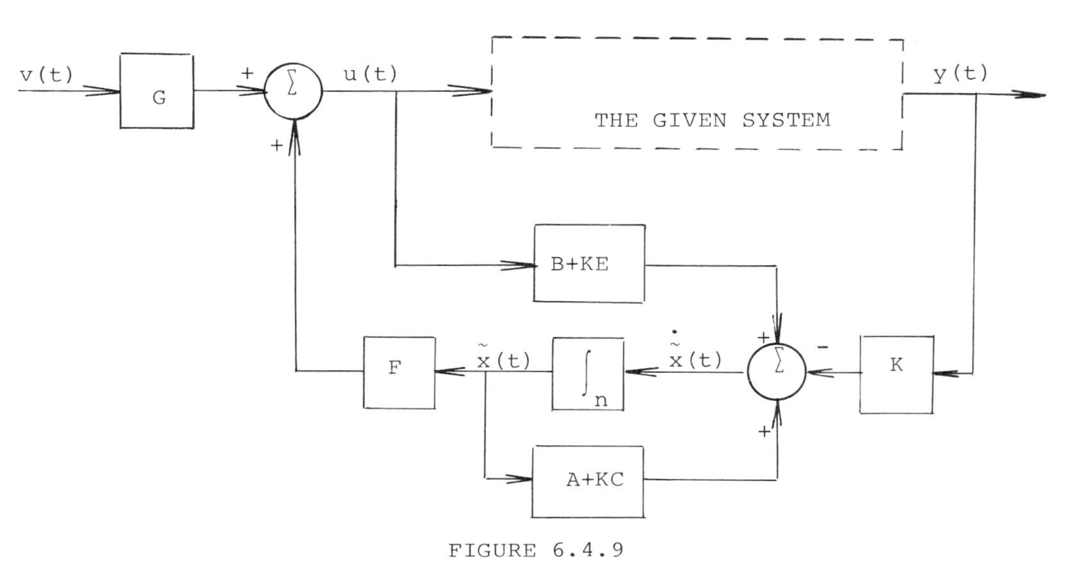

FIGURE 6.4.9

CLOSED LOOP EXPONENTIAL ESTIMATION

(LUENBERGER OBSERVATION)

$$\begin{bmatrix} \dot{x}(t) \\ \dot{\tilde{x}}(t) \end{bmatrix} = \begin{bmatrix} A & BF \\ -KC & A+KC+BF \end{bmatrix} \begin{bmatrix} x(t) \\ \tilde{x}(t) \end{bmatrix} + \begin{bmatrix} BG \\ BG \end{bmatrix} v(t); \qquad 6.4.10a$$

$$\begin{bmatrix} y(t) \\ \tilde{y}(t) \end{bmatrix} = \begin{bmatrix} C & EF \\ 0 & I \end{bmatrix} \begin{bmatrix} x(t) \\ \tilde{x}(t) \end{bmatrix} + \begin{bmatrix} EG \\ 0 \end{bmatrix} v(t) \qquad 6.4.10b$$

It is now convenient to transform the above via the equivalence transformation $Q = \begin{bmatrix} I & 0 \\ I & -I \end{bmatrix} = Q^{-1}$, in order to obtain the equivalent system:

$$\begin{bmatrix} \dot{x}(t) \\ \dot{x}(t) - \dot{\tilde{x}}(t) \end{bmatrix} = \begin{bmatrix} A+BF & -BF \\ 0 & A+KC \end{bmatrix} \begin{bmatrix} x(t) \\ x(t) - \tilde{x}(t) \end{bmatrix} + \begin{bmatrix} BG \\ 0 \end{bmatrix} v(t); \qquad 6.4.11a$$

$$\begin{bmatrix} y(t) \\ \tilde{y}(t) \end{bmatrix} = \begin{bmatrix} C+EF & -EF \\ I & -I \end{bmatrix} \begin{bmatrix} x(t) \\ x(t) - \tilde{x}(t) \end{bmatrix} + \begin{bmatrix} EG \\ 0 \end{bmatrix} v(t). \qquad 6.4.11b$$

It is now immediately apparent that the entire n-dimensional "state", $x(t) - \tilde{x}(t)$, is not controllable, which we note should be the case since we have forced $x(t) - \tilde{x}(t)$ to exponentially approach the origin of the state space regardless of the input $u(t)$. We further note that

$$|\lambda I - \begin{bmatrix} A+BF & -BF \\ 0 & A+KC \end{bmatrix}| = |\lambda I-A-BF| \times |\lambda I-A-KC|, \qquad 6.4.12$$

or that the (2n) eigenvalues of the composite closed loop system are equal to the individual eigenvalues of both $A + BF$ and $A + KC$. Therefore, by Theorem 6.3.10 and its dual, Theorem 6.4.1, we note that <u>given the system 6.3.1, if the pair $\{A,B\}$ is controllable and the pair $\{A,C\}$ is observable, a pair $\{F,K\}$ of gain matrices can be chosen to insure complete and arbitrary pole assignment and, therefore, the asymptotic stability of the 2n-dimensional closed loop system consisting of the given system, 6.3.1, compensated by an exponential estimator</u> (as depicted in Figure 6.4.9).

We further note that the closed loop transfer matrix, $T_{F,G,K}(s)$, of the composite 2n-dimensional system can be found via 4.2.2 using the system matrices in 6.4.11; i.e.

$$T_{F,G,K}(s) = [C+EF, \quad -EF]\begin{bmatrix} sI-A-BF & BF \\ 0 & sI-A-KC \end{bmatrix}^{-1}\begin{bmatrix} BG \\ 0 \end{bmatrix} + [EG], \qquad 6.4.13$$

or

$$T_{F,G,K}(s) = \{(C+EF)(sI-A-BF)^{-1}BG + EG\} \frac{|sI-A-KC|}{|sI-A-KC|}. \qquad 6.4.14$$

The reader can verify that with the exception of the polynomial $|sI-A-KC|$, which appears as a stable (uncontrollable) cancellable factor in $T_{F,G,K}(s)$, this expression, 6.4.14, for the closed loop transfer matrix is identical to the one which would be obtained if the actual state feedback control law 6.2.7 were used instead of 6.4.8. We thus conclude that <u>exponential estimator compensation yields the same closed loop transfer matrix as actual l.s.v.f.</u>

We have thus far said nothing about the specific locations of the (n) eigenvalues of the state estimator (the zeros of $|\lambda I-A-KC|$). As we have just shown, insofar as the closed loop transfer matrix is concerned, it does not appear to make any difference what K is employed as long as it produces a stable observer. We note, however, that by 6.4.5, the eigenvalues of $A + KC$ represent a measure of the speed with which $\tilde{x}(t)$ approaches $x(t)$; the more negative the eigenvalues, the faster $\tilde{x}(t)$ and $x(t)$ converge. Excessively fast convergence times usually require large gains in the system which may, in turn, cause certain signals to saturate, resulting in unpredictable nonlinear performance. Improperly located high gains can also excessively amplify any noise which may be present in the system, also causing unsatisfactory performance. On the other hand, slow convergence times can cause poor transient behavior, especially if the initial conditions on $x(t)$ and $\tilde{x}(t)$ differ significantly, and should therefore be avoided if transient performance is important. The term "tradeoff" is often used to describe the ad hoc trial and error procedure by which a control system designer often reaches a final acceptable design configuration, and this same term perhaps best describes the process which might be employed here to obtain a "good" gain matrix K in the case of completely deterministic (noiseless) systems. If the system is subjected to "certain types" of noise whose statistical properties are either known or can be approximated, a "best" or "optimum" gain matrix K^* can be explicitly determined by solving essentially the dual of the output regulator problem (Section 6.2), thereby producing a special exponential estimator which is commonly called the "Kalman filter". We merely note this here for general information, since stochastic systems and the Kalman filter are subjects which go beyond the intended scope of this text.

The actual implementation of 6.4.2 as an exponential estimator of the entire state of 6.3.1 involves the construction of an n-dimensional dynamical system, whose (n) state components, $\tilde{x}_i(t)$, approximate the dynamical behavior of corresponding components, $x_i(t)$, of the given system's state. We note, however, that we already have p output variables, $y(t)$, available and directly measureable, assuming that each of the individual components, $y_i(t)$, of $y(t)$ represents a known, independent linear combination of the actual state variables. Therefore, one might logically question the need to construct an <u>n-dimensional</u> observer; i.e. is it possible to employ an observer of reduced dimension $(n-p)$ to reconstruct only those linear combinations of the state which are not present in $Cx(t) = y(t) - Eu(t)$? The answer to this question is an affirmative one as we will now demonstrate.

In particular, we will find it convenient to consider the scalar output $(p = 1)$ case first, and to employ the observable companion form introduced in Section 3.6. More specifically, let us consider an observable system in the state form 6.3.1 with C equal to an n-dimensional row vector. Since the pair $\{A,C\}$ is assumed observable, we can transform the given system, via $\bar{Q}^{-T}$ (Section 3.6), to an equivalent observable companion form, where $\{\hat{A},\hat{B},\hat{C},E\} = \{\bar{Q}^{-T}A\bar{Q}^{T},\bar{Q}^{-T}B,C\bar{Q}^{T},E\}$,

$$\hat{A} = \begin{bmatrix} 0 & 0 & \cdots & 0 & -a_o \\ 1 & 0 & \cdots & 0 & -a_1 \\ 0 & 1 & & & \vdots \\ \vdots & \vdots & \ddots & & \\ 0 & 0 & \cdots & 1 & -a_{n-1} \end{bmatrix}, \text{ and } \hat{C} = [0 \;\; 0 \;\; \cdots \;\; 0 \;\; 1] \qquad 6.4.15$$

In view of the above (scalar) observable companion form, it is clear that all (n) eigenvalues of $\hat{A} + \hat{K}\hat{C}$ can be completely and arbitrarily assigned via the n-dimensional column vector $\hat{K} = [\hat{k}_1,\hat{k}_2,\ldots,\hat{k}_n]^T = \bar{Q}^{-T}K$, since

$$\hat{A} + \hat{K}\hat{C} = \begin{bmatrix} 0 & 0 & \dots & 0 & \hat{k}_1 - a_o \\ 1 & 0 & \dots & 0 & \hat{k}_2 - a_1 \\ 0 & 1 & & \vdots & \vdots \\ \vdots & \vdots & \ddots & & \\ 0 & 0 & \dots & 1 & \hat{k}_n - a_{n-1} \end{bmatrix}, \qquad 6.4.16$$

which clearly implies that

$$|\lambda I - \hat{A} - \hat{K}\hat{C}| = \lambda^n + (a_{n-1} - \hat{k}_n)\lambda^{n-1} + \dots + (a_1 - \hat{k}_2)\lambda + a_o - \hat{k}_1 \qquad 6.4.17$$

We can therefore construct an n-dimensional state observer for this system using 6.4.2, as noted earlier. However, it is also possible to construct an observer of reduced order $(n-p) = (n-1)$ whose state, $\tilde{x}(t) = [\tilde{x}_1(t), \tilde{x}_2(t), \dots, \tilde{x}_{n-1}(t)]^T$ exponentially approaches the $(n-1)$ state variables $\hat{x}_1(t), \hat{x}_2(t), \dots, \hat{x}_{n-1}(t)$, (excluding the externally measurable signal $\hat{x}_n(t) = y(t) - Eu(t)$) of the single output, observable, companion form system:

$$\dot{\hat{x}}(t) = \hat{A}\hat{x}(t) + \hat{B}u(t); \qquad 6.4.18a$$

$$y(t) = \hat{C}\hat{x}(t) + Eu(t) \qquad 6.4.18b$$

To show this, consider the equivalence transformation

$$P = \begin{bmatrix} 1 & 0 & \dots & 0 & -p_o \\ 0 & 1 & \dots & 0 & -p_1 \\ \vdots & \vdots & \ddots & \vdots & \\ & & & & -p_{n-2} \\ 0 & 0 & \dots & 0 & 1 \end{bmatrix}, \qquad 6.4.19$$

where the p_i are, as yet, unspecified real numbers. If we now set $\bar{x}(t) = P\hat{x}(t)$, or $\hat{x}(t) = P^{-1}\bar{x}(t)$, it follows that the system:

$$\dot{\bar{x}}(t) = \bar{A}\bar{x}(t) + \bar{B}u(t); \qquad 6.4.20a$$

$$y(t) = \bar{C}\bar{x}(t) + Eu(t), \qquad 6.4.20b$$

where $\bar{A} = P\hat{A}P^{-1}$, $\bar{B} = P\hat{B}$, and $\bar{C} = \hat{C}P^{-1}$, is equivalent to 6.4.18 and, therefore, to 6.3.1 as well. Furthermore, the system 6.4.20 now assumes a rather useful form. In particular, since

$$P^{-1} = \begin{bmatrix} 1 & 0 & \cdots & 0 & p_o \\ 0 & 1 & \cdots & 0 & p_1 \\ \vdots & \vdots & \ddots & \vdots & \\ & & & & p_{n-2} \\ 0 & 0 & \cdots & 0 & 1 \end{bmatrix} \tag{6.4.21}$$

it follows that

$$P\hat{A}P^{-1} = \bar{A} = \left[\begin{array}{cccc|c} 0 & 0 & \cdots & -p_o & -p_o p_{n-2} - a_o + p_o a_{n-1} \\ 1 & 0 & \cdots & -p_1 & p_o - p_1 p_{n-2} - a_1 + p_1 a_{n-1} \\ 0 & 1 & \cdots & -p_2 & p_1 - p_2 p_{n-2} - a_2 + p_2 a_{n-1} \\ \vdots & \vdots & \ddots & \vdots & \vdots \\ & & & -p_{n-2} & p_{n-3} - p_{n-2}^2 - a_{n-2} + p_{n-2} a_{n-1} \\ 0 & 0 & \cdots & 1 & p_{n-2} - a_{n-1} \end{array}\right], \tag{6.4.22}$$

where the final (n-th) column of $\bar{A}$ has been separated from the remainder of the matrix by dotted lines for clarity. We further note that

$$P\hat{B} = \bar{B} = \begin{bmatrix} b_1 - p_o b_n \\ b_2 - p_1 b_n \\ \vdots \\ b_{n-1} - p_{n-2} b_n \\ b_n \end{bmatrix} \tag{6.4.23}$$

The matrices $\hat{C}$ and E are clearly unaffected by the equivalence transformation P. Let us denote the first (n-1) components of $\bar{x}(t)$

by $\bar{x}_{\bar{n}}(t) = \begin{bmatrix} \bar{x}_1(t) \\ \bar{x}_2(t) \\ \vdots \\ \bar{x}_{n-1}(t) \end{bmatrix}$ and define $\bar{B}_{\bar{n}}$ as the first (n-1) rows of $\bar{B}$.

If we now define $\bar{A}_{n-1}$ as the (n-1) dimensional companion matrix obtained by eliminating both the n-th row and the n-th column of $\bar{A}$ and let $\bar{A}_{\bar{n}}$ represent the column vector consisting of the first (n-1) elements of the last column of $\bar{A}$, we can obtain a concise state form dynamical equation for the (n-1)-dimensional system with state $\bar{x}_{\bar{n}}(t)$. In particular, from 6.4.20a, 6.4.22, and 6.4.23, it now follows that

$$\dot{\bar{x}}_{\bar{n}}(t) = \bar{A}_{n-1}\bar{x}_{\bar{n}}(t) + \bar{A}_{\bar{n}}\bar{x}_n(t) + \bar{B}_{\bar{n}}u(t), \qquad 6.4.24$$

and since

$$y(t) = \bar{x}_n(t) + Eu(t), \qquad 6.4.25$$

we obtain, by substituting $y(t) - Eu(t)$ for $\bar{x}_n(t)$ in 6.4.24, the relation:

$$\dot{\bar{x}}_{\bar{n}}(t) = \bar{A}_{n-1}\bar{x}_{\bar{n}}(t) + \bar{A}_{\bar{n}}y(t) + [\bar{B}_n - \bar{A}_{\bar{n}}E]u(t) \qquad 6.4.26$$

We now claim that the following (n-1)-dimensional system is an exponential estimator of 6.4.26; i.e. if

$$\dot{\tilde{x}}(t) = \bar{A}_{n-1}\tilde{x}(t) + \bar{A}_{\bar{n}}y(t) + [\bar{B}_{\bar{n}} - \bar{A}_{\bar{n}}E]u(t), \qquad 6.4.27$$

then

$$\dot{\bar{x}}_{\bar{n}}(t) - \dot{\tilde{x}}(t) = \bar{A}_{n-1}[\bar{x}_{\bar{n}}(t) - \tilde{x}(t)], \qquad 6.4.28$$

and

$$\bar{x}_{\bar{n}}(t) - \tilde{x}(t) = e^{\bar{A}_{n-1}(t-t_o)}[\bar{x}_{\bar{n}}(t_o) - \tilde{x}(t_o)]. \qquad 6.4.29$$

Therefore, if the (n-1) real scalars $p_o, p_1, \ldots, p_{n-2}$ are chosen such that the (n-1) eigenvalues of $\bar{A}_{n-1}$, which are equal to the zeros of $\lambda^{n-1} + p_{n-2}\lambda^{n-2} + \ldots + p_1\lambda + p_o$, lie in the half-plane $Re(\lambda) < 0$, then $\tilde{x}(t)$ approaches $\bar{x}_{\bar{n}}(t)$ exponentially. Since $\bar{x}(t) = P\hat{x}(t)$ and

$\hat{x}(t) = \overline{Q}^{-T} x(t)$, it follows that $\overline{x}(t) = P\overline{Q}^{-T} x(t)$, or that an exponential estimate of the actual state $x(t)$ is given by $\overline{Q}^{T-1} P \begin{bmatrix} \tilde{x}(t) \\ y(t)-Eu(t) \end{bmatrix}$, since the n-th component, $\overline{x}_n(t)$, of $\overline{x}(t)$ is given by $y(t)-Eu(t)$ in accordance with 6.4.25; i.e.

$$x(t) = \overline{Q}^{T-1} P\, \overline{x}(t) = \overline{Q}^{T-1} P \begin{bmatrix} \tilde{x}(t) \\ y(t)-Eu(t) \end{bmatrix} \qquad 6.4.30$$

Our construction of a reduced order observer in the scalar case is thus complete.

This scalar output result can now be extended to include the more general multivariable case. In particular, we first recall from Section 3.6 that if the system is observable and C is of full rank $p \leq n$, then the system 3.6.1 can be reduced via $\overline{Q}^{-T}$ to an observable companion form $\{\hat{A},\hat{B},\hat{C},E\}$ where $\hat{A}$ and $\hat{C}$ assume the special form represented by 3.6.19; i.e.

$$\overline{Q}^{-T} A \overline{Q}^{T} = \hat{A} = \begin{bmatrix} \hat{A}_{11} & \hat{A}_{12} & \cdots & \hat{A}_{1p} \\ \hat{A}_{21} & \hat{A}_{22} & \cdots & \hat{A}_{2p} \\ \vdots & & & \vdots \\ \hat{A}_{p1} & & \cdots & \hat{A}_{pp} \end{bmatrix}, \qquad 6.4.31$$

where each $\hat{A}_{ii}$ is a (lower left identity) companion matrix of dimension $\overline{d}_i$, and for each $j \neq i$, $\hat{A}_{ij}$ is a $\overline{d}_i \times \overline{d}_j$ matrix which is identically zero except perhaps for its final ($\overline{d}_j$-th) column. Also

$$C\overline{Q}^{T} = \hat{C} = \begin{bmatrix} 0 & 0 & \cdots & 1 & 0 & \cdots & 0 & \cdots & 0 \\ 0 & 0 & \cdots & x & 0 & \cdots & 1 & \cdots & 0 \\ \vdots & \vdots & & \vdots & \vdots & & \vdots & & \vdots \\ 0 & 0 & \cdots & x & 0 & \cdots & x & \cdots & 1 \end{bmatrix}, \qquad 6.4.32$$

and is identically zero except for its (p) $\overline{\sigma}_k$ $(= \sum_1^k \overline{d}_i)$ columns.

If we now consider any $\bar{d}_i$-dimensional subsystem of the above; i.e. a subsystem with state $\hat{x}_{Ti}(t) = \begin{bmatrix} \hat{x}_{\bar{\sigma}_{i-1}+1}(t) \\ \vdots \\ \hat{x}_{\bar{\sigma}_i}(t) \end{bmatrix}$, it follows in view of the special companion form representation for the observable pair $\{\hat{A},\hat{C}\}$, that

$$\dot{\hat{x}}_{Ti}(t) = \hat{A}_{ii}\hat{x}_{Ti}(t) + \sum_{\substack{j=1 \\ j\neq i}}^{p} \hat{A}_{ij}\hat{x}_{Tj}(t) + \hat{B}_{Ti}u(t), \qquad 6.4.33$$

where $\hat{B}_{Ti}$ represents the ordered $\bar{d}_i$ rows of $\hat{B}$ numbered from $\bar{\sigma}_{i-1}+1$ to $\bar{\sigma}_i$. Since the first $(\bar{d}_i-1)$ columns of $\hat{A}_{ij}$ are identically zero, the second term on the right side of 6.4.33 can be expressed in terms of the (p) components, $\hat{x}_{\bar{\sigma}_j}(t)$, of the total state $\hat{x}(t)$ which, together with $Eu(t)$, completely determine the output $y(t)$; i.e.

$$\hat{A}_{ij}\hat{x}_{Tj}(t) = \hat{A}_{i\bar{\sigma}_j}\hat{x}_{\bar{\sigma}_j}(t), \qquad 6.4.34$$

where $\hat{A}_{i\bar{\sigma}_j}$ represents the final ($\bar{d}_j$-th) column of $\hat{A}_{ij}$, while $\hat{x}_{\bar{\sigma}_j}(t)$ denotes the single state variable so numbered. We also note, in view of 6.4.32, that

$$y(t) = \hat{C}_p\hat{x}_{T\bar{\sigma}}(t) + Eu(t), \qquad 6.4.35$$

where $\hat{C}_p$ denotes the $p \times p$ lower left triangular matrix with 1's along the diagonal obtained by eliminating all but the (p) nonzero $\bar{\sigma}_k$ rows of $\hat{C}$, and $\hat{x}_{T\bar{\sigma}}(t) = [\hat{x}_{\bar{\sigma}_1}(t), \hat{x}_{\bar{\sigma}_2}(t), \ldots, \hat{x}_{\bar{\sigma}_p}(t)]^T$. If we now substitute 6.4.34 for the second term in 6.4.33, noting that in view of 6.4.35

$$\hat{x}_{\bar{\sigma}_j}(t) = \hat{C}^{-1}_{pj}[y(t) - Eu(t)], \qquad 6.4.36$$

where $\hat{C}^{-1}_{pj}$ denotes the j-th row of $\hat{C}^{-1}_p$, we see that

$$\dot{\hat{x}}_{Ti}(t) = \hat{A}_{ii}\hat{x}_{Ti}(t) + \sum_{\substack{j=1 \\ j\neq i}}^{p} \hat{A}_{i\bar{\sigma}_j}\hat{C}_{pj}^{-1}[y(t) - Eu(t)] + \hat{B}_{Ti}u(t) \qquad 6.4.37$$

At this point we have reduced the general multivariable system to (p) individual scalar output systems of the form 6.4.37, each driven by the directly measurable system signals $u(t)$ and $y(t)$, and each characterized by a known measurable final state component $\hat{x}_{\bar{\sigma}_i}(t) = \hat{C}_{pi}^{-1}[y(t) - Eu(t)]$. Therefore, the technique which was first presented for exponentially estimating the entire state of a scalar output system via an observer of total dimension $n-1$ can now be employed here p times for each of these scalar output subsystems, each employment requiring an observer of total dimension $\bar{d}_i - 1$. Consequently, <u>if $\rho[C] = p \leq n$ and $\{A,C\}$ is observable, an observer of total dimension $\sum_1^p(\bar{d}_i-1) = \sum_1^p \bar{d}_i-p = n-p$ can be employed to estimate the entire n-dimensional state of a state-space system</u>. It might finally be noted that an alternative, frequency domain approach to exponential state estimation will be presented in the next chapter.

EXAMPLE 6.4.38: To illustrate the procedures outlined above for constructing state observers of total dimension $n-p$, we consider the following system $\{\hat{A},\hat{B},\hat{C},E\}$ which has already been reduced to observable companion form:

$$\hat{A} = \left[\begin{array}{ccc|cc} 0 & 0 & -2 & 0 & 0 \\ 1 & 0 & 4 & 0 & -1 \\ 0 & 1 & 0 & 0 & 3 \\ \hline 0 & 0 & 1 & 0 & 2 \\ 0 & 0 & -1 & 1 & 0 \end{array}\right], \quad \hat{B} = \left[\begin{array}{c} 1 \\ 0 \\ 1 \\ \hline -1 \\ 0 \end{array}\right], \quad \hat{C} = \left[\begin{array}{ccc|cc} 0 & 0 & 1 & 0 & 0 \\ 0 & 0 & -1 & 0 & 1 \end{array}\right], \text{ and } E = \begin{bmatrix} 1 \\ 2 \end{bmatrix}$$

This system is clearly observable, with $m = 1$, $p = 2$, $\bar{d}_1 = 3$, $\bar{d}_2 = 2$, $\bar{\sigma}_1 = 3$, and $\bar{\sigma}_2 = 5 = n$. We will only illustrate the exponential estimation of the subsystem with state $\hat{x}_{T1}(t) = \begin{bmatrix} \hat{x}_1(t) \\ \hat{x}_2(t) \\ \hat{x}_3(t) \end{bmatrix}$, since the estimation of $\hat{x}_{T2} = \begin{bmatrix} \hat{x}_4(t) \\ \hat{x}_5(t) \end{bmatrix}$ is completely analogous. In view of 6.4.33, we first note that

$$\dot{\hat{x}}_{T1}(t) = \underbrace{\begin{bmatrix} 0 & 0 & -2 \\ 1 & 0 & 4 \\ 0 & 1 & 0 \end{bmatrix}}_{\hat{A}_{11}} \hat{x}_{T1}(t) + \underbrace{\begin{bmatrix} 0 & 0 \\ 0 & -1 \\ 0 & 3 \end{bmatrix}}_{\hat{A}_{12}} \hat{x}_{T2}(t) + \underbrace{\begin{bmatrix} 1 \\ 0 \\ 1 \end{bmatrix}}_{\hat{B}_{T1}} u(t)$$

Equation 6.4.34 is now employed in order to simplify the expression for $\hat{A}_{12}\hat{x}_{T2}(t)$; i.e. we note that $\hat{A}_{12}\hat{x}_{T2}(t) = \hat{A}_{1\bar{\sigma}_2}\hat{x}_{\bar{\sigma}_2}(t) = \begin{bmatrix} 0 \\ -1 \\ 3 \end{bmatrix} \hat{x}_5(t)$. By 6.4.35, it also follows that $y(t) = \underbrace{\begin{bmatrix} 1 & 0 \\ -1 & 1 \end{bmatrix}}_{\hat{C}_p} \underbrace{\begin{bmatrix} \hat{x}_3(t) \\ \hat{x}_5(t) \end{bmatrix}}_{\hat{x}_{T\bar{\sigma}}(t)} + \underbrace{\begin{bmatrix} 1 \\ 2 \end{bmatrix}}_{E} u(t)$, or since $\hat{C}_p$ is nonsingular, that

$$\begin{bmatrix} \hat{x}_3(t) \\ \hat{x}_5(t) \end{bmatrix} = \underbrace{\begin{bmatrix} 1 & 0 \\ 1 & 1 \end{bmatrix}}_{C_p^{-1}} \left\{ y(t) - \begin{bmatrix} 1 \\ 2 \end{bmatrix} u(t) \right\}$$

It is therefore clear that $\hat{x}_5(t) = [1 \;\; 1] \left\{ y(t) - \begin{bmatrix} 1 \\ 2 \end{bmatrix} u(t) \right\}$, in accordance with 6.4.36 and, consequently, in view of 6.4.37, that

$$\dot{\hat{x}}_{T1}(t) = \underbrace{\begin{bmatrix} 0 & 0 & -2 \\ 1 & 0 & 4 \\ 0 & 1 & 0 \end{bmatrix}}_{\hat{A}_{11}} \hat{x}_{T1}(t) + \underbrace{\begin{bmatrix} 0 \\ -1 \\ 3 \end{bmatrix}}_{\hat{A}_{1\bar{\sigma}_2}} \underbrace{[1 \;\; 1]}_{\hat{C}_{p2}^{-1}} \left\{ y(t) - \underbrace{\begin{bmatrix} 1 \\ 2 \end{bmatrix}}_{E} u(t) \right\} + \underbrace{\begin{bmatrix} 1 \\ 0 \\ 1 \end{bmatrix}}_{\hat{B}_{T1}} u(t).$$

We can now simplify this expression for $\dot{\hat{x}}_{T1}$ by performing the matrix operations indicated above and then combining the appropriate terms; i.e.

$\dot{\hat{x}}_{T1}(t) = \begin{bmatrix} 0 & 0 & -2 \\ 1 & 0 & 4 \\ 0 & 1 & 0 \end{bmatrix} \hat{x}_{T1}(t) + \begin{bmatrix} 0 & 0 \\ -1 & -1 \\ 3 & 3 \end{bmatrix} y(t) + \begin{bmatrix} 1 \\ 3 \\ -8 \end{bmatrix} u(t)$, where $\hat{x}_{\bar{\sigma}_1}(t) = \hat{x}_3(t) = [1 \quad 0]y(t) - u(t)$, a known measurable signal. We have therefore reduced the first subsystem of the given multivariable system to an equivalent scalar output system which is driven by both $u(t)$ and $y(t)$. Since $\hat{x}_3(t) = y_1(t) - u(t)$, we need only estimate $\hat{x}_1(t)$ and $\hat{x}_2(t)$, the first $2(= \bar{d}_1 - 1)$ states of this subsystem. We do this by first employing the equivalence transformation $P = \begin{bmatrix} 1 & 0 & -2 \\ 0 & 1 & -3 \\ 0 & 0 & 1 \end{bmatrix}$, which implies that the eigenvalues of the exponential estimator of $\hat{x}_1(t)$ and $\hat{x}_2(t)$ will equal the zeros of $\lambda^2 + 3\lambda + 2$, or be given by $\lambda_1 = -1$ and $\lambda_2 = -2$. The reader can readily verify that the substitution $\bar{x}_{T1}(t) = P\hat{x}_{T1}(t)$ yields the equivalent system:

$\dot{\bar{x}}_{T1}(t) = \begin{bmatrix} 0 & -2 & -8 \\ 1 & -3 & -5 \\ 0 & 1 & 3 \end{bmatrix} \bar{x}_{T1}(t) + \begin{bmatrix} -6 & -6 \\ -10 & -10 \\ 3 & 3 \end{bmatrix} y(t) + \begin{bmatrix} 17 \\ 27 \\ -8 \end{bmatrix} u(t)$. Since the third component of $\bar{x}_{T1}(t)$, namely $\bar{x}_{T13}(t) = \hat{x}_{\bar{\sigma}_1}(t) = \hat{x}_3(t) = y_1(t) + u(t)$, a directly measurable signal, we need estimate only the first two states of this subsystem which are defined by a second $(\bar{d}_1 - 1)$ order system driven by both $y(t)$ and $u(t)$; i.e.

$$\begin{bmatrix} \dot{\bar{x}}_1(t) \\ \dot{\bar{x}}_2(t) \end{bmatrix} = \begin{bmatrix} 0 & -2 \\ 1 & -3 \end{bmatrix} \begin{bmatrix} \bar{x}_1(t) \\ \bar{x}_2(t) \end{bmatrix} + \begin{bmatrix} -14 & -6 \\ -15 & -10 \end{bmatrix} \begin{bmatrix} y_1(t) \\ y_2(t) \end{bmatrix} + \begin{bmatrix} 9 \\ 22 \end{bmatrix} u(t).$$

It therefore follows that the system:

$\begin{bmatrix} \dot{\tilde{x}}_1(t) \\ \dot{\tilde{x}}_2(t) \end{bmatrix} = \begin{bmatrix} 0 & -2 \\ 1 & -3 \end{bmatrix} \begin{bmatrix} \tilde{x}_1(t) \\ \tilde{x}_2(t) \end{bmatrix} + \begin{bmatrix} -14 & -6 \\ -15 & -10 \end{bmatrix} \begin{bmatrix} y_1(t) \\ y_2(t) \end{bmatrix} + \begin{bmatrix} 9 \\ 22 \end{bmatrix} u(t)$, which can readily be constructed, represents an exponential estimator of $\bar{x}_1(t)$ and $\bar{x}_2(t)$, since

$$\begin{bmatrix} \dot{\bar{x}}_1(t) - \dot{\tilde{x}}_1(t) \\ \dot{\bar{x}}_2(t) - \dot{\tilde{x}}_2(t) \end{bmatrix} = \begin{bmatrix} 0 & -2 \\ 1 & -3 \end{bmatrix} \begin{bmatrix} \bar{x}_1(t) - \tilde{x}_1(t) \\ \bar{x}_2(t) - \tilde{x}_2(t) \end{bmatrix}.$$

An exponential estimate of the complete subsystem state, $\hat{x}_{T1}(t)$, can now be obtained from $\tilde{x}_1(t)$ and $\tilde{x}_2(t)$ via the equivalence transformation P: i.e.

$$\hat{x}_{T1}(t) = P^{-1}\bar{x}_{T1} \approx P^{-1}\begin{bmatrix} \tilde{x}_1(t) \\ \tilde{x}_2(t) \\ \bar{x}_3(t) \end{bmatrix},$$

where $\tilde{x}_1(t)$ and $\tilde{x}_2(t)$ are the states of the $(\bar{d}_1-1)$ dimensional observer, and $\bar{x}_3(t) = \hat{x}_3(t) = y_1(t) + u(t)$ is a known measurable signal which need not be estimated.

6.5 CONCLUDING REMARKS AND REFERENCES

This chapter has served to introduce a control technique which is one of the most common methods employed to compensate state form dynamical sytems, namely linear state variable feedback (l.s.v.f.). We motivated the general employment of this synthesis procedure here via linear quadratic optimal control and, in particular, by formulating the solution to the classical output regulator problem which, in one form or another, dates back to some of Weiner's work [W8] in the late 1940's. Although a somewhat heuristic frequency domain characterization and solution to this problem was given in [N1], it was not until 1960 when Kalman [K1], employing the then revolutionary concepts of controllability and observability, rigorously formulated and solved the general linear quadratic optimization problem. His development included, as a special case, the output regulator problem presented here, thus explicitly establishing the importance of l.s.v.f. as a compensation technique.

The solution to the quadratic optimization problem suggested the possibility that l.s.v.f. might be used for other purposes as well, such as complete and arbitrary closed loop pole placement, and it was first Morgan [M4] and then Brockett [B3], the latter employing a scalar companion form system matrix, similar to the approach used here, who constructively established the fact that l.s.v.f. can indeed be used to completely and arbitrarily position all (n) poles of a closed loop controllable system in the single-input case. Wonham [W6] later extended this result to include the more general multivariable case, using a rather involved constructive procedure, and since then a number of other investigators [D3][H4][W7] have developed alternative constructive methods for arbitrarily assigning the (n) poles of a controllable multivariable system via l.s.v.f. It might again be noted here that there are a number of other important applications of l.s.v.f. compensation which will be presented in this text, but not until after we have discussed certain frequency domain implications of this compensation technique from the point of view of the structure theorem, which we will do in the next chapter.

The development of exponential estimators or Luenberger observers for approximating the state of deterministic linear systems is generally credited to Luenberger [L4], as noted in Section 6.4, although the existence of such state reconstruction schemes was undoubtedly recognized earlier by Kalman, in view of his results on system observability and its implications. Luenberger [L5] also extended his observer theory to include reduced order observers, although by that time, Bass and Gura [B4] had also discovered the equivalent of reduced order observers and displayed their utility in closed loop pole assignment.

PROBLEMS - CHAPTER 6

6-1 Show that the eigenvalues of the system matrix $A + BF$ of 6.3.3a are identical to those associated with the matrix $\hat{A} + \hat{B}\hat{F}$ of system 6.3.8a.

6-2 Consider a state-space system of the form 6.3.1, with

$A = \begin{bmatrix} 0 & 1 \\ 1 & 0 \end{bmatrix}$, $B = \begin{bmatrix} 0 \\ 1 \end{bmatrix}$, $C = [0 \quad 1]$, and $E = [0]$. Find the (optimal) control, $u^*(t)$, which minimizes $J = \int_{t_o}^{\infty} [y^2(t)+2\,u^2(t)]dt$.

Determine the (minimum) value of J if $x_1(t_o) = x_2(t_o) = -2$.

6-3 Determine all of the solutions to the matrix Riccati equation of Example 6.2.8 and verify that only one of the solutions is positive definite.

6-4 Show that the l.s.v.f. gain vector, $fx(t)$, of a scalar, controllable, state-space system is uniquely specified by any set of (n) desired closed loop poles.

6-5 Show that the eigenvalues of a square matrix A are generally altered if two rows of A are interchanged. If, however, the same two numbered columns of A are <u>also</u> interchanged, show that the eigenvalues of A will remain unaltered.

6-6 Show that if G is nonsingular, l.s.v.f. cannot affect system controllability, although it can affect system observability; i.e. show that l.s.v.f. can alter the rank of the observability matrix $\mathcal{O}$. Can l.s.v.f. increase or decrease (or both) $\rho[\mathcal{O}]$? (Hint: Consider a scalar example.)

6-7 What difficulties, if any, are encountered in the construction of a Luenberger state estimator if $B = -KE$ in Equation 6.4.2?

6-8 Can an exponential state estimator of a system be constructed if the system is not (completely state) observable, but all of its unobservable modes are asymptotically stable? unstable? Explain your answers.

6-9 Show that it is possible to construct an exponential state estimator of a given system with $K = 0$ if all of the eigenvalues of A lie in the asymptotically stable half-plane, $\text{Re}(\lambda) < 0$.

6-10 Outline the dual of the pole assignment algorithm used to constructively establish Theorem 6.3.10.

6-11 Find a state feedback gain matrix, $Fx(t)$, which places the closed loop poles of the system considered in Example 3.6.10 at $s = -1 \pm j$ and $-2 \pm j$.

6-12 Consider the state-space system 6.3.1. If <u>linear output feedback</u> is defined as the control law: $u(t) = Hy(t) + \hat{G}v(t)$, where H and $\hat{G}$ are constant $(m \times p)$ and $(m \times q)$ matrices respectively, under what conditions will linear output feedback be completely equivalent to l.s.v.f.?

6-13 Construct a second order state observer, with poles at $-1 \pm j$, of the (scalar) state-space system 6.3.1 with $A = \begin{bmatrix} -1 & 1 & 2 \\ 1 & 0 & -1 \\ 0 & 1 & 0 \end{bmatrix}$, $b = \begin{bmatrix} 1 \\ 0 \\ 1 \end{bmatrix}$, $c = [1 \quad 0 \quad -1]$, and $e = 0$.

6-14 A state-space system is said to be <u>single input controllable</u> if there exists a column vector g such that the pair $\{A, Bg\}$ is controllable. Show that if a system is single input controllable, then a single linear function of the state, $fx(t)$, can be employed to completely and arbitrarily assign all (n) closed loop poles.

6-15 Show that if a state-space system is both observable and single input controllable, then a single state estimator of total order $\nu-1$, where $\nu = \max \bar{d}_i$ is the observability index of the system, can be employed in the feedback path to completely and arbitrarily assign all $(n+\nu-1)$ closed loop poles of the compensated system. (Hint: Consider the effect of employing the same observer poles to estimate the state of each subsystem of the multivariable system in conjunction with the results obtained in Problem 6-14.)

6-16 Verify that l.s.v.f. can sometimes be used to alter the zeros of the transfer matrix elements of a linear multivariable system without affecting the poles. In particular, show that if the state system, $\dot{x}(t) = Ax(t) + Bu(t)$; $y(t) = Cx(t)$, with

$$A = \begin{bmatrix} 0 & 1 & 0 & 0 \\ 0 & 0 & 0 & 0 \\ 0 & 0 & 0 & 1 \\ 0 & 0 & 1 & 0 \end{bmatrix}, \quad B = \begin{bmatrix} 0 & 0 \\ 1 & 0 \\ 0 & 0 \\ 0 & 1 \end{bmatrix}, \text{ and } C = [1 \quad 1 \quad 0 \quad 1]$$

is compensated via the linear state feedback control law, $u(t) = Fx(t) + Gv(t)$, with

$$F = \begin{bmatrix} 1/5 & 0 & -2/5 & 0 \\ -2/5 & 0 & -1/5 & 0 \end{bmatrix} \text{ and } G = \begin{bmatrix} 2 & -1 \\ 1 & 2 \end{bmatrix},$$

then the open loop transfer matrix,

$$T(s) = C(sI-A)^{-1}B = \left[\frac{s+1}{s^2}, \quad \frac{s}{s^2-1}\right], \text{ becomes } T_{F,G}(s) =$$

$$C(sI-A-BF)^{-1}BG = \left[\frac{3s+2}{s^2}, \quad \frac{s-1}{s^2-1}\right].$$

(Hint: Theorem 4.3.3 can be employed to facilitate the computations.)

CHAPTER 7
FREQUENCY DOMAIN COMPENSATION

7.1 INTRODUCTION

The purpose of this chapter is to present an in-depth study of certain frequency domain implications of linear state variable feedback (l.s.v.f.) from the point of view of the structure theorem in order to develop some rather general procedures for the compensation of linear multivariable systems. We begin, in Section 7.2, by representing the transfer matrix of a linear system in such a way that those parts or characteristics which are totally unaffected by l.s.v.f. are "separated" from those which can be completely and arbitrarily altered.

In Section 7.3 we employ the results developed in Section 7.2 in order to outline a frequency domain procedure for realizing the equivalent of a l.s.v.f. control law implemented via a Luenberger observer. The technique employed relies on the ability to represent a proper transfer matrix, $T(s)$, as the product $R(s)P(s)^{-1}$, where $R(s)$ and $P(s)$ are relatively right prime polynomial matrices, an assumption which implies the (complete state) observability of any equivalent time domain realization and, therefore, the ability to construct a Luenberger observer for exponentially estimating the entire state of the system. As we show, a complete frequency domain analog of l.s.v.f. and its implementation via Luenberger estimation can be accomplished without any reference whatsoever to the time domain notion of state. The technique employed is a constructive one and relies on an extension of the classical "eliminant matrix" of two polynomials to include polynomial matrices as well.

We begin Section 7.4 by illustrating that l.s.v.f. compensation can be considered as merely a special case of a more general type of compensation. In particular, we show that any l.s.v.f. design can be implemented via an "equivalent" frequency domain compensation scheme consisting of an appropriate proper transfer matrix placed in the feedforward path, and then show that a design based on the employment of <u>any</u> proper transfer matrix in the feedforward path can be realized via an appropriate "equivalent" design which involves dynamic feedforward compensation in conjunction with l.s.v.f. The more general compensation scheme is constructively developed by employing the results given in Section 7.3 as well as a polynomial matrix version of the scalar "division algorithm" which was introduced in Section 2.5. Some concluding remarks and references are then given in Section 7.5.

7.2 FREQUENCY DOMAIN IMPLICATIONS OF STATE FEEDBACK

In the previous chapter we discussed the employment of linear state variable feedback (l.s.v.f.) as a means of compensating dynamical systems in the state form:

$$\dot{x}(t) = Ax(t) + Bu(t); \qquad 7.2.1a$$

$$y(t) = Cx(t) + Eu(t), \qquad 7.2.1b$$

with $\rho[B] = m \leq n$. In this section we will discuss certain frequency domain implications associated with the compensation of this class of systems via the general l.s.v.f. control law 6.2.7, namely

$$u(t) = Fx(t) + Gv(t) \qquad 7.2.2$$

from the point of view of the structure theorem, which was introduced in Section 4.3. The motivation for this discussion will become apparent in the subsequent sections as we evolve a general frequency domain approach to linear system compensation.

To begin, we recall from our work in Section 6.3 that if 7.2.1 is controllable, the effect of l.s.v.f. on the closed loop system can be clarified considerably once the system has been transformed to an equivalent controllable companion form $\{\hat{A},\hat{B},\hat{C},E\} = \{QAQ^{-1},QB,CQ^{-1},E\}$ via an appropriate equivalence transformation Q (see Section 3.6). In particular, as we demonstrated in Section 6.3, every one of the (mn) entries which comprise $\hat{A}_m$, the (m) ordered σ_k rows of $\hat{A}$, as well as $\hat{B}_m$, the (m) ordered σ_k rows of $\hat{B}$, can be completely and arbitrarily altered via the feedback pair $\{\hat{F},G\}$. More specifically, $\hat{B}_m\hat{F} = \hat{B}_m FQ^{-1}$ is added to $\hat{A}_m$, $\hat{B}_m$ is postmultiplied by G, and the companion form structure, 3.6.9, of the pair $\{\hat{A}+\hat{B}\hat{F},\hat{B}G\}$ is preserved, although the (m) ordered σ_k rows of $\hat{B}G$ do not necessarily retain a nonsingular, upper right triangular form.

Since the controllable companion form structure of the system is preserved under l.s.v.f., we can directly employ the controllable version of the structure theorem (Section 4.3) in order to obtain an expression for $T_{F,G}(s)$, the closed loop transfer matrix of the system; i.e. in view of 4.2.3,

$$T_{F,G}(s) = (C+EF)(sI-A-BF)^{-1}BG + EG \qquad 7.2.3a$$

$$= (\hat{C}+E\hat{F})(sI-\hat{A}-\hat{B}\hat{F})^{-1}\hat{B}G + EG, \qquad 7.2.3b$$

and, therefore, by Theorem 4.3.3,

$$T_{F,G}(s) = (\hat{C}+E\hat{F})S(s)\delta_F^{-1}(s)\hat{B}_mG + EG \qquad 7.2.3c$$

$$= [(\hat{C}+E\hat{F})S(s)+E\hat{B}_m^{-1}\delta_F(s)][\hat{B}_m^{-1}\delta_F(s)]^{-1}G, \qquad 7.2.3d$$

with

$$\delta_F(s) = \begin{bmatrix} s^{d_1} & & & 0 \\ & s^{d_2} & & \\ & & \ddots & \\ 0 & & & s^{d_m} \end{bmatrix} - (\hat{A}_m+\hat{B}_m\hat{F})S(s) = \delta(s) - \hat{B}_m\hat{F}S(s) \qquad 7.2.4$$

In view of 7.2.4, we note that $E\hat{B}_m^{-1}\delta_F(s) = E\hat{B}_m^{-1}\delta(s) - E\hat{F}S(s)$, which allows us to simplify the final expression, 7.2.3d, for $T_{F,G}(s)$. In particular, in view of the above,

$$T_{F,G}(s) = [\hat{C}S(s) + E\hat{B}_m^{-1}\delta(s)][\hat{B}_m^{-1}\delta_F(s)]^{-1}G \qquad 7.2.5$$

As in the open loop, uncompensated case which was discussed in Section 4.3, we can again make some interesting observations and comparisons regarding this final expression for the closed loop transfer matrix of the system. In particular, by equating the n-th degree (monic) denominator polynomials of 7.2.3a and 7.2.3c, we see that

$$|sI-A-BF| = |sI-\hat{A}-\hat{B}\hat{F}| = |\delta_F(s)| \stackrel{\Delta}{=} \Delta_F(s), \qquad 7.2.6$$

or that the closed loop poles of the controllable system 7.2.1 are the zeros of $|\delta_F(s)|$. Furthermore, if we recall 4.3.8 and 4.3.9 and define

$$P_F(s) \stackrel{\Delta}{=} \hat{B}_m^{-1}\delta_F(s) = \hat{B}_m^{-1}[\delta(s) - \hat{B}_m\hat{F}S(s)], \qquad 7.2.7$$

it follows, in view of 4.3.9, that

$$P_F(s) = P(s) - \hat{F}S(s), \qquad 7.2.8$$

or, in view of 4.3.8 and the above, that the expression for $T_{F,G}(s)$ can be written more succinctly as:

$$T_{F,G}(s) = R(s)P_F^{-1}(s)G \qquad 7.2.9$$

If we now compare this expression for the closed loop transfer matrix to 4.3.10, an analogous expression for the open loop transfer matrix, $T_{0,I}(s) = T(s)$, we can, loosely speaking, separate those portions of the transfer matrix which are completely and arbitrarily altered by l.s.v.f. from the l.s.v.f. invariant parts. In particular, it is immediately obvious that R(s) is unaltered by l.s.v.f. and thus represents a l.s.v.f. invariant part of $T_{F,G}(s)$. Furthermore, $P_F(s)$ remains column proper and, more specifically

$$\Gamma_c[P_F(s)] = \Gamma_c[P(s)] = \hat{B}_m^{-1}, \qquad 7.2.10$$

with

$$\partial_{ci}[P_F(s)] = \partial_{ci}[P(s)] = d_i, \tag{7.2.11}$$

which, together, represent the only other l.s.v.f. invariant relations; i.e. other than satisfying 7.2.10 and 7.2.11, $P_F(s)$ can be completely altered via F (actually $\hat{F} = FQ^{-1}$). It is also clear that G, the postmultiplier of the closed loop transfer matrix, is also arbitrarily assignable.

We now recall from Section 6.3 that any uncontrollable poles of the system are unaffected by l.s.v.f. and appear as cancellable factors in both the numerator and denominator of the transfer matrix. Therefore, if the system 7.2.1 is not (completely state) controllable, we can still apply the preceding results to the "controllable part" of the system; i.e. to the quadruple $\{\hat{A}_c, \hat{B}_c, \hat{C}_c, E\}$ (see Section 3.6). Of course $|sI-\hat{A}_{\bar{c}}| = \Delta_{\bar{c}}(s)$ would represent an additional l.s.v.f. invariant term.

Perhaps the most interesting and important fact that should be derived from the results presented in this section is that the effect of l.s.v.f., a time domain compensation scheme, on the closed loop system can be quite readily observed in the frequency domain via the structure theorem "factorization", 7.2.9, of the closed loop transfer matrix. We finally note that our observations regarding the l.s.v.f. variable and invariant parts of a system's transfer matrix are completely consistent with a well known fact regarding scalar systems, namely that l.s.v.f. can be used to completely and arbitrarily assign all of the controllable closed loop poles of a scalar system (the zeros of the denominator of its rational transfer function), although l.s.v.f. has no effect whatsoever on the zeros of the system (the numerator of its transfer function). We have now established a generalization of this fact in the multivariable case.

7.3 FREQUENCY DOMAIN STATE ESTIMATION AND FEEDBACK

The major emphasis on system compensation thus far has been placed on linear state variable feedback (l.s.v.f.) and various implications regarding its employment in the time domain via the feedback pair $\{F,G\}$. It should be realized, however, that as often as not, the dynamical behavior of a linear system is represented by a transfer matrix; i.e. its input/output frequency domain characteristics, which are independent of any internal state representation. Furthermore, as we will show, a number of important multivariable system design objectives can best be stated in terms of a particular input/output relationship (e.g. a diagonal transfer matrix in the case of "decoupling" --Chapter 8). In view of these observations, it is not unusual for an actual design based on the employment of l.s.v.f. to begin with a transfer matrix description for the dynamical behavior of the system, followed by a transfer, via realization theory, to some equivalent time domain realization, followed by the development of an appropriate state feedback control law, followed by the implementation of the l.s.v.f. control law via exponential state estimation. It might be noted that each of these individual design steps has already been presented in this text, and an overall design employing all of these steps could involve a substantial amount of computation, much of which can be avoided if we employ a more direct approach to the problem which we will now illustrate. More specifically, the objective of this section is to develop a direct frequency domain approach to the synthesis of linear multivariable systems, one which is completely analogous to the employment of l.s.v.f. and its implementation via exponential state estimation in the time domain.

The Scalar Case

To begin, we will introduce the compensation scheme in its most basic form in order to motivate and simplify the design and analysis of the compensator in the more general multivariable case. In particular, let us consider the dynamical behavior of a single input/output (scalar) n-th order system, which is represented by the frequency domain relationship:

$$y(s) = t(s)u(s), \tag{7.3.1}$$

where $t(s)$ is a transfer function expressible as the ratio of two relatively prime polynomials in the Laplace operator s; i.e.

$$t(s) = r(s)p(s)^{-1} \tag{7.3.2}$$

We assume that the transfer function is a proper one, or that the degree of $r(s)$ is less than or equal to n, the degree of $p(s)$; i.e. $\partial[r(s)] \leq \partial[p(s)] = n$. By introducing a scalar variable, $z(s)$, 7.3.1 can be represented, in view of 7.3.2, by the following two equations in Laplace transformed differential operator form:

$$p(s)z(s) = u(s); \tag{7.3.3a}$$

$$y(s) = r(s)z(s), \tag{7.3.3b}$$

since by 7.3.3a, $z(s) = p(s)^{-1}u(s)$, and by 7.3.3b, $y(s) = r(s)p(s)^{-1}u(s) = t(s)u(s)$. In view of 7.3.3, the open loop system can be represented by the following block diagram:

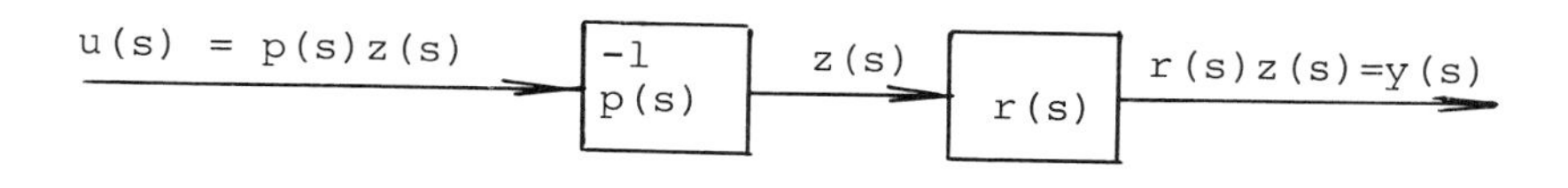

FIGURE 7.3.4

THE OPEN LOOP SCALAR SYSTEM

We now recall from the previous section that in the case of controllable scalar systems, linear state variable feedback can be used to completely and arbitrarily specify all (n) closed loop poles of the system, although the zeros of the transfer function are unaffected by l.s.v.f. In view of these observations, the objective of the frequency domain compensation scheme which will now be developed is to completely and arbitrarily specify all (n) closed loop poles of the given system without any "apparent increase" in system order. More specifically, the objective will be to design a linear feedback compensator, which is driven by the two available signals, $y(s)$ and $u(s)$, such that the closed loop transfer function $t_{f,g}(s)$, relating $y(s)$ to the external input, $v(s)$, is "equivalent to" $r(s)p_f^{-1}(s)g$, where $g^{-1}p_f(s)$ is any desired n-th degree polynomial.

Let us now consider the compensator, depicted in Figure 7.3.5, which we will employ to achieve this objective, noting that the closed

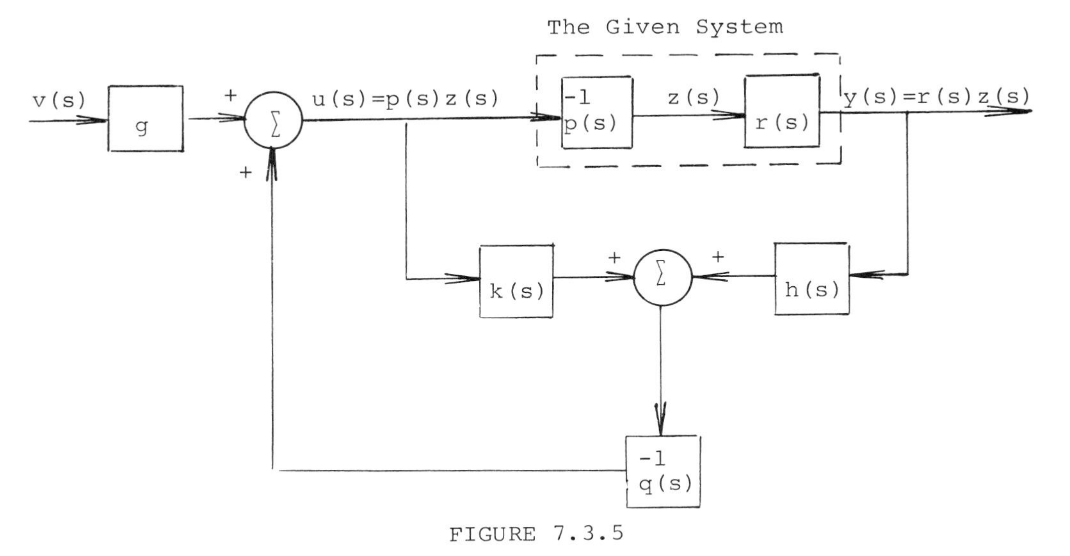

FIGURE 7.3.5

THE SCALAR COMPENSATION SCHEME

loop dynamical behavior of this system can readily be determined if we equate signals at the first summation junction; i.e.

$$u(s) = p(s)z(s) = g\,v(s) + q^{-1}(s)[k(s)p(s)z(s) + h(s)r(s)z(s)], \quad 7.3.6$$

or

$$[q(s)p(s) - k(s)p(s) - h(s)r(s)]z(s) = q(s)g\ v(s). \qquad 7.3.7$$

Let us now note that if we can choose $k(s)$, $h(s)$, and $q(s)$ such that for any arbitrary polynomial $f(s)$ of degree no greater than $n-1$,

(i) $q(s)$ is a stable polynomial

(ii) $k(s)p(s) + h(s)r(s) = q(s)f(s)$, 7.3.8

and (iii) both $q^{-1}(s)k(s)$ and $q^{-1}(s)h(s)$ are (stable) proper transfer functions, then it follows that this scalar compensation scheme would be both realizable without the need to differentiate signals and would also yield the desired closed loop transfer function, $r(s)p_f^{-1}(s)g$. In particular, if we substitute 7.3.8 into 7.3.7 we obtain

$$q(s)[p(s) - f(s)]z(s) = q(s)g\ v(s), \qquad 7.3.9$$

and since $y(s) = r(s)z(s)$ and $p_f(s) = p(s) - f(s)$ is invertible, it clearly follows that

$$y(s) = r(s)p_f^{-1}(s)q^{-1}(s)q(s)g\ v(s) = r(s)p_f^{-1}(s)g\ v(s), \qquad 7.3.10$$

the desired closed loop transfer function (after the stable pole-zero cancellations corresponding to $q(s)$ and its inverse have been made).

We thus note that in order to practically realize this desired closed loop transfer function we must formally establish that the three conditions noted above pertaining to the selection of $k(s)$, $h(s)$, and $q(s)$ can be satisfied. A major step in establishing these relations is now provided by the following:

DEFINITION 7.3.11: The <u>eliminant (matrix)</u>, M_e, of two nonzero polynomials $r(s) = r_o+r_1s+\ldots+r_ns^n$ and $p(s) = p_o+p_1s+\ldots+p_ns^n$, with $p_n \neq 0$ is defined as the following $(2n \times 2n)$ constant matrix consisting of the "shifted" polynomial coefficients:

$$M_e \triangleq \begin{bmatrix} r_o & r_1 & \cdots & & r_n & 0 & 0 & \cdots & 0 \\ 0 & r_o & \cdots & & r_{n-1} & r_n & 0 & \cdots & 0 \\ \vdots & & & & \vdots & & & & \vdots \\ 0 & 0 & \cdots & r_o & r_1 & r_2 & r_3 & \cdots & r_n \\ p_o & p_1 & \cdots & & p_n & 0 & 0 & \cdots & 0 \\ 0 & p_o & \cdots & & p_{n-1} & p_n & 0 & \cdots & 0 \\ \vdots & & & & \vdots & & & & \vdots \\ 0 & 0 & \cdots & p_o & p_1 & p_2 & p_3 & \cdots & p_n \end{bmatrix} \qquad 7.3.11$$

THEOREM 7.3.12 (Sylvester): <u>The polynomials</u> $r(s) = r_o+r_1s+\ldots+r_ns^n$ <u>and</u> $p(s) = p_o+p_1s+\ldots+p_ns^n$, <u>with</u> $p_n \neq 0$, <u>are relatively prime if and only if their eliminant is nonsingular</u>.

<u>Proof</u>: We first establish necessity by contradiction. In particular, suppose that $r(s)$ and $p(s)$ are relatively prime but that the determinant of their eliminant, M_e, which is sometimes called the <u>resultant</u> of $p(s)$ and $r(s)$, is zero; i.e. $|M_e| = 0$. This latter assumption implies the existence of some nonzero, 2n-dimensional row vector $\alpha = [\alpha_1,\alpha_2] = [\alpha_{10},\alpha_{11},\ldots,\alpha_{1,n-1},\alpha_{20},\ldots,\alpha_{2,n-1}]$ such that $\alpha M_e = 0$, or if we define $S_e(s) \triangleq [1,s,s^2,\ldots,s^{2n-1}]^T$, it follows that

$$\alpha M_e S_e(s) = [\alpha_1,\alpha_2]\begin{bmatrix} r(s) \\ sr(s) \\ \vdots \\ s^{n-1}r(s) \\ p(s) \\ \vdots \\ s^{n-1}p(s) \end{bmatrix} = \alpha_1(s)r(s) + \alpha_2(s)p(s) \qquad 7.3.13$$

equals zero for some $\alpha_1(s) = \alpha_{10}+\alpha_{11}s+\ldots+\alpha_{1,n-1}s^{n-1}$ and $\alpha_2(s) = \alpha_{20}+\alpha_{21}s+\ldots+\alpha_{2,n-1}s^{n-1}$, with at least one of the $\alpha_{ij} \neq 0$. It is clear, however, that both $\alpha_1(s)$ and $\alpha_2(s)$ must be nonzero since (in view of 7.3.13) the converse would imply that either $r(s)$ or $p(s)$ would be identically zero. Therefore, in view of 7.3.13

$$p(s) = -\frac{\alpha_1(s)r(s)}{\alpha_2(s)}, \qquad 7.3.14$$

with both $\alpha_1(s) \neq 0$ and $\alpha_2(s) \neq 0$. If we now assume, for convenience, that $\alpha_1(s)$ and $\alpha_2(s)$ are relatively prime, it then follows, in view of the fact that $\partial[\alpha_1(s)] < \partial[p(s)]$, that $\frac{r(s)}{\alpha_2(s)} = -\frac{p(s)}{\alpha_1(s)}$ represents a nontrivial polynomial divisor of both $p(s)$ and $r(s)$ and, therefore, that $r(s)$ and $p(s)$ are not relatively prime which is contrary to our initial assumption. Necessity is thus established by contradiction.

We now establish sufficiency. In particular, to show that a nonzero resultant implies a relatively prime polynomial pair $\{r(s),p(s)\}$, we let $\alpha = [1,0,\ldots,0]M_e^{-1}$, which immediately implies that

$$\alpha M_e S_e(s) = \alpha_1(s)r(s) + \alpha_2(s)p(s) = 1, \qquad 7.3.15$$

for some pair $\{\alpha_1(s),\alpha_2(s)\}$ of polynomials. Therefore 1 is a greatest common divisor of $r(s)$ and $p(s)$; i.e. $r(s)$ and $p(s)$ are relatively prime.

As a consequence of the sufficiency proof of this theorem, we now have:

COROLLARY 7.3.16: <u>If $r(s)$ and $p(s)$ are relatively prime, with $\partial[p(s)] = n$ and $\partial[r(s)] \leq n$, then any arbitrary polynomial of degree $2n-1$ can be obtained as the sum, $\alpha_1(s)r(s) + \alpha_2(s)p(s)$ for some appropriate choice of $\alpha_1(s)$ and $\alpha_2(s)$.</u>

<u>Proof</u>: We simply note that if the polynomial $\beta(s) =$

$\beta_o+\beta_1 s+\ldots+\beta_{2n-1}s^{2n-1} = [\beta_o,\beta_1,\ldots,\beta_{2n-1}]S_e(s) = S_e(s)$ is desired, it can be obtained by setting

$$\alpha = M_e^{-1} \qquad 7.3.17$$

and then solving 7.3.13 for $\alpha_1(s)$ and $\alpha_2(s)$, both of which would be of degree no greater than $n-1$.

In view of Corollary 7.3.16, it is now relatively easy to establish the fact that if $r(s)$ and $p(s)$ are relatively prime then $k(s)$, $h(s)$, and $q(s)$ can be chosen to satisfy the three conditions which enable us to realize the desired stable closed loop transfer function, $r(s)p_f^{-1}(s)g$. In particular, if we set $q(s) = q_o+q_1 s+\ldots+q_{n-1}s^{n-1}$ equal to any desired stable polynomial of degree $n-1$, and let $f(s) = f_o+f_1 s+\ldots+f_{n-1}s^{n-1}$ equal any desired polynomial of degree no greater than $n-1$, then the product of the two, a polynomial of degree $\leq 2n-2$, can be expressed as the product of some constant vector β and $S_e(s)$; i.e.

$$q(s)f(s) = \beta S_e(s) \qquad 7.3.18$$

If we now employ 7.3.17 and 7.3.13, we can write

$$q(s)f(s) = \alpha M_e S_e(s) = [\alpha_1(s),\alpha_2(s)]\begin{bmatrix} r(s) \\ p(s) \end{bmatrix} = \alpha_1(s)r(s)+\alpha_2(s)p(s), \qquad 7.3.19$$

where $\partial[\alpha_1(s)] \leq n-1$ and $\partial[\alpha_2(s)] \leq n-1$. Since $\partial[q(s)] = n-1$, it is clear that if we set $h(s) = \alpha_1(s)$ and $k(s) = \alpha_2(s)$, then both $q(s)^{-1}h(s)$ and $q(s)^{-1}k(s)$ will be (stable) proper transfer functions. Furthermore, in view of 7.3.19, it also follows that 7.3.8 is satisfied. We can now summarize the preceding by noting that <u>if</u> $t(s) = r(s)p(s)^{-1}$ <u>is a proper transfer function with $r(s)$ and $p(s)$ relatively prime and $\partial[p(s)] = n$, then one can achieve any desired (stable) closed loop transfer function</u> $t_{f,g}(s) = r(s)p_f^{-1}(s)g = r(s)[g^{-1}p_f(s)]^{-1}$ <u>via the compensation scheme depicted in Figure 7.3.5</u>,

<u>where the only requirement on</u> $\bar{g}^{-1}p_f(s)$ <u>is that it also be of degree n.</u>

It might be noted at this point that the frequency domain design just outlined is completely independent of the notion of "state"; i.e. the design is based solely on the input/output dynamical behavior of the system and relies only on the assumption that the numerator and denominator polynomials of the transfer function are relatively prime. We further remark that this compensation scheme represents a frequency domain analog of the combined time domain notions of linear state variable feedback for complete and arbitrary closed loop pole assignment and the implementation of the state feedback design via Luenberger observer feedback. More will be said regarding this observation after we have presented a more general multivariable version of this compensation technique, which we now do.

<u>The Multivariable Case</u>:

The scalar results which were just presented will now serve to simplify the extension of the compensation scheme to the multivariable case. In particular, it might first be noted that, as in the scalar case, the dynamical behavior of an m-input, p-output, linear, time-invariant system may be represented by a proper $p \times m$ transfer matrix, T(s), where

$$y(s) = T(s)u(s). \tag{7.3.20}$$

Furthermore, as in the scalar case, a proper rational transfer matrix can always be factored as the product,

$$T(s) = R(s)P^{-1}(s), \tag{7.3.21}$$

where R(s) and P(s) are relatively right prime polynomial matrices of dimensions $p \times m$ and $m \times m$ respectively, with P(s) column proper, $\partial[|P(s)|] = n$, and

$$\partial_{cj}[R(s)] \leq \partial_{cj}[P(s)] = d_j. \tag{7.3.22}$$

Sections 2.6 and 5.4 contain the details regarding the above observations. In view of the results presented in the previous section, it follows that given a controllable companion form realization of $T(s)$, l.s.v.f. can be used to completely and arbitrarily alter the (m^2) polynomials associated with $P_F(s) = P(s) - \hat{F}S(s)$. The (m) ordered controllability indices, d_j, as well as the nonsingularity of $\Gamma_c[P_F(s)] = \Gamma_c[P(s)]$ would, however, remain unaltered by l.s.v.f. In view of these observations, the objective of the frequency domain compensation scheme which we will now develop is to achieve a closed loop transfer matrix of the form $R(s)P_F^{-1}(s)G$, with $\Gamma_c[P_F(s)] = \Gamma_c[P(s)]$ and $\partial_{cj}[P_F(s)] = d_j$ for all $j = 1,2,\ldots,m$, but $P_F(s)$ otherwise arbitrary and completely assignable. We begin by assuming complete knowledge of $R(s)$ and $P(s)$, a polynomial matrix pair which specifies the open loop transfer matrix of a given dynamical system.

THEOREM 7.3.23: Consider the $(p \times m)$ open loop transfer matrix, $T(s) = R(s)P(s)^{-1}$, of a dynamical system where $\partial_c[R(s)] \leq \partial_c[P(s)]$ and $P(s)$ is column proper with $\partial_{cj}[P(s)] = d_j \geq 1$ for all $j = 1,2,\ldots m$. If $R(s)$ and $P(s)$ are relatively right prime, then for any arbitrary $(m \times m)$ polynomial matrix $F(s)$ which satisfies the relation: $\partial_c[F(s)] < \partial_c[P(s)]$, there exists a triple, $\{H(s),K(s),Q(s)\}$, of polynomial matrices which satisfies the following relations:

(i) The zeros of $|Q(s)|$ lie in the stable half-plane $Re(s) < 0$,

(ii) $K(s)P(s) + H(s)R(s) = Q(s)F(s)$, 7.3.24

and

(iii) $Q(s)^{-1}K(s)$ and $Q(s)^{-1}H(s)$ are both (stable) proper transfer matrices.

Rather than first establishing this theorem, it might prove enlightening to discuss its significance from the point of view of the frequency domain compensation scheme depicted in Figure 7.3.25.

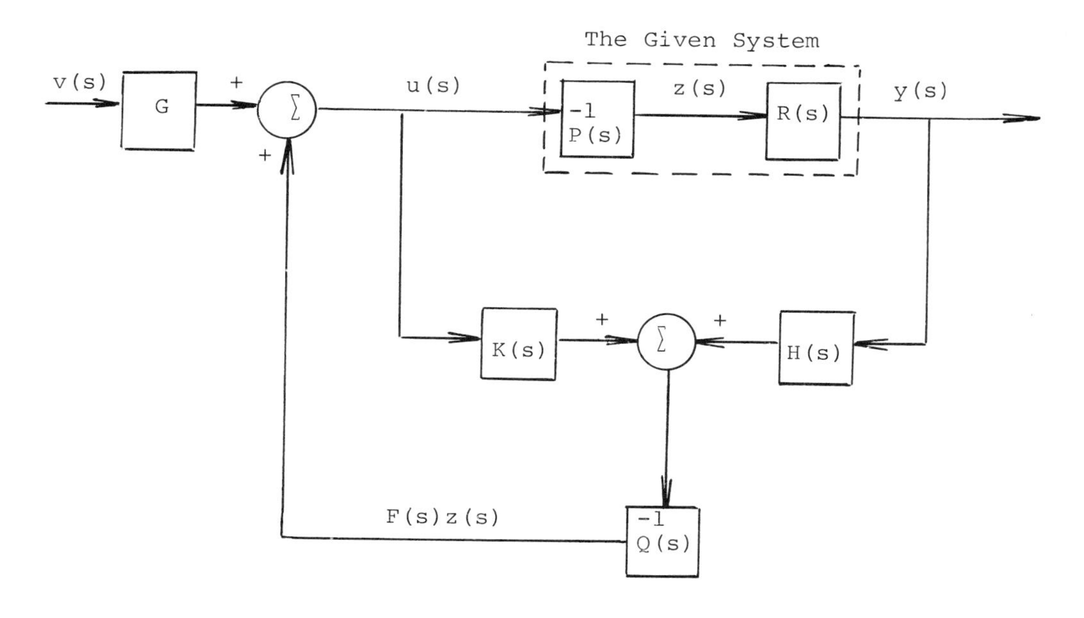

FIGURE 7.3.25

THE COMPENSATOR

In particular, we first note that the original open loop system with $T(s) = R(s)P(s)^{-1}$ is enclosed by the dotted lines and, through the introduction of the m-vector partial state $z(s)$, can be represented via the equations:

$$P(s)z(s) = u(s); \quad y(s) = R(s)z(s), \qquad 7.3.26$$

since $z(s) = P(s)^{-1}u(s)$ and $y(s) = R(s)z(s) = R(s)P(s)^{-1}u(s) = T(s)u(s)$. If we now equate signals at the first summation junction, we readily obtain an expression for the closed loop transfer matrix of the system; i.e.

$$u(s) = P(s)z(s) = Gv(s) + Q(s)^{-1}[K(s)P(s)+H(s)R(s)]z(s), \quad 7.3.27$$

and substituting the right side of 7.3.24 for $K(s)P(s) + H(s)R(s)$ in 7.3.27, we obtain after some rudimentary manipulations:

$$y(s) = R(s)[P(s) - F(s)]^{-1}Q^{-1}(s)Q(s)Gv(s) = R(s)P_F^{-1}(s)Gv(s), \quad 7.3.28$$

where $P_F(s) = P(s) - F(s)$, with $F(s)$ completely analogous to $\hat{F}S(s)$ in 7.2.8. This latter relationship thus represents the desired closed loop transfer matrix (after the stable pole-zero cancellations represented by $Q(s)$ and its inverse have been made). We therefore conclude that if Theorem 7.3.23 holds, then <u>the stable, physically realizable compensation scheme depicted in Figure 7.3.25 can be used to achieve any desired closed loop transfer matrix of the form 7.3.28</u>, which is equivalent to the transfer matrix which would be obtained under actual l.s.v.f. We now establish Theorem 7.3.23.

<u>Proof</u>: In order to establish this theorem, we will first extend the notions of the eliminant matrix of two polynomials and its determinant, the resultant, which were developed earlier in this section, to the matrix case. In particular, let us first consider any pair, $\{R(s),P(s)\}$, of polynomial matrices of dimensions $p \times m$ and $m \times m$ respectively with $\partial_c[R(s)] \leq \partial_c[P(s)]$, $P(s)$ column proper, and $\partial_{cj}[P(s)] = d_j \geq 1$ for all $j = 1,2,\ldots m$. In view of these assumptions, it follows that for $k = 1,2,\ldots,$ the $k(m + p) \times m$ polynomial matrix: $[R^T(s),sR^T(s),\ldots,s^{k-1}R^T(s),P^T(s),\ldots,s^{k-1}P^T(s)]^T$ can be expressed as the product of a constant $k(m + p) \times (n + mk)$ matrix, M_{ek}, and an $(n + mk) \times m$ matrix, $S_{ek}(s)$, consisting of monic single term polynomial elements; i.e.

$$\begin{bmatrix} R(s) \\ sR(s) \\ \vdots \\ s^{k-1}R(s) \\ P(s) \\ \vdots \\ s^{k-1}P(s) \end{bmatrix} = M_{ek}S_{ek}(s) = M_{ek} \begin{bmatrix} 1 & 0 & \cdots & 0 \\ s & 0 & & 0 \\ \vdots & \vdots & & \vdots \\ s^{d_1+k-1} & 0 & & 0 \\ 0 & 1 & \cdots & 0 \\ 0 & s & & 0 \\ \vdots & \vdots & & \vdots \\ 0 & s^{d_2+k-1} & & 0 \\ 0 & 0 & & 0 \\ \vdots & \vdots & & \vdots \\ 0 & 0 & & 1 \\ 0 & 0 & & s \\ \vdots & \vdots & & \vdots \\ 0 & 0 & \cdots & s^{d_m+k-1} \end{bmatrix} \qquad 7.3.29$$

for some constant matrix M_{ek}. It also follows that for all $k \geq \frac{n}{p}$, $pk \geq n$, and hence that once $k \geq \frac{n}{p}$, M_{ek} will have as many or more rows $(pk + mk)$ than columns $(n + mk)$. The <u>eliminant (matrix)</u>, M_e, of the matrix pair $\{R(s),P(s)\}$ with $P(s)$ column proper and $\partial_c[R(s)] \leq \partial[P(s)]$ is now defined as $M_{e\nu}$, where ν is the least integer k in 7.3.29 for which $n+mk - \rho[M_{ek}]$ is a minimum.† $S_e(s)$ is then defined as $S_{e\nu}(s)$.

Now that the eliminant matrix has been defined to include polynomial matrices, we can establish an important result which generalizes the constant matrix rank test for determining whether or not two polynomials are relatively prime to the more general polynomial matrix case. In particular, in order to establish Theorem 7.3.23, we will first establish:

---

†It should be noted that this generalized definition of the eliminant is consistent with the classical definition (7.3.11) in the scalar case only when $r(s)$ and $p(s)$ are relatively prime; i.e. $k = \nu$, and when $r(s)$ and $p(s)$ are polynomials, $\nu = n$ only if they are relatively prime.

THEOREM 7.3.30: <u>The pair of polynomial matrices $\{R(s),P(s)\}$ employed in the definition of M_e are relatively right prime if and only if their eliminant has full (column) rank $n + m\nu$.</u>

<u>Proof</u>: We will first establish necessity. In particular, if $R(s)$ and $P(s)$ are relatively right prime, it follows that the differential operator representation implied by 7.3.26, namely: $P(D)z(t) = u(t)$; $y(t) = R(D)z(t)$ is observable (see Section 5.3). This observation directly implies the ability to reconstruct the entire "state" of the system by repeated differentiation of $y(t)$ and its derivatives or, equivalently, that for some $\nu \leq n$, the $p\nu$ functions of $z(t)$ defined by $y(t) = R(D)z(t)$, $Dy(t) = DR(D)z(t),\ldots,D^{\nu-1}y(t) = D^{\nu-1}R(D)z(t)$ must yield n linearly independent functions of $z(t)$ and its derivatives in addition to the $m\nu$ linearly independent functions: $P(D)z(t)$, $DP(D)z(t),\ldots,D^{\nu-1}P(D)z(t)$, which comprise the final $m\nu$ rows of $M_eS_e(D)z(t)$. This is clearly possible only if for some ν, $\rho[M_e] = n + m\nu$, thus establishing necessity.†

We next establish sufficiency by constructively showing that if $\rho[M_e] = n + m\nu$, then an $m \times \nu(m + p)$ constant gain matrix $[H,K]$ can be chosen such that $[H,K]M_eS_e(s) = H(s)R(s) + K(s)P(s)$ equals any arbitrary desired $(m \times m)$ polynomial matrix $\beta(s)$, including I_m, which satisfies the relation:

$$\partial_{cj}[\beta(s)] \leq d_j + \nu - 1, \qquad 7.3.31$$

for $j = 1,2,\ldots,m$. To begin, let us first define $\hat{M}_e$ as the nonsingular matrix consisting of the first $(n + m\nu)$ linearly independent

---

† By considering a controllable companion form realization, $\{A,B,C,E\}$, of $T(s) = R(s)P(s)^{-1}$, which is also observable whenever $R(s)$ and $P(s)$ are relatively right prime, it can be shown that the integer ν, which we have used in defining M_e, corresponds to the observability index (see Section 3.6) of the equivalent state-space realization.

rows of M_e and $[\hat{H},\hat{K}]$ as the $m \times (n + m\nu)$ matrix obtained from $[H,K]$ by deleting those $(\nu p-n)$ columns of $[H,K]$ which correspond to the same numbered $(\nu p-n)$ rows of M_e which were eliminated to form $\hat{M}_e$. It is then clear in view of 7.3.29 that any arbitrary $m \times m$ polynomial matrix, $\beta(s) = \beta S_e(s)$ which satisfies 7.3.31 can be obtained by simply solving

$$[\hat{H},\hat{K}]\hat{M}_e S_e(s) = \beta S_e(s) \qquad 7.3.32$$

for $[\hat{H},\hat{K}]$; i.e.

$$[\hat{H},\hat{K}] = \hat{M}_e^{-1}$$

In order to find an appropriate $[H,K]$ we can now insert $(\nu p-n)$ identically zero columns in $[\hat{H},\hat{K}]$ corresponding to those numbered $(\nu p-n)$ rows of M_e which were eliminated to form $\hat{M}_e$. If we do so, it follows that $[\hat{H},\hat{K}]\hat{M}_e = [H,K]M_e = \beta$ and therefore, in view of 7.3.29 and 7.3.32, that

$$[H,K]M_e S_e(s) = [H,K]\begin{bmatrix} R(s) \\ sR(s) \\ \vdots \\ s^{\nu-1}R(s) \\ P(s) \\ \vdots \\ s^{\nu-1}P(s) \end{bmatrix} = H(s)R(s) + K(s)P(s) = \beta S_e(s) \qquad 7.3.34$$

for some appropriate polynomial matrix pair $\{H(s),K(s)\}$. If we now choose $\beta S_e(s) = \beta(s) = I_m$, then in view of the above,

$$H_I(s)R(s) + K_I(s)P(s) = I_m \qquad 7.3.35$$

for some pair, $\{H_I(s),K_I(s)\}$, of polynomial matrices. It is thus clear that I_m is a g.c.r.d. of $R(s)$ and $P(s)$ (see Section 2.5) and, therefore, that $R(s)$ and $P(s)$ are relatively right prime.

In view of this constructive proof of sufficiency, we can also readily establish the fact that the polynomial matrices $H(s)$, $K(s)$, and $Q(s)$ can be chosen to satisfy the three conditions of Theorem 7.3.23. In particular, by setting

$$Q(s) = \begin{bmatrix} s^{\nu-1} & 0 & \cdots & 0 & q_{1m}(s) \\ -1 & s^{\nu-1} & \cdots & 0 & q_{2m}(s) \\ 0 & -1 & & & \\ \vdots & \vdots & \ddots & \vdots & \vdots \\ 0 & 0 & \cdots & -1 & s^{\nu-1}+q_{mm}(s) \end{bmatrix}, \qquad 7.3.36$$

where

$$q_{im}(s) = \sum_{k=0}^{\nu-2} a_{(i-1)(\nu-1)+k}\, s^k \qquad 7.3.37$$

for all $i = 1,2,\ldots,m$, and evaluating $|Q(s)|$ by last column minors, it follows that

$$|Q(s)| = a_o + a_1 s + \ldots + a_{m\nu-m-1} s^{m\nu-m-1} + s^{m\nu-m} \qquad 7.3.38$$

and, therefore, that any arbitrary polynomial of degree $m\nu - m$ can be chosen as $|Q(s)|$ by an appropriate choice of the $q_{im}(s)$. If $F(s)$ is any arbitrary $m \times m$ polynomial matrix which satisfies the relation: $\partial_c[F(s)] < \partial_c[P(s)]$, it is also clear that the product of $Q(s)$ and $F(s)$ is a polynomial matrix of column (j) degree $< d_j + \nu - 1$. Therefore,

$$Q(s)F(s) = \beta(s) = \beta S_e(s) \qquad 7.3.39$$

for some constant matrix β. If $[H,K]$ is now chosen such that

$$[H,K]M_e S_e(s) = Q(s)F(s) = \beta S_e(s), \qquad 7.3.40$$

for this particular choice (7.3.39) for β, it follows in view of 7.3.34 that

$$H(s)R(s) + K(s)P(s) = Q(s)F(s) \qquad 7.3.41$$

where both $\partial_{ri}[H(s)] \leq \nu-1$ and $\partial_{ri}[K(s)] \leq \nu-1$ for all $i = 1,2,\ldots,m$. Since $\partial_{ri}[Q(s)] = \nu-1$, it is clear that if all of the zeros of $|Q(s)|$ are chosen to lie in the stable half-plane $Re(s) < 0$ then both $Q^{-1}(s)K(s)$ and $Q^{-1}(s)H(s)$ will be stable proper transfer matrices. Theorem 7.3.23 is therefore established.

As in the scalar case, it should be noted that the "cancellation" represented by $Q(s)$ and its inverse in 7.3.28 is a valid practical one in view of the assumption that $|Q(s)|$ is a stable polynomial. It should also be noted that this general compensation scheme requires no knowledge of state-space methods, although it represents a frequency domain analog of the state-space notion of linear state variable feedback and its implementation via Luenberger state estimation. To illustrate this point further, without a detailed comparison with these state-space methods, it might again be noted that ν, a measure of the required additional dynamics, is identical to the observability index of any equivalent state-space system. The fact that any equivalent state-space system is completely observable, and hence has an observability index, follows directly from the assumption that $R(s)$ and $P(s)$ are relatively right prime, and was formally established in Section 5.3. Once ν has been found, a physically realizable system of total order $(\nu-1)m$, represented by the relation:

$$F(s)z(s) = Q^{-1}(s)[K(s)u(s) + H(s)y(s)], \qquad 7.3.42$$

is then required to implement the compensation scheme. While this number $(\nu m-m)$ can equal, exceed, or be exceeded by the increase in system order $(n-p)$ required to realize a Luenberger observer (see Section 6.4), it should be noted that if $m = p$ and $\nu = \frac{n}{p}$, then both l.s.v.f. equivalent schemes require the same increase in system order. Furthermore, if a given observable system is single input controllable or, equivalently, if $m = 1$, then both schemes will yield any desired set of $(n+\nu-1)$ closed loop poles via a feedback compensator of total order $\nu-1$ (see Problem 6-15). We finally note that the increase in system order associated with this compensation scheme is not apparent from (7.3.28) the transfer matrix of the closed loop system, a result which also holds when Luenberger observer feedback is used in lieu of actual l.s.v.f. (see 6.4.14).

EXAMPLE 7.3.43: To illustrate this compensation scheme, let us now consider the particular system whose transfer matrix

$$T(s) = \frac{\begin{bmatrix} -s^2+1, & 0 \\ -s, & s^2 \end{bmatrix}}{-s^3+s^2} = \underbrace{\begin{bmatrix} s+1, & 0 \\ 1, & 1 \end{bmatrix}}_{R(s)} \underbrace{\begin{bmatrix} s^2, & 0 \\ 1, & -s+1 \end{bmatrix}^{-1}}_{P^{-1}(s)}.$$

For this system, it is clear that $n = 3$, $m = p = 2$, $d_1 = 2$, and $d_2 = 1$. Furthermore

$$\Gamma_c[P(s)] = \begin{bmatrix} 1 & 0 \\ 0 & -1 \end{bmatrix},$$

which implies that $P(s)$ is column proper. Suppose we are now required to compensate the system so as to place the $(n=3)$ closed loop poles at $s = -1$ and $-1\pm j$. One way to achieve this goal is to require that G by nonsingular, with

$$G^{-1}P_F(s) = \begin{bmatrix} s^2+2s+2, & 0 \\ s+1, & s+1 \end{bmatrix},$$

a column proper polynomial matrix which shares the same ordered d_j as $P(s)$. It might be noted that this particular choice for $G^{-1}P_F(s)$ also results in a diagonal (decoupled) --see Section 8.3 closed loop transfer matrix, namely

$$R(s)[G^{-1}P_F(s)]^{-1} = R(s)P_F^{-1}(s)G = \begin{bmatrix} \dfrac{s+1}{s^2+2s+2}, & 0 \\ 0, & \dfrac{1}{s+1} \end{bmatrix}.$$

To find $F(s)$ and G, we first note that $\Gamma_c[G^{-1}P_F(s)] = G^{-1}\Gamma_c[P_F(s)] = G^{-1}\Gamma_c[P(s)]$, or that

$$G = \Gamma_c[P(s)]\{\Gamma_c[G^{-1}P_F(s)]\}^{-1}. \qquad 7.3.44$$

For this example, therefore, $G = \begin{bmatrix} 1 & 0 \\ 0 & -1 \end{bmatrix}\begin{bmatrix} 1 & 0 \\ 0 & 1 \end{bmatrix} = \begin{bmatrix} 1 & 0 \\ 0 & -1 \end{bmatrix} = G^{-1}$. Furthermore,

$$F(s) = P(s) - P_F(s), \qquad 7.3.45$$

and in this example $F(s) = \begin{bmatrix} -2s-2, & 0 \\ s+2, & 2 \end{bmatrix}$, which is consistent with the requirement that $\partial_c[F(s)] < \partial_c[P(s)]$. Once $F(s)$ and G have been determined, we can implement the compensation scheme depicted in Figure 7.3.25 by now obtaining a feedback triple $\{H(s),K(s),Q(s)\}$ which satisfies the conditions of Theorem 7.3.23.

To begin, we note that since $\nu \geq \frac{n}{p} = \frac{3}{2}$, an initial value of $k = 2$ is assumed; i.e. if $k = 2$,

$$\begin{bmatrix} R(s) \\ sR(s) \\ P(s) \\ sP(s) \end{bmatrix} = \begin{bmatrix} s+1 & 0 \\ 1 & 1 \\ s^2+s & 0 \\ s & s \\ s^2 & 0 \\ 1 & -s+1 \\ s^3 & 0 \\ s & -s^2+s \end{bmatrix} = \underbrace{\begin{bmatrix} 1 & 1 & 0 & 0 & 0 & 0 & 0 \\ 1 & 0 & 0 & 0 & 1 & 0 & 0 \\ 0 & 1 & 1 & 0 & 0 & 0 & 0 \\ 0 & 1 & 0 & 0 & 0 & 1 & 0 \\ 0 & 0 & 1 & 0 & 0 & 0 & 0 \\ 1 & 0 & 0 & 0 & 1 & -1 & 0 \\ 0 & 0 & 0 & 1 & 0 & 0 & 0 \\ 0 & 1 & 0 & 0 & 0 & 1 & -1 \end{bmatrix}}_{M_{e2}} \underbrace{\begin{bmatrix} 1 & 0 \\ s & 0 \\ s^2 & 0 \\ s^3 & 0 \\ 0 & 1 \\ 0 & s \\ 0 & s^2 \end{bmatrix}}_{S_{e2}(s)}.$$

The reader can now verify that $\rho[M_{e2}] = 7$ and, therefore, that there cannot be any further minimization of the difference between $n+mk$ $(=7)$ and $\rho[M_{ek}]$ $(= \rho[M_{e2}] = 7)$. $M_{e2} = M_e$, the eliminant of $R(s)$ and $P(s)$, while $S_{e2}(s) = S_e(s)$. It might be noted that the determination that M_{e2} has full column rank immediately implies that $R(s)$ and $P(s)$ are relatively right prime and, therefore, that our compensation scheme can be employed to yield the desired transfer matrix. The sixth row of M_e is now found to be the first linearly dependent one and is consequently eliminated to form the nonsingular matrix

$$\hat{M}_e = \begin{bmatrix} 1 & 1 & 0 & 0 & 0 & 0 & 0 \\ 1 & 0 & 0 & 0 & 1 & 0 & 0 \\ 0 & 1 & 1 & 0 & 0 & 0 & 0 \\ 0 & 1 & 0 & 0 & 0 & 1 & 0 \\ 0 & 0 & 1 & 0 & 0 & 0 & 0 \\ 0 & 0 & 0 & 1 & 0 & 0 & 0 \\ 0 & 1 & 0 & 0 & 0 & 1 & -1 \end{bmatrix}.$$

Since $\nu = 2$ and $m = 2$, we can now choose $Q(s)$ in accordance with 7.3.36; i.e. $Q(s) = \begin{bmatrix} s & q_{12} \\ -1 & s+q_{22} \end{bmatrix}$, and if we (arbitrarily) require that the poles of $Q(s)^{-1}$ lie at $s = -1 \pm j2$, or that $|Q(s)| = s^2+2s+5$, then $q_{12} = 5$ and $q_{22} = 2$. We can now solve for $\beta S_e(s)$ in accordance with 7.3.39; i.e. $Q(s)F(s) = \begin{bmatrix} -2s^2+3s+10, & 10 \\ s^2+6s+6\ , & 2s+4 \end{bmatrix} = \beta S_e(s)$, where $\beta = \begin{bmatrix} 10 & 3 & -2 & 0 & 10 & 0 & 0 \\ 6 & 6 & 1 & 0 & 4 & 2 & 0 \end{bmatrix}$. Since

$$\hat{M}_e^{-1} = \begin{bmatrix} 1 & 0 & -1 & 0 & 1 & 0 & 0 \\ 0 & 0 & 1 & 0 & -1 & 0 & 0 \\ 0 & 0 & 0 & 0 & 1 & 0 & 0 \\ 0 & 0 & 0 & 0 & 0 & 1 & 0 \\ -1 & 1 & 1 & 0 & -1 & 0 & 0 \\ 0 & 0 & -1 & 1 & 1 & 0 & 0 \\ 0 & 0 & 0 & 1 & 0 & 0 & -1 \end{bmatrix},$$ it follows that

$$[\hat{H},\hat{K}] = \beta\hat{M}_e^{-1} = \begin{bmatrix} 0 & 10 & 3 & 0 & -5 & 0 & 0 \\ 2 & 4 & 2 & 2 & -1 & 0 & 0 \end{bmatrix}$$ and, therefore, that

$$[H,K] = \begin{bmatrix} 0 & 10 & 3 & 0 & -5 & 0 & 0 & 0 \\ 2 & 4 & 2 & 2 & -1 & 0 & 0 & 0 \end{bmatrix},$$ which we obtain by adding a zero (sixth) column to $[\hat{H},\hat{K}]$ corresponding to the (sixth) row of M_e

which we eliminated to form $\hat{M}_e$. If we now employ 7.3.34 we readily determine that $H(s) = \begin{bmatrix} 3s, & 10 \\ 2s+2, & 2s+4 \end{bmatrix}$ and $K(s) = \begin{bmatrix} -5, & 0 \\ -1, & 0 \end{bmatrix}$. If this triple, $\{H(s), K(s), Q(s)\}$, is now employed in the feedback scheme depicted in Figure 7.3.25, the desired (decoupled) closed loop transfer matrix with apparent poles at $s = -1$ and $-1 \pm j$, and cancelled, uncontrollable poles at $-1 \pm j2$ is obtained.

7.4 A GENERAL COMPENSATION TECHNIQUE

As we illustrated in the previous section, compensation via linear state variable feedback, as implemented by a Luenberger observer can be accomplished entirely in the frequency domain with no reference whatsoever to the time domain notion of state. We will now show that this result represents only a special case of a more general result.

To begin, we first note that since $\partial_c[P_F(s)] = \partial_c[P(s)]$, and both are column proper, $P(s)P_F^{-1}(s)$ is a proper transfer matrix (see Section 4.3). If we therefore compensate a system, whose open loop transfer matrix is $R(s)P^{-1}(s)$, with a feedforward compensator whose transfer matrix is $P(s)P_F^{-1}(s)G$, as depicted in Figure 7.4.1, it is

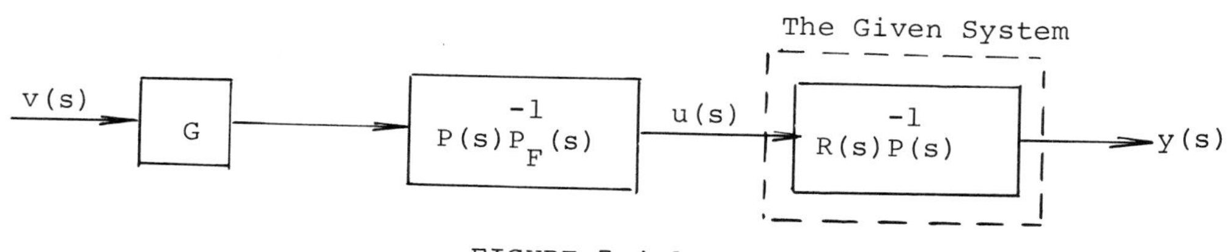

FIGURE 7.4.1

A FEEDFORWARD REPRESENTATION OF LINEAR STATE VARIABLE FEEDBACK

clear that the overall input/output behavior of the compensated system is given by $T_{F,G}(s)$; i.e.

$$y(s) = R(s)P^{-1}(s)P(s)P_F^{-1}(s)Gv(s) = R(s)P_F^{-1}(s)Gv(s) = T_{F,G}(s)v(s) \qquad 7.4.2$$

In view of this observation, it is clear that <u>any l.s.v.f. compensation scheme can be represented via feedforward compensation through a system with a proper transfer matrix</u>. We realize, of course, that there are a number of disadvantages associated with compensating a system via the open loop scheme depicted in Figure 7.4.1. In particular, exact cancellation of $P(s)$ and its inverse is virtually impossible in practical cases and, therefore, if any of the zeros of $|P(s)|$ lie in the half-plane $Re(s) > 0$, instability results. Furthermore, none of the advantages of a feedback design, such as reduction in the sensitivity of the system to plant parameter variations, are obtained via open loop compensation.

This feedforward scheme is useful, however, since it serves to motivate a more general closed loop compensation scheme which is based on the observation that if a <u>dynamic compensator</u>, which we now define as any system with a proper transfer matrix, is placed in series with the given system, then it can also be represented in the same way that we are able to represent l.s.v.f. In view of this observation, it therefore follows that <u>any compensator scheme which employs a system with a proper transfer matrix</u>, $T_c(s)$, <u>in the feedforward path can be realized by an "equivalent" compensation scheme which employs dynamic compensation in combination with l.s.v.f.</u>, and the remainder of this section will focus on the development of a rather general synthesis algorithm based on this fact. As we will show, the synthesis algorithm essentially "separates" any dynamic precompensator, $T_c(s)$, into two mutually exclusive parts, a "low order" feedforward compensator and a state feedback compensator as developed in the previous section.

In order to constructively outline the synthesis algorithm, we will require a fundamental result involving the "division" of two polynomial matrices, namely:

THEOREM 7.4.3: Consider the $(p \times m)$ and $(q \times m)$ polynomial matrices $J(s)$ and $\bar{J}(s)$, respectively. If $\rho[J(s)] = m$, there exists a pair of polynomial matrices, $\{M(s),N(s)\}$, of dimension $(q \times p)$ and $(q \times m)$ respectively, such that

$$\bar{J}(s) = M(s)J(s) + N(s) \qquad 7.4.4$$

with

$$\partial_c[N(s)] < \partial_c[J(s)] \qquad 7.4.5$$

Proof: We first define $\hat{J}(s)$ as the nonsingular m-dimensional matrix obtained from $J(s)$ by eliminating all but its first m linearly independent rows.[†] It is then clear that $\bar{J}(s)\hat{J}^{-1}(s)$ represents a rational transfer matrix, $\left[\frac{c_{ij}(s)}{d_{ij}(s)}\right]$, which is not necessarily proper. However, if we divide the numerator of each element by its denominator, we can uniquely express each element as $\frac{c^*_{ij}(s)}{d_{ij}(s)} + m^*_{ij}(s)$, where $\partial[c^*_{ij}(s)] < \partial[d_{ij}(s)]$ if $\partial[d_{ij}(s)] \neq 0$ or $c^*_{ij}(s) = 0$ otherwise. This procedure now enables us to represent the rational transfer matrix $\bar{J}(s)\hat{J}^{-1}(s)$ as the sum of its quotient, $M^*(s) = [m^*_{ij}(s)]$, and its strictly proper part, $N(s)\hat{J}^{-1}(s)$; (see Problem 4-20); i.e.

$$\bar{J}(s)\hat{J}^{-1}(s) = M^*(s) + N(s)\hat{J}^{-1}(s), \qquad 7.4.6$$

which immediately implies that

$$\bar{J}(s) = M^*(s)\hat{J}(s) + N(s) \qquad 7.4.7$$

with $\partial_c[N(s)] < \partial_c[\hat{J}(s)]$ (see Problem 5-8). If we now define $M(s)$ as the $(q \times p)$ matrix obtained by simply inserting $(p-m)$ identically

---

† This choice for $\hat{J}(s)$ is made only for convenience in establishing the theorem. In general, therefore, $M(s)$ and $N(s)$ will not be unique. However, if $J(s)$ is nonsingular, both $M(s)$ and $N(s)$ will be unique (as in Step 3 of the synthesis algorithm).

zero columns in $M^*(s)$ corresponding to those numbered rows of $J(s)$ which were eliminated to form $\hat{J}(s)$, it follows that 7.4.4 holds, thus establishing the theorem.

Let us now consider any open loop system with a proper, rational $(p \times m)$ transfer matrix $T(s) = R(s)P(s)^{-1}$, where $R(s)$ and $P(s)$ are relatively right prime and $P(s)$ is column proper. If we compensate this system by a system with a proper transfer matrix, $T_c(s)$, placed in the feedforward path as illustrated in Figure 7.4.8, it is clear

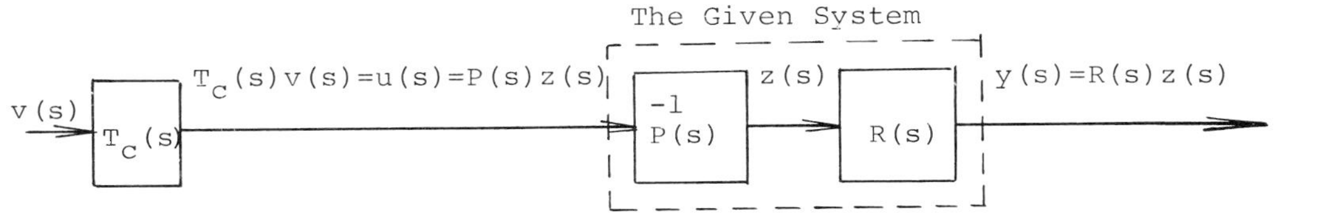

FIGURE 7.4.8

OPEN LOOP COMPENSATION

that the input/output dynamical behavior of the composite system is given by the desired transfer matrix, $T_d(s) = R(s)P(s)^{-1}T_c(s)$; i.e.

$$y(s) = R(s)P(s)^{-1}T_c(s)v(s) = T_d(s)v(s) \qquad 7.4.9$$

The remainder of this section will now focus on the development of a synthesis algorithm for realizing this desired transfer matrix via "low order" dynamic feedforward compensation in combination with l.s.v.f.

The Synthesis Algorithm

Step 1: Premultiply $T_c(s)$ by $P(s)^{-1}$ and express the resulting proper transfer matrix as $P_c(s)^{-1}L_c(s)$, where $P_c(s)$ and $L_c(s)$ are relatively left prime polynomial matrices with $L_c(s)$ in column proper, upper right triangular form (see Theorem 2.5.11); i.e.

$$P(s)T_c^{-1}(s) = P_c^{-1}(s)L_c(s), \qquad 7.4.10$$

with $P_c(s)$ and $L_c(s)$ of the appropriate form.

Step 2: If $P(s)P_c^{-1}(s)$ is a proper transfer matrix, define $\bar{P}(s) = P_c(s)$ and $L(s) = L_c(s)$ and proceed to the next step. If $P(s)P_c^{-1}(s)$ is not a proper transfer matrix, postmultiply $P_c^{-1}(s)$ by the inverse of any nonsingular, stable diagonal polynomial matrix, $G_L(s) = \mathrm{diag}[g_i(s)]$ of row degree "just sufficient enough" to insure that $P(s)P_c^{-1}(s)G_L^{-1}(s) = P(s)[G_L(s)P_c(s)]^{-1}$ is proper. Define $\bar{P}(s) = G_L(s)P_c(s)$ and $L(s) = G_L(s)L_c(s)$, noting that in view of 7.4.10,

$$P(s)T_c^{-1}(s) = \bar{P}^{-1}(s)L(s), \qquad 7.4.11$$

with $P(s)\bar{P}^{-1}(s)$ a proper transfer matrix.

Step 3: In view of Theorem 7.4.3, it now follows that since $P(s)$ is nonsingular, we can uniquely express $\bar{P}(s)$ as:

$$\bar{P}(s) = G(s)P(s) - F(s), \qquad 7.4.12$$

where both $G(s)$ and $F(s)$ are $(m \times m)$ polynomial matrices with

$$\partial_c[F(s)] < \partial_c[P(s)] \qquad 7.4.13$$

Step 4: We now construct the system depicted in Figure 7.4.14 using Theorem 7.3.23 to insure that $K(s)P(s) + H(s)R(s) = Q(s)F(s)$, with both $Q^{-1}(s)K(s)$ and $Q^{-1}(s)H(s)$ proper and stable.

This fourth step completes the actual computations involved in representing $T_c(s)$ as combined l.s.v.f./dynamic feedforward compensation. Before we display the utility of this synthesis algorithm, however, a number of remarks are in order.

R 1: We first show that $G^{-1}(s)L(s)$ is a well defined and proper transfer matrix, and therefore, that the dynamic precompensator defined by

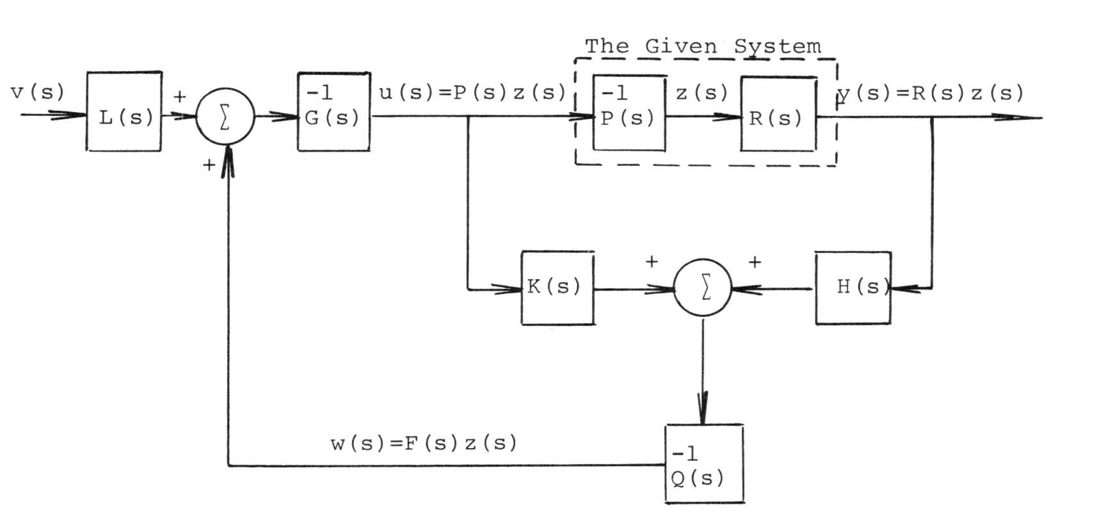

FIGURE 7.4.14

THE GENERAL COMPENSATION SCHEME

the differential operator representation:

$$G(D)u(t) = L(D)v(t) + w(t), \qquad 7.4.15$$

where $w(t) = F(D)z(t)$, can be implemented without differentiators.

In particular, we now recall from Step 2 of the synthesis algorithm that since $P(s)\bar{P}^{-1}(s)$ is proper, $\partial_c[P(s)] \le \partial_c[\bar{P}(s)]$ (see Problem 5-8). Therefore, in view of 7.4.13, it is clear that

$$\partial_c[F(s)] < \partial_c[\bar{P}(s)] \qquad 7.4.16$$

Since $T_c(s) = P(s)\bar{P}^{-1}(s)L(s)$ is proper, it follows in view of 7.4.12 that

$$P(s)[\bar{P}(s) + F(s)]^{-1}L(s) = P(s)[G(s)P(s)]^{-1}L(s) = G^{-1}(s)L(s) \qquad 7.4.17$$

is also proper, thus establishing the desired result.

R 2: We now show that the closed loop transfer matrix of the system depicted in Figure 7.4.14 is equal to $T(s)T_c(s) = T_d(s)$, as desired. In particular, if we sum signals at the first summation junction, we readily obtain an expression for the closed loop behavior of the system; i.e.

$$G(s)P(s)z(s) = L(s)v(s) + F(s)z(s), \tag{7.4.18}$$

or, in view of 7.4.12,

$$\bar{P}(s)z(s) = L(s)v(s) \tag{7.4.19}$$

Since $P(s)T_c^{-1}(s) = \bar{P}^{-1}(s)L(s)$, it now follows that

$$\begin{aligned} y(s) = R(s)z(s) = R(s)P^{-1}(s)T_c(s)v(s) &= T(s)T_c(s)v(s) \\ &= T_d(s)v(s), \end{aligned} \tag{7.4.20}$$

the desired transfer matrix. It is thus clear that <u>the compensation scheme depicted in Figure 7.4.14 can be employed to achieve any desired transfer matrix which can be specified via feedforward compensation through a proper transfer matrix</u>, $T_c(s)$.

<u>R 3</u>: We next note that once a $T_c(s)$ has been specified, the synthesis algorithm "separates" $T_c(s)$ in a way which attempts to minimize the feedforward dynamics, as represented by $G^{-1}(s)L(s)$, while placing "as much of $T_c(s)$ as possible" in the state feedback path. More specifically, the requirements that $L_c(s)$ be in column proper, upper right triangular form and that $P_c(s)$ and $L_c(s)$ be relatively left prime were imposed in the first step of the algorithm in order to minimize the number of differentiations of $v(t)$ in 7.4.15 and, therefore, the number of integrators required to insure proper feedforward compensation. In fact, <u>if</u> $T_c(s)$ <u>is unique and its rank is m, then this algorithm will minimize the order of any dynamic feedforward compensation which may be required</u> and consequently, will produce a stable l.s.v.f. control law whenever such a design is feasible (see Problem 7-15). While no such guarantee can be given in the more

general case, the uniqueness of the polynomial matrix division relationship 7.4.12 in combination with a minimization of the degree of $L(s)$ act together to produce "low order" input dynamics.

R 4: It should be noted that the poles of the final, closed loop compensated system are equal to the zeros of $|Q(s)|$, an arbitrarily assignable polynomial, and the zeros of $|\bar{P}(s)|$, which correspond to the uncancelled poles of $P(s)T_c^{-1}(s)$. It is therefore important that $T_c(s)$ always be chosen so that all of the uncancelled poles of $P(s)T_c^{-1}(s)$ lie in the stable half-plane $Re(s) < 0$.

R 5: We again note that the compensation scheme outlined in this section actually represents a combination of both linear state variable feedback and feedforward compensation. The feedforward part of the design is often referred to as added dynamics or input dynamics as well as dynamic compensation in the control literature, although we realize that the feedback portion of a practical design might also require the employment of added dynamics via Luenberger state estimation whenever the entire state of a system is not directly measurable.

R 6: One of the primary advantages of the compensation algorithm which has just been outlined is that it represents a complete and rather general design procedure; i.e. exponential state estimation is included as part of the design, but it can be eliminated if the entire state is directly measurable. We further note that state feedback is required from the original system only and, therefore, we need not be concerned about first adding "sufficient" input dynamics to "precondition" the system and then feeding back from the entire state of the composite system.

R 7: Finally, this synthesis algorithm is rather general in that it can readily be applied to a number of synthesis questions (e.g.

arbitrary pole placement, "decoupling", and "exact model matching") by first determining whether or not the particular design objective can be achieved via a dynamic feedforward compensator, $T_c(s)$, a fact which will be rather thoroughly exploited in the next chapter.

EXAMPLE 7.4.21: To illustrate the synthesis algorithm, consider a system with the open loop transfer matrix: $T(s) = \begin{bmatrix} \frac{s+1}{s^2} & \frac{s+2}{s^2+1} \\ \frac{2}{s} & \frac{2s+3}{s^2+1} \end{bmatrix} =$

$$\underbrace{\begin{bmatrix} s+1, & 1 \\ 2s, & 3 \end{bmatrix}}_{R(s)} \underbrace{\begin{bmatrix} s^2 & -s^2 \\ 0 & s^2+1 \end{bmatrix}^{-1}}_{P^{-1}(s)}$$

, with $R(s)$ and $P(s)$ relatively right prime.

The reader can readily verify that it is possible to achieve the diagonal or "decoupled" closed loop transfer matrix,

$$T_d(s) = \begin{bmatrix} \frac{1}{s^2+3s+2}, & 0 \\ 0, & \frac{1}{s^2+3s+2} \end{bmatrix}$$

via feedforward compensation (of $T(s)$)

by $$T_c(s) = \begin{bmatrix} \frac{2s^3+3s^2}{s^3+6s^2+11s+6}, & \frac{-s^3-2s^2}{s^3+6s^2+11s+6} \\ \frac{-2s^3-2s}{s^3+6s^2+11s+6}, & \frac{s^3+s^2+s+1}{s^3+6s^2+11s+6} \end{bmatrix} = P(s)R^{-1}(s)T_d(s)$$ (Section 8.3 contains a rather thorough treatment of "decoupling"). Since $T_c(s)$ is proper, it can be realized by the compensation scheme depicted in Figure 7.4.14, and we can therefore apply the synthesis algorithm to now determine $G(s)$, $L(s)$, and $F(s)$.

<u>Step 1</u>: Since $T_c(s) = P(s)R^{-1}(s)T_d(s)$, $P^{-1}(s)T_c(s) = R^{-1}(s)T_d(s)$ which is equal to $\dfrac{\begin{bmatrix} 3, & -1 \\ -2s, & s+1 \end{bmatrix}}{s^3+6s^2+11s+6}$. In view of the results given in Sections

2.5 and 5.4, we now express this transfer matrix as the product:

$$\underbrace{\begin{bmatrix} s^3+6s^2+11s+6, & 0 \\ 2s^3+6s^2+4s\ , & 3s^2+9s+6 \end{bmatrix}^{-1}}_{P_c^{-1}(s)} \underbrace{\begin{bmatrix} 3, & -1 \\ 0, & 1 \end{bmatrix}}_{L_c(s)}$$

with $P_c(s)$ and $L_c(s)$ relatively left prime and $L_c(s)$ in upper right triangular form.

Step 2: We next note that $P(s)P_c^{-1}(s)$ is a proper transfer matrix and, therefore, that $\overline{P}(s) = P_c(s)$ and $L(s) = L_c(s)$. It might be noted that this must be the case in this example, since $T_c(s)$ is both nonsingular and unique--see Problem 7-15.

Step 3: Employing the constructive proof of Theorem 7.4.3, we now determine that

$$\overline{P}(s)P^{-1}(s) = \underbrace{\begin{bmatrix} s+6, & s+6 \\ 2s+6, & 2s+9 \end{bmatrix}}_{G(s)} - \underbrace{\begin{bmatrix} \frac{-11s-6}{s^2} & \frac{-10s}{s^2+1} \\ -\frac{4}{s} & \frac{-11s+3}{s^2+1} \end{bmatrix}}_{F(s)P^{-1}(s)}$$

and, therefore, that

$$F(s) = F(s)P^{-1}(s)P(s) = \begin{bmatrix} -11s-6, & s+6 \\ -4s\ , & -7s+3 \end{bmatrix},$$

which is consistent with 7.4.13; i.e. $\partial_c[F(s)] < \partial_c[P(s)]$.

Step 4: Since $R(s)$ and $P(s)$ were determined earlier, we can now employ the algorithm given in the previous section in order to obtain a feedback triple, $\{H(s), K(s), Q(s)\}$, which produces this $F(s)$. We merely note that a feedback system of total order $2 (= \partial[|Q(s)|])$ would be required in this example, a fact which the reader can readily verify.

It is also worth noting that since

$$G^{-1}(s) = \begin{bmatrix} \frac{2/3\ s + 3}{s + 6}\ , & -1/3 \\ \frac{-2/3\ s - 2}{s + 6}\ , & 1/3 \end{bmatrix}$$

and $L(s) = L = \begin{bmatrix} 3, & -1 \\ 0, & 1 \end{bmatrix}$ in this example, the feedforward portion of our compensation scheme can be realized by a first ("low") order dynamic compensator. Furthermore, since $T_c(s)$ is both nonsingular and unique, it is clear that this compensation algorithm does minimize the order of the required feedforward dynamics, which is consistent with an earlier observation--see R3.

7.5. CONCLUDING REMARKS AND REFERENCES

A number of new and potentially useful results have been presented in this chapter. We first showed that the time domain notion of l.s.v.f. has a rather useful frequency domain interpretation from the point of view of the structure theorem; one which enables us to separate those "portions" of the transfer matrix which are unaffected or invariant under l.s.v.f. from those which can be arbitrarily altered. It is interesting to note that one of the easiest ways of illustrating the effect which l.s.v.f. has on the closed loop performance of a system is through this frequency domain interpretation. Our results also generalize, to the multivariable case, the well known scalar result that only the poles of a transfer function can be altered via l.s.v.f. [M4][B3]. As we will later show in Chapter 8, our frequency domain interpretation of l.s.v.f. can also be employed to determine whether or not certain specific design objectives can be attained via l.s.v.f. alone.

In Section 7.3 we presented a complete frequency domain compensation scheme analogous to the time domain implementation of a l.s.v.f. control law via exponential state estimation. This constructive compensation procedure was based on the ability to express the open loop transfer matrix of a system as the product $R(s)P(s)^{-1}$, where $R(s)$ and $P(s)$ are relatively right prime, and directly employed an extension of the classical eliminant matrix of two polynomials to the more general polynomial matrix case. Sylvester is usually credited with first defining the eliminant of two polynomials and establishing the fact that it is nonsingular, if and only if the polynomials are relatively prime, a result which can be found in a number of earlier mathematical texts; e.g. [T1]. There have also been some other partially successful attempts to generalize the eliminant to include the matrix case [R4][C5]. It might be noted that the definition employed here in Section 7.3 is somewhat restrictive, in that $P(s)$ must be column proper. However, this restriction can be lessened somewhat (see Problem 7-13). In view of the results presented in Section 7.3, it is clear that any l.s.v.f. compensation scheme can be implemented in the frequency domain with no reference whatsoever to the time domain notion of state. Furthermore, the observability index of a multivariable system can also be defined in the frequency domain with no explicit reference to the time domain notion of state.

The first point we illustrated in Section 7.4 was that any l.s.v.f. compensation scheme can be implemented via an "equivalent" procedure which employs an appropriate proper transfer matrix placed in the feedforward path. While this method of system compensation was shown to possess certain practical disadvantages in comparison to a feedback scheme, it did serve to motivate the development of a rather general form of feedback compensation. In particular, we showed that any control scheme which can be specified and implemented by <u>any</u> proper transfer matrix, $T_c(s)$, placed in the feedforward path can also be implemented via an "equivalent" feedforward/feedback scheme. The reasons for developing a compensation scheme more general than l.s.v.f. will become apparent in the next chapter when we discuss certain specific design objectives and show that l.s.v.f. alone cannot always yield a suitable closed loop system.

We finally note that the results given in this chapter were first introduced in [W3] and [W13] and are partially related to those of Chen [C1] [C3] [C4], who first illustrated the scalar algorithm of Section 7.3 for achieving the frequency domain equivalent of an exponentially estimated l.s.v.f. design. Several other investigators have developed a variety of multivariable compensation techniques which involve the application of l.s.v.f. as well as feedback and internal feedforward dynamics. Unlike the general frequency domain scheme developed here in Section 7.4, however, their schemes, for the most part, involve time domain dynamics, usually with some specific design objective in mind, such as complete and arbitrary closed loop pole placement [P4][B5], noninteraction [F1][G3][W9][M5][S5], or exact model matching [E1]. We will say more regarding the similarities and differences between their respective approaches and ours in the next chapter when we discuss these specific design objectives.

PROBLEMS - CHAPTER 7

7-1 For what values of F and G does $T_{F,G}(s) = T(s)$ when $\rho[T(s)] = m$?

7-2 Consider the open loop state-space system, $\{A,B,C,E\}$, where

$$A = \begin{bmatrix} 0 & 0 & 0 & 0 \\ 1 & 0 & 0 & 0 \\ 0 & 0 & 0 & 0 \\ 1 & 0 & 1 & 0 \end{bmatrix}, \quad B = \begin{bmatrix} 1 & 0 \\ 0 & 0 \\ 0 & 1 \\ 0 & 0 \end{bmatrix}, \quad C = \begin{bmatrix} 1 & 1 & 1 & 0 \\ 0 & 0 & 0 & 1 \end{bmatrix}, \text{ and } \quad E = \begin{bmatrix} 1 & 1 \\ 1 & 1 \end{bmatrix}$$

and the l.s.v.f. control law, $\{F,G\}$, where $F = \begin{bmatrix} 1 & 0 & 1 & 0 \\ 0 & -1 & 0 & 0 \end{bmatrix}$ and $G = \begin{bmatrix} 1 & -1 \\ 0 & 1 \end{bmatrix}$. Determine an $R(s)$, $P(s)$, $\hat{B}_m$, and $P_F(s)$ for this example and verify that both 7.2.10 and 7.2.11 hold.

7-3 Consider the expression, 7.2.9, for the closed loop transfer matrix of a controllable system compensated by l.s.v.f. Show that if G is nonsingular and $P_{F,G}(s) \overset{\Delta}{=} G^{-1} P_F(s)$, then $T_{F,G}(s) = R(s) P_{F,G}^{-1}(s)$ with $P_{F,G}(s)$ column proper and $\partial_{ci}[P_{F,G}(s)] = \partial_{ci}[P(s)] = d_i$. In view of the above, verify that (i) $R(s)$, (ii) the (m) ordered d_i, and (iii) the fact that $P_{F,G}(s)$ remains column proper represent a complete set of l.s.v.f. invariants whenever $|G| \neq 0$.

7-4 In view of Problem 7-3, show that whenever G is nonsingular, $\Gamma_c[P_{F,G}(s)]$ can be completely and arbitrarily altered via G. Discuss the possibility of generalizing this result to include cases where G is either nonsquare or singular.

7-5 Verify that the eliminant of $r(s) = s-1$ and $p(s) = s^2-1$ is singular.

7-6 Consider a (scalar) system with the open loop transfer function

$t(s) = \frac{s-1}{s^2}$. Determine the eliminant matrix of $s-1$ and s^2 and use it to find an $h(s)$, $k(s)$, $q(s)$, and g so that the compensation scheme depicted in Figure 7.3.5 yields the desired closed loop transfer function: $t_{f,g}(s) = \frac{(s-1)(s+2)}{(s+1)(s+2)(s+3)} = \frac{s-1}{s^2+4s+3}$.

7-7 State the dual of Theorem 7.4.3.

7-8 Show that if $t(s) = \frac{r(s)}{p(s)}$, where $p(s) = r(s)\bar{p}(s)$, then the compensation scheme depicted in Figure 7.3.5 can still be used to achieve $t_{f,g}(s) = \frac{r(s)}{p_f(s)} g$ as the closed loop transfer function, where $p_f(s) = r(s)\bar{p}_f(s)$, with $\bar{p}_f(s)$ any arbitrary polynomial of the same degree as $\bar{p}(s)$. How do you interpret this result from the viewpoint of system controllability and observability? Can it be generalized to the multivariable case?

7-9 Consider the transfer function $t(s) = \frac{r(s)}{p(s)}$ where $q = \partial[p(s)] - \partial[r(s)]$. Show that compensation via l.s.v.f. alone cannot alter q, although compensation via l.s.v.f. in combination with (proper) input dynamics can increase q arbitrarily. How might this result be generalized to the multivariable case?

7-10 If we approximate the transfer function $t(s) = \frac{1}{(s+1)(2s-1)(.01s+1)}$ by its "dominant poles" transfer function $\tilde{t}(s) = \frac{1}{(s+1)(2s-1)}$ and then employ the compensation scheme depicted in Figure 7.3.5 to arbitrarily position the $n(=2)$ closed loop poles of the "dominant poles" system at $s = -1 \pm j$, show that the effect of this compensation scheme on the actual system is "approximately equivalent" to its effect on $\tilde{t}(s)$.

7-11 Consider a system with the proper transfer function $t(s) = \frac{r(s)}{p(s)}$, with $r(s)$ and $p(s)$ relatively prime and $r(0) \neq 0$. Show that if we first compensate this system by the scheme depicted in Figure 7.3.5 in order to obtain $t_{f,g}(s) = \frac{r(s)}{p_f(s)} g$ as the resulting transfer function, we can then achieve a "zero position error" unity feedback scheme (as depicted below) which has any desired closed loop pole configuration.

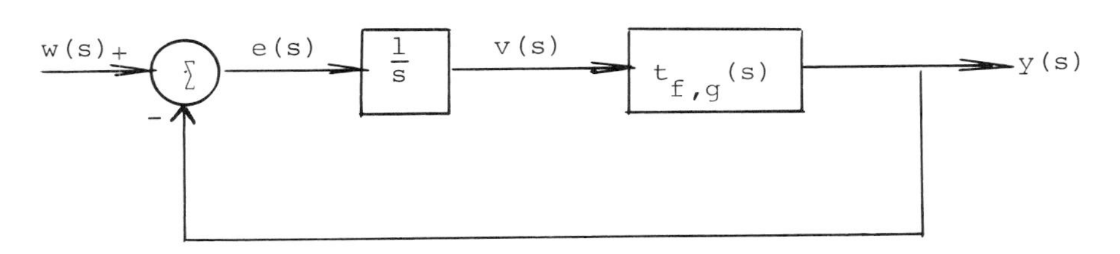

A "Zero Position Error" System

7-12 In view of the fact that any rational transfer matrix, $T(s)$, can be "factored" as the product, $R(s)P(s)^{-1}$, and also uniquely expressed as the sum of its quotient, $E(s)$, and its strictly proper part, $\hat{R}(s)P(s)^{-1}$, show that $R(s)$ and $P(s)$ are relatively right prime if and only if $\hat{R}(s)$ and $P(s)$ are relatively right prime.

7-13 Show that whenever the rank of $\begin{bmatrix} R(s) \\ \hline P(s) \end{bmatrix}$ is equal to the number (m) of columns of the composite matrix, and more than m rows of $\begin{bmatrix} R(s) \\ \hline P(s) \end{bmatrix}$ are nonzero, then an eliminant matrix can be defined and utilized to determine whether or not $R(s)$ and $P(s)$ are relatively right prime. Discuss the extension of this result to include more than two polynomial matrices.

7-14 Given the relatively right prime pair: $P(s) = \begin{bmatrix} s^2 & 0 \\ 1 & s^2 \end{bmatrix}$ and $R(s) = \begin{bmatrix} s & 0 \\ 0 & 1 \end{bmatrix}$, find an $H(s)$ and a $K(s)$ such that $H(s)R(s)+K(s)P(s) = I_2$. Prove that there does not exist an $\overline{H}(s)$ and a $\overline{K}(s)$ such that $R(s)\overline{H}(s)+P(s)\overline{K}(s) = I_2$.

7-15 Prove that the synthesis algorithm of Section 7.4 does minimize $\partial[|G(s)|]$ whenever $T_c(s)$ is unique and its rank is m. (Hint: First show that $P(s)P_c^{-1}(s)$ is proper whenever $\rho[T_c(s)] = m$ and, therefore, that $\overline{P}(s) = P_c(s)$ and $L(s) = L_c(s)$ in Step 2 of the algorithm.) Show that dynamic feedforward compensation is not required and, therefore, that l.s.v.f. alone can be employed for compensation if and only if $\partial[|\overline{P}(s)|] = \partial[|P(s)|]=n$.

7-16 Employ duality to define the eliminant matrix of the pair, $\{P(s),Q(s)\}$, with $P(s)$ a $(p \times p)$ row proper polynomial matrix and $Q(s)$ a $(p \times m)$ polynomial matrix which satisfies the condition: $\partial_r[Q(s)] \leq \partial_r[P(s)]$.

7-17 Verify (by direct computation) that the determinant of

$$P(s) = \begin{bmatrix} s^2 & -1 & 0 \\ 0 & s^2 & -1 \\ a_o+a_1 s & a_2+a_3 s & a_4+a_5 s+s^2 \end{bmatrix} \text{ is equal to}$$

$s^6+a_5 s^5+\ldots+a_1 s+a_o$.

7-18 Consider a system with the open loop transfer matrix $T(s) = R(s)P(s)^{-1}$, where $R(s)$ and $P(s)$ are as given in Problem 7-14. Find a triple, $\{H(s),K(s),Q(s)\}$, of polynomial matrices with all poles of $Q^{-1}(s)$ at $s = -1$ so that the feedback compensation scheme depicted in Figure 7.3.25 yields a closed loop transfer matrix with $n(=4)$ closed loop poles at $s = -1, -2,$

and $-1\pm j$. Verify that both $Q^{-1}(s)H(s)$ and $Q^{-1}(s)K(s)$ are proper. Show that such a design is not possible if $R(s)$ remains unaltered but $P(s) = \begin{bmatrix} s^2 & 1 \\ 0 & s^2 \end{bmatrix}$.

7-19 Employ the synthesis algorithm of Section 7.4 in order to realize the "decoupled" closed loop transfer matrix:

$$T_d(s) = \begin{bmatrix} \frac{1}{s^2+3s+2}, & 0 \\ 0, & \frac{1}{s^3+4s^2+6s+4} \end{bmatrix},$$ given a system with the open loop transfer matrix $$T(s) = \frac{\begin{bmatrix} s+1, & s \\ 1, & 1 \end{bmatrix}}{s^2}$$

7-20 Consider the synthesis algorithm of Section 7.4. If $T_c(s) = \tilde{P}^{-1}(s)$ for some polynomial matrix $\tilde{P}(s)$, show that the order of any dynamic feedforward compensator (the degree of $|G(s)|$) must equal the order of a minimal realization of $T_c(s)$ (the degree of $|\tilde{P}(s)|$).

7-21 In view of the results given in Sections 7.2 and 7.3, show that if the controllable state-space system: $\dot{x}(t) = Ax(t) + Bu(t)$; $y(t) = Cx(t) + Eu(t)$, and the controllable differential operator system: $P(D)z(t) = u(t)$; $y(t) = R(D)z(t)$ are equivalent, then it is possible to uniquely associate with every state-space feedback control law of the form $u(t) = Fx(t) + Gv(t)$, an equivalent differential operator feedback control law of the form: $u(t) = F(D)z(t) + Gv(t)$. Show that $F(D)$ will satisfy the relation: $\partial_c[F(D)] < \partial_c[P(D)]$.

7-22 Consider a linear multivariable system with the proper, rational transfer matrix, $T(s)$, compensated by a feedback (or a feedforward/unity feedback) system with the proper, rational transfer matrix, $T_f(s)$, as depicted below in Figure 7-22a (or Figure 7-22b). Show that if the $\lim_{s \to \infty} (I-T_F(s)T(s))$ is nonsingular, then there exists an "equivalent" feedforward compensator with a proper, rational transfer matrix, $T_c(s)$, as depicted in Figure 7-22c, such that the compensated transfer matrices of the systems depicted in Figures a (or b) and c are equal. Explicitly determine $T_c(s)$ in terms of $T(s)$ and $T_f(s)$ in both cases. In view of the general synthesis algorithm of Section 7.4, show that the compensation schemes in Figures a and b can be achieved via an "equivalent" scheme which employs l.s.v.f. in combination with input dynamics.

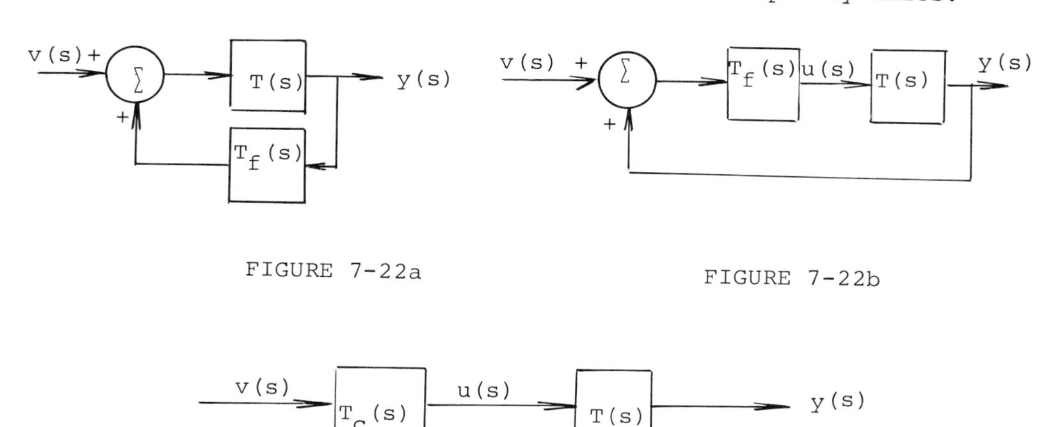

FIGURE 7-22a

FIGURE 7-22b

FIGURE 7-22c

CHAPTER 8
DESIGN OBJECTIVES

8.1 INTRODUCTION

In this chapter we employ many of the techniques and results developed in previous chapters to resolve a number of questions pertaining to the compensation of linear, time-invariant, multivariable systems. This chapter therefore represents a natural culmination of the general methods developed earlier for the analysis and design of this important class of systems. It should be emphasized that the synthesis procedures which will be presented in this chapter do not represent the "final word" in multivariable system compensation nor do they constitute a complete set of design objectives. Rather, they represent a fairly thorough compilation of some of the more recent control techniques which have been devised for compensating multivariable systems, and it is felt that the results given here are as complete and current as can be found anywhere. Furthermore, the procedures developed earlier and applied in this chapter can be enlisted to resolve a number of allied synthesis questions, a point which will be illustrated in the problem section of this chapter.

In Section 8.2 the question of arbitrary pole placement is considered, first from the point of view of linear output feedback, which involves no increase in system order, and then through the use of "modified forms" of linear state variable feedback. More specifically, the notion of a cyclic matrix is introduced and employed in the construction of "low order" single-input and single-output pole placement compensation schemes based on the frequency domain methods developed in Chapter 7. We show that for arbitrary pole placement, these latter schemes represent a compromise (in terms of the required added dynamics) between linear output feedback and the l.s.v.f. pole placement scheme developed earlier in Section 6.3.

In Section 8.3 the question of (dynamic) decoupling or diagonalizing the transfer matrix of a multivariable system is resolved. In particular, it is shown that any right invertible system can be decoupled by the general synthesis algorithm outlined in Section 7.4. In view of this observation, a general decoupling algorithm for invertible systems is developed by appropriately extending the general synthesis algorithm. A number of practical questions related to the decoupling question are then resolved, including a detailed, separate resolution of the question of decoupling via l.s.v.f. alone.

In Section 8.4, we consider the question of static decoupling, or the attainment of a compensated system in which a step change in the static, steady-state level of each (i-th) input is reflected by a corresponding change in the steady-state level of the corresponding (i-th) output and only that output. We show that static decoupling is a less restrictive design objective than dynamic decoupling, and one which can often be achieved via linear output feedback alone, with no increase in the dynamic order of the system.

The question of exact model matching, or achieving a compensated system whose transfer matrix equals (exactly matches) that of some desired model is then discussed in Section 8.5. Since this design objective is more restrictive than any of the others, we focus our attention almost exclusively on the question of exact model matching via combined l.s.v.f. and input dynamics as implemented by the synthesis algorithm of Section 7.4. In particular, we show that whenever the given system is left invertible, then provided it exists, a proper feedforward compensator, $T_c(s)$, which yields the model transfer matrix will be unique. In the other cases, we develop a procedure for characterizing the entire class of $T_c(s)$ which yield the model transfer matrix. This characterization then allows us to not only resolve the model matching question but also to gain considerable insight into related questions concerning the stability and the increase in dynamic order of the required compensator. Some concluding remarks and references are then given in Section 8.6.

8.2 ARBITRARY POLE PLACEMENT

The question of complete and arbitrary closed loop pole placement via l.s.v.f. was discussed in Section 6.3, and this same question of pole placement via state estimation feedback was covered in Section 6.4. We formally established in Section 6.3 that all of the controllable poles of a linear system can be arbitrarily assigned via l.s.v.f. and in Section 6.4 we demonstrated that Luenberger state estimator feedback can be used in lieu of actual l.s.v.f. with similar results, provided the system is observable. These results were then extended to the frequency domain in Section 7.3, where we presented a frequency domain analog of l.s.v.f. and its implementation via Luenberger state estimation.

While all of these pole assignment techniques serve to illustrate that complete and arbitrary placement of the controllable and observable poles of a given system is possible in both the time and frequency domains whether or not the entire state of the system is directly measurable, they are not, in general, the most "efficient" means of achieving a desired closed loop pole configuration, a fact which will form the basis of our discussions in this section. In particular, our objective here will be to present certain pole placement schemes which do not require as large an increase in system order as those developed earlier.

To begin, let us consider a system with a known proper $(p \times m)$ transfer matrix, $T(s)$. In view of Theorem 5.4.1, it is clear that $T(s)$ can be factored as the product: $R(s)P(s)^{-1}$, with $P(s)$ column proper and $\partial_c[R(s)] \leq \partial_c[P(s)]$. By introducing the m-vector partial state, $z(s)$, the dynamical behavior of the system can therefore be represented by the equations:

$$P(s)z(s) = u(s); \quad y(s) = R(s)z(s), \qquad 8.2.1$$

where $y(s)$ and $u(s)$ are the Laplace transforms of the output and input of the system, respectively. If $R(s)$ and $P(s)$ are also assumed to be relatively right prime, the synthesis procedure outlined in Section 7.3 can be used to achieve a closed loop transfer matrix, $R(s)P_F^{-1}(s)G$, with any arbitrary set of $n = \partial[|P(s)|] = \partial[|P_F(s)|]$ closed loop poles. However, such a procedure would require an increase in system order equal to $m(\nu-1)$, where ν is the observability index of the system, and we now ask whether or not some alternative technique for arbitrary pole placement can be found which would not require this large an increase in system order.

Since it involves <u>no increase</u> in system order (no added dynamics), we begin our investigation of alternative schemes by considering the employment of <u>linear output feedback</u> (l.o.f.), which is defined as the control law:

$$u(s) = Hy(s) + Gv(s), \qquad 8.2.2.a$$

or in the time domain as

$$u(t) = Hy(t) + Gv(t), \qquad 8.2.2b$$

where H is a constant $(m \times p)$ gain matrix and $v(t)$ is a q-vector external input. If we substitute the control law 8.2.2a for $u(s)$ in 8.2.1, we obtain the l.o.f. closed loop system:

$$[P(s)-HR(s)]z(s) \triangleq P_H(s)z(s) = Gv(s); \quad y(s) = R(s)z(s), \qquad 8.2.3$$

with transfer matrix:

$$T_{H,G}(s) = R(s)[P(s)-HR(s)]^{-1}G = R(s)P_H^{-1}(s)G, \qquad 8.2.4$$

provided the inverse of $P_H(s) = P(s)-HR(s)$ exists. Since H is completely arbitrary, however, it can always be selected such that (i) $P_H^{-1}(s)$ exists, and (ii) $\partial[|P_H(s)|] = \partial[|P(s)|] = n$, and we will restrict our attention in the remainder of this section to these cases. We therefore note that the closed loop poles of the system are equal to the (n) zeros of

$$\Delta_H(s) \triangleq |P_H(s)| = |P(s)-HR(s)| \qquad 8.2.5$$

The question of closed loop pole placement via l.o.f. thus reduces to the determination of the effect which the (pm) individual elements (feedback gains) of H have on the zeros of $\Delta_H(s)$. This point is an important one, and we will now illustrate that 8.2.5 can often be directly employed to resolve this question.

EXAMPLE 8.2.6: Let us consider the system whose transfer matrix

$$T(s) = R(s)P(s)^{-1} = \begin{bmatrix} s-1, & 0 \\ 0, & 2 \end{bmatrix} \begin{bmatrix} s^2+s-1, & 0 \\ 1, & s^2-4 \end{bmatrix}^{-1}.$$ If we let $H = \begin{bmatrix} h_1 & h_2 \\ h_3 & h_4 \end{bmatrix}$,

the l.o.f. closed loop poles of the system are given by the zeros of $\Delta_H(s)$, where $\Delta_H(s) = |P(s)-HR(s)| = \begin{vmatrix} s^2+(1-h_1)s+h_1-1, & -2h_2 \\ -h_3s+h_3+1, & s^2-4-2h_4 \end{vmatrix}$, or

$$\Delta_H(s) = s^4+(1-h_1)s^3+(-5+h_1-2h_4)s^2+(-4+4h_1-2h_4+2h_1h_4-2h_2h_3)s$$
$$+ (4-4h_1+2h_2+2h_4-2h_1h_4+2h_2h_3).$$

It is therefore clear in this example that given some desired $\Delta_H(s) = a_0+a_1s+a_2s^2+a_3s^3+s^4$, h_1 can first be chosen to achieve any desired a_3, and h_4 can then be selected so that any desired a_2 is obtained. The product h_2h_3 is then specified by a_1, and h_2 by a_0 once h_2h_3 has been determined. Provided $h_2 \neq 0$ (or $h_2h_3 = 0$ if $h_2 = 0$) therefore, the four gains which comprise H can be selected so that any desired l.o.f. closed loop pole configuration is achieved.

We should perhaps remark here that, in general, the determination of an appropriate H is dependent on the solution of n nonlinear algebraic equations which involve the (pm) free parameters of H, and there is no guarantee that a general solution exists even when $pm > n$.† Furthermore, a solution to the set of nonlinear equations is not always as transparent as this example might lead one to believe. The question of pole placement via l.o.f. therefore remains one of the most important unresolved questions in linear system theory.‡ We also note that our "brute force" resolution of the question of pole placement via l.o.f. is not confined solely to systems whose dynamical behavior is given in terms of a known transfer matrix; i.e. if we began our investigation in the time domain with some known state-space representation $\{A,B,C,E\}$, we could employ the controllable version of the structure theorem in order to find an equivalent representation of the form 8.2.1 for the controllable part of the system. Equation 8.2.5 could then be used to determine the effect of the l.o.f. control law 8.2.2b on the closed loop poles of the system.

While the employment of l.o.f. for closed loop pole placement is a question which remains largely unresolved, there are systematic techniques for complete and arbitrary closed loop pole assignment which require "relatively low" increases in system order, and we will devote the remainder of this section to the discussion of certain of these procedures.

In order to motivate one such scheme, we note that the transfer

---

† If $pm < n$, it is clear that no general solution can exist since there would be more equations than unknowns and, therefore, complete and arbitrary closed loop pole placement via l.o.f. would be impossible.

‡ In the scalar (single input/output) case this question is resolved by the now classical root locus, which was introduced by Evans [E2] over two decades ago.

matrix of a single input, multiple (p) output, controllable and observable system can be represented by the p-dimensional (column vector) transfer matrix:

$$T(s) = \tilde{R}(s)p(s)^{-1} = \frac{\begin{bmatrix} \tilde{r}_1(s) \\ \tilde{r}_2(s) \\ \vdots \\ \tilde{r}_p(s) \end{bmatrix}}{p(s)}, \qquad 8.2.7$$

where $\tilde{R}(s)$ and $p(s)$ are relatively right prime. If we also assume that $T(s)$ is proper, then $\partial_c[\tilde{R}(s)] \leq \partial[p(s)] = n$. In view of the results presented in Section 7.3, it would require an increase in system order equal to $m(\nu-1)$, or $\nu-1$ in this case since $m = 1$, to achieve any desired closed loop transfer matrix of the form:

$$T_f(s) = \tilde{R}(s)p_f(s)^{-1}, \qquad 8.2.8$$

where $p_f(s) = p(s)-f(s)$ for any arbitrary polynomial $f(s)$ of degree less than n.

In view of the above, we now note that we can always reduce a multiple input system, which we assume for convenience is both controllable and observable, to a scalar input system by linearly combining the m inputs into one; i.e. by simply setting

$$u(t) = gv(t) + w(t) = \begin{bmatrix} g_1 \\ g_2 \\ \vdots \\ g_m \end{bmatrix} v(t) + w(t) \qquad 8.2.9a$$

in the time domain, or

$$u(s) = gv(s) + w(s) \qquad 8.2.9b$$

in the frequency domain for any constant $(m \times 1)$ gain vector g. If we take care to choose g (if possible) so that the resulting system remains controllable, then all n closed loop poles can be arbitrarily assigned via the procedure given in Section 7.3 with a net increase in system order equal to $\nu-1$, where ν is the observability index of the system. The final closed loop system would have apparent order n, however, since the poles associated with the $(\nu-1)$-dimensional exponential estimator would appear as cancellable pole-zero pairs in the final closed loop transfer matrix (see 7.3.28).

To illustrate this procedure, consider the compensation scheme depicted in Figure 8.2.10. If g is chosen such that the controllable

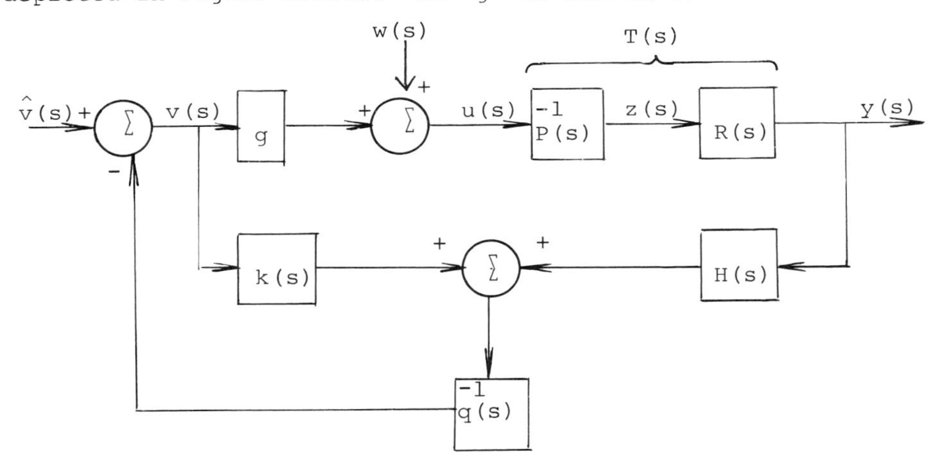

FIGURE 8.2.10

A SINGLE INPUT POLE PLACEMENT COMPENSATOR

and observable system with transfer matrix $T(s) = R(s)P(s)^{-1}$ remains (completely state) controllable from the scalar input $v(s)$, then for $w(s) = 0$,

$$y(s) = R(s)P(s)^{-1}g\, v(s) = \tilde{R}(s)p(s)^{-1}v(s), \qquad 8.2.11$$

for some polynomial vector $\tilde{R}(s)$ and polynomial $p(s)$. Since system observability is unaffected by the control law 8.2.9, it follows that $p(s) = \alpha|P(s)|$ for some nonzero scalar α with $\tilde{R}(s)$ and $p(s)$ relatively right prime. In view of 8.2.11, a system equivalent to the one depicted in Figure 8.2.10 (with $w(s) = 0$) can now be represented via a new partial state, $\hat{z}(s)$; i.e. if we set $y(s) = \tilde{R}(s)\hat{z}(s)$ and $p(s)\hat{z}(s) = u(s)$, we can represent the original system by the equivalent one depicted in Figure 8.2.12. The closed loop performance of

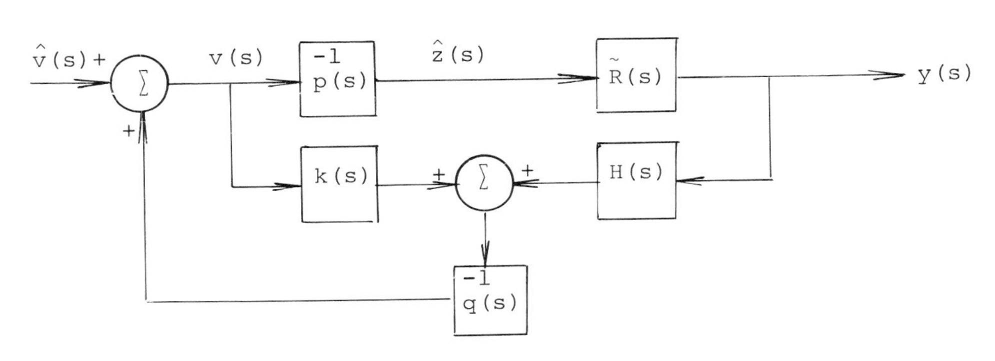

FIGURE 8.2.12

AN EQUIVALENT "SINGLE INPUT" COMPENSATOR

this equivalent compensator can now be readily determined by equating signals at the first summation junction. In particular,

$$p(s)\hat{z}(s) = v(s) = \hat{v}(s) + q^{-1}(s)[H(s)y(s) + k(s)v(s)], \qquad 8.2.13$$

and since $y(s) = \tilde{R}(s)\hat{z}(s)$, we find that

$$[q(s)p(s) - H(s)\tilde{R}(s) - k(s)p(s)]\hat{z}(s) = q(s)\hat{v}(s) \qquad 8.2.14$$

Since $\tilde{R}(s)$ and $p(s)$ are relatively right prime we can now choose $H(s)$ and $k(s)$ so that

$$H(s)\tilde{R}(s) + k(s)p(s) = q(s)f(s), \qquad 8.2.15$$

where $q(s)$ is any arbitrary (stable) polynomial of degree $\nu-1$, and

$f(s)$ is any arbitrary polynomial of degree less than $n = \partial[p(s)]$ (see Theorem 7.3.23). If we then employ 8.2.15 in 8.2.14, we readily obtain an expression for the closed loop transfer matrix of the system; i.e. $q(s)[p(s) - f(s)]\hat{z}(s) = q(s)\hat{v}(s)$, or

$$y(s) = \tilde{R}(s)p_f^{-1}(s)q^{-1}(s)q(s)\hat{v}(s) = \tilde{R}(s)p_f^{-1}(s)\hat{v}(s), \qquad 8.2.16$$

where $p_f(s) = p(s) - f(s)$, an n-th degree polynomial whose zeros are completely and arbitrarily assignable via $f(s)$. If this same triple, $\{H(s),k(s),q(s)\}$, is now employed in the compensator depicted in Figure 8.2.10, it follows that the $n+\nu-1$ closed loop poles of that (equivalent) system would also equal the zeros of $q(s)$ and $p_f(s)$ and, therefore, that the transfer matrix between $y(s)$ and $w(s)$ would also display these same poles. It might be noted that the entire compensator depicted in Figure 8.2.10, including g, can be placed in the feedback path, with $\hat{v}(s) \equiv 0$ and $w(s)$ the primary external input.

A major question regarding this pole placement scheme which still remains unresolved is whether or not g can be chosen in 8.2.9 so that (complete state) controllability is preserved, and we now resolve this question via the notion of "cyclic matrices".

DEFINITION 8.2.17: Any $(n \times n)$ matrix A with elements in $\mathscr{R}$ is said to be cyclic if and only if there exists an $(n \times 1)$ cyclic vector b with elements in $\mathscr{R}$ such that $\rho[b,Ab,\ldots,A^{n-1}b] = n$. We merely note here that A is cyclic if and only if its characteristic and minimal polynomials are identical; i.e. if and only if $\Delta = \Delta_m$ (see Section 2.4), a well known result in linear algebra which can be found in a number of mathematical texts (e.g. [H1]). In view of this definition, we now state:

THEOREM 8.2.18: Consider any state-space representation, $\{A,B,C,E\}$, of a dynamical system. If the pair $\{A,B\}$ is controllable and the matrix A is cyclic, then there exists an m-vector g such that the scalar input pair $\{A,Bg\}$ is controllable.

Proof: A formal constructive proof of this theorem is given in [W6] and will not be repeated here since it is rather involved and, as it turns out, if the conditions of the theorem are satisfied then "almost any" g will work; i.e. in mathematical terms, all $(m \times 1)$ gain vectors, g, except those belonging to "a set of measure zero" will produce a controllable pair $\{A,Bg\}$ (see Example 8.2.20).

In view of Theorem 8.2.18, it is thus clear that whenever A is cyclic, it is relatively easy to reduce a controllable multi-input state representation to one which is single input controllable; i.e. to find a g such that $\{A,Bg,C,Eg\}$ is (completely state) controllable via the scalar input $v(t)$, defined by the control law 8.2.9a.

We could, at this time, present a time domain compensation scheme for completely and arbitrarily assigning all (n) closed loop poles of any controllable and observable state-space system characterized by a cyclic A matrix through the employment of a single Luenberger observer of total order $\nu-1$. Rather than presenting this result from a time domain point of view, however, it is perhaps easier and more natural to illustrate this procedure from a frequency domain viewpoint, employing the results outlined earlier in this section. In order to do this, however, we must first present a frequency domain analog of the notion of cyclic matrices. Therefore, let us consider any $(p \times m)$ proper, rational transfer matrix, $T(s)$, which has a minimal state-space realization, $\{A,B,C,E\}$. $T(s)$ will be called a cyclic transfer matrix if and only if A is cyclic; i.e. if and only if $\Delta = \Delta_m$. In view of this definition, we can now state and establish the following:

THEOREM 8.2.19: <u>Consider any $(p \times m)$ proper rational transfer matrix $T(s)$. There exists an m-vector g such that $T(s)g = \tilde{R}(s)p(s)^{-1}$, with $\tilde{R}(s)$ and $p(s)$ relatively right prime and $p(s) = \Delta(s)$ if and only if $T(s)$ is cyclic.</u>

<u>Proof</u>: We first establish sufficiency. In particular, if $T(s)$ is cyclic with minimal realization $\{A,B,C,E\}$, then, by definition A will also be cyclic. Therefore, by Theorem 8.2.12 we can readily find an m-vector g such that $\{A,Bg\}$ is controllable; i.e. such that Bg is a cyclic vector of A. Since $T(s) = C(sI-A)^{-1}B+E$, it is clear that $T(s)g = C(sI-A)^{-1}Bg+Eg$, which we can express as the product: $\tilde{R}(s)p(s)^{-1}$ of some relatively right prime pair, $\{\tilde{R}(s),p(s)\}$, with $\tilde{R}(s)$ a p-dimensional polynomial vector and $p(s)$ a monic polynomial. In view of the results presented in Section 5.3, we further note that since $\{A,C\}$ is observable and $\{A,Bg\}$ is controllable $p(s) = |sI-A| = \Delta(s) = \Delta_m(s)$.

To establish necessity, we assume that $T(s)g = \tilde{R}(s)p(s)^{-1}$, with $\tilde{R}(s)$ and $p(s)$ relatively right prime and $p(s) = \Delta(s) = |sI-A|$, where $\{A,B,C,E\}$ is a minimal realization of $T(s)$. Since $T(s)g = C(sI-A)^{-1}Bg+Eg = \tilde{R}(s)p(s)^{-1}$, $\{A,Bg,C,Eg\}$ is a realization of $T(s)g$. Furthermore, since $\tilde{R}(s)$ and $p(s)$ are relatively right prime with $p(s) = |sI-A|$, it follows (see Section 5.3) that $\{A,Bg,C,Eg\}$ is a minimal realization of $T(s)g$. Therefore, $\{A,Bg\}$ must be controllable, which implies that A must be cyclic. By definition, it now follows that $T(s)$ must be cyclic, thus establishing the theorem.

In view of Theorem 8.2.19, it is thus clear that <u>whenever the transfer matrix, $T(s)$, of a system is cyclic with $n = \partial[\Delta]$, the system can be made single input controllable via some m-vector g, after which the single input compensation scheme outlined earlier in this section can be employed to arbitrarily assign all $(n+\nu-1)$ of its</u>

closed loop poles.

EXAMPLE 8.2.20: To illustrate the above procedure for arbitrary closed loop pole assignment, let us consider the same system employed earlier in Example 8.2.6; i.e.

$$T(s) = R(s)P^{-1}(s) = \begin{bmatrix} s-1 & 0 \\ 0 & 2 \end{bmatrix} \begin{bmatrix} s^2+s-1, & 0 \\ 1, & s^2-4 \end{bmatrix}^{-1} = \begin{bmatrix} \dfrac{1}{s^2+s-1}, & 0 \\ \dfrac{-1}{(s^2-4)(s^2+s-1)}, & \dfrac{1}{s^2-4} \end{bmatrix}$$

Since $R(s)$ and $P(s)$ are relatively right prime and all four zeros of $|P(s)|$ differ, it is clear in this example that $T(s)$ is cyclic (see Problem 8-15) and furthermore, that any nonzero g other than $g = \alpha \begin{bmatrix} 0 \\ 1 \end{bmatrix}$, which we note represents only one of an infinite number of "directions" in $\mathscr{R}^2$ (a "set of measure zero"), will produce a single input transfer matrix of the form 8.2.7 with $p(s) = |P(s)| = (s^2-4)(s^2+s-1)$. If, for example, we set $g = \begin{bmatrix} 1 \\ 0 \end{bmatrix}$, then

$$T(s)g = \frac{\begin{bmatrix} s^2-4 \\ -1 \end{bmatrix}}{s^4+s^3-5s^2-4s+4} = \tilde{R}(s)p^{-1}(s),$$ a scalar input, two (=p) output system. The reader can now verify that $\nu = 2$ in this example, and therefore that the eliminant matrix associated with the relatively right prime pair $\{\tilde{R}(s),p(s)\}$ is equal to M_{e2} (see 7.3.29); i.e.

$$\begin{bmatrix} \tilde{R}(s) \\ s\tilde{R}(s) \\ p(s) \\ sp(s) \end{bmatrix} = M_{e2}S_{e2}(s) = M_e(s)S_e(s) = \underbrace{\begin{bmatrix} -4 & 0 & 1 & 0 & 0 & 0 \\ -1 & 0 & 0 & 0 & 0 & 0 \\ 0 & -4 & 0 & 1 & 0 & 0 \\ 0 & -1 & 0 & 0 & 0 & 0 \\ 4 & -4 & -5 & 1 & 1 & 0 \\ 0 & 4 & -4 & -5 & 1 & 1 \end{bmatrix}}_{M_{e2} = M_e} \underbrace{\begin{bmatrix} 1 \\ s \\ s^2 \\ s^3 \\ s^4 \\ s^5 \end{bmatrix}}_{S_{e2}(s)=S_e(s)}$$

If we now employ the synthesis algorithm of Section 7.3, it is clear that a triple $\{H(s),k(s),q(s)\}$ can be found such that the compensation scheme depicted in Figure 8.2.10 (or Figure 8.2.12) yields any desired closed loop transfer matrix (between $y(s)$ and $\hat{v}(s)$) of the form $\tilde{R}(s)p_f^{-1}(s)$, where $p_f(s)$ has any arbitrary, desired set of (n=4) zeros. Furthermore, $q(s)$ can be chosen to be any (stable) first degree polynomial.

Some remarks are in order at this point. In particular, we first note that in order to achieve complete and arbitrary closed loop pole placement via this scheme, we must add a dynamical system of total order $\nu-1 = 1$ in this example. As noted earlier, however, l.o.f. can often be used to achieve virtually the same result with no increase in system order (see Example 8.2.6). Furthermore, if we employed the feedback compensation scheme of Section 7.3, we would have to increase the system order by an amount equal to $m(\nu-1) = 2$ in this example. Therefore, while the single input scheme which we have just outlined for pole placement requires an increase in system order equal to $\nu-1$, which is more efficient for arbitrary pole assignment than the more general l.s.v.f. scheme of Section 7.3, it is not as efficient as l.o.f. provided, of course, complete and arbitrary pole placement via l.o.f. is possible.

We finally note that we do not require explicit knowledge of $R(s)$ and $P(s)$, a "factorization" of $T(s)$, in order to implement this single input scheme; i.e. it should now be clear that given any $(p \times m)$ proper transfer matrix $T(s)$, and any constant $m \times 1$ vector g, $T(s)g$ can be expressed as $\tilde{R}(s)p^{-1}(s)$, where $\tilde{R}(s)$ and $p(s)$ are relatively right prime. Furthermore, all of the zeros of $p(s)$ (thus defined) can be arbitrarily assigned via this compensation scheme, although $p(s)$ may not satisfy the condition of Theorem 8.2.19. The relatively right prime factorization, $R(s)P^{-1}(s)$, of $T(s)$ was employed in Example 8.2.20 only to illustrate that $p(s) = \alpha|P(s)|$ when controllability is preserved by g; i.e. when $T(s)$ is cyclic and g is properly chosen.

If we are not particularly concerned about achieving a closed loop scheme with the same apparent order (n) as the given system, we can simplify the single input compensator somewhat, and still achieve a closed loop system of total order $n+\nu-1$, all of whose poles are completely and arbitrarily assignable whenever $T(s)$ is cyclic. In particular, let us now consider the compensation scheme depicted in Figure 8.2.12, with $k(s) = 0$, which we have redrawn for convenience in Figure 8.2.21. If we equate signals at the input junction, we

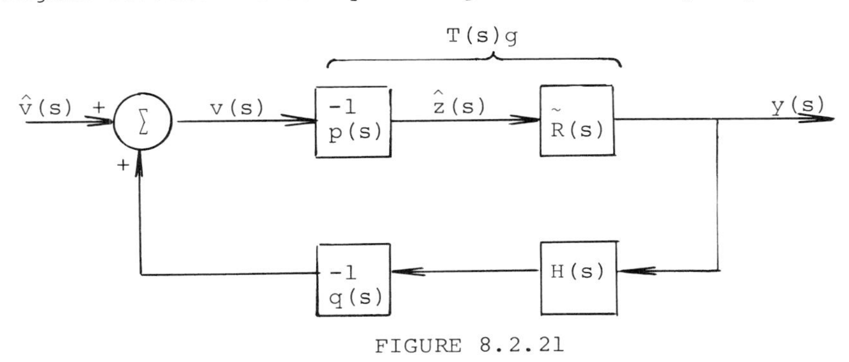

FIGURE 8.2.21

THE SINGLE INPUT COMPENSATOR WITH $k(s) = 0$

readily obtain an expression for the closed loop performance of the system, namely

$$[q(s)p(s) - H(s)\tilde{R}(s)]\hat{z}(s) = q(s)\hat{v}(s) \qquad 8.2.22$$

We now note that since $\tilde{R}(s)$ and $p(s)$ are relatively right prime, $q(s)$ and $H(s)$ can be chosen such that $q(s)p(s) - H(s)\tilde{R}(s)$ is any desired polynomial, $\bar{p}(s)$, of degree $\leq n+\nu-1$ (see Theorem 7.3.23). In order to insure a proper $q(s)^{-1}H(s)$, let us therefore choose $q(s)$ and $H(s)$ so that

$$q(s)p(s) - H(s)\tilde{R}(s) = \bar{p}(s), \qquad 8.2.23$$

with $\partial[q(s)] = \nu-1$ and $\bar{p}(s)$ equal to any desired polynomial of degree $n+\nu-1$. Substituting 8.2.23 into 8.2.22, and solving for $y(s)$ in terms of $\hat{v}(s)$, we readily obtain an expression for the closed loop transfer matrix of this system, namely

$$y(s)=\tilde{R}(s)\bar{p}^{-1}(s)q(s)\hat{v}(s) \qquad 8.2.24$$

Since we are no longer free to choose $q(s)$, $\bar{p}(s)$ and $q(s)$ will generally be relatively prime, and the system depicted in Figure 8.2.21 will therefore have apparent order $n+\nu-1$ with its poles equal to the zeros of $\bar{p}(s)$. It also follows that if $k(s) = 0$ and this same $q(s)$ and $H(s)$ are employed in the feedback scheme depicted in Figure 8.2.10, then the apparent order of that system will also be $n+\nu-1$, in general, with its poles equal to the $(n+\nu-1)$ zeros of $\bar{p}(s)$.

We will now show that there is a dual to this latter pole placement scheme which involves single output systems (instead of single input systems) and the controllability index μ (instead of the observability index ν). More specifically, by duality it follows that if $T(s)$ is a cyclic, proper $(p \times m)$ rational transfer matrix with minimal realization $\{A,B,C,E\}$, then there exists a $(1 \times p)$ vector $\tilde{g}$ such that $\tilde{g}T(s) = p^{-1}(s)\tilde{R}(s)$, where $p(s)$ and $\tilde{R}(s)$ are relatively left prime, and $p(s)$ is both the characteristic and minimal polynomial of A (see Problem 8-14). As in the single input case, if $T(s)$ is cyclic, "almost any" $\tilde{g}$ will work [W6].

Let us now consider the compensation scheme depicted in Figure 8.2.25, which is dual to the scheme depicted in Figure 8.2.21. If we equate signals at the input summation junction, we readily obtain an expression for the closed loop transfer matrix of the system; i.e. since $u(s) = v(s) + \tilde{H}(s)\tilde{q}^{-1}(s)\hat{y}(s)$, where $\hat{y}(s) = p^{-1}(s)\tilde{R}(s)u(s)$, we have

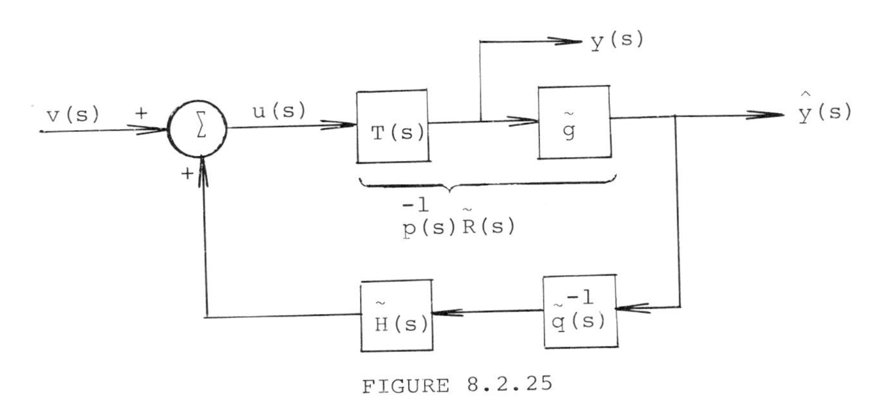

FIGURE 8.2.25

A SINGLE OUTPUT POLE PLACEMENT COMPENSATOR

$$\hat{y}(s) = p^{-1}(s)\tilde{R}(s)[v(s) + \tilde{H}(s)\tilde{q}^{-1}(s)\hat{y}(s)], \tag{8.2.26}$$

or

$$[p(s)\tilde{q}(s) - \tilde{R}(s)\tilde{H}(s)]\hat{y}(s) = \tilde{q}(s)\tilde{R}(s)v(s) \tag{8.2.27}$$

It now follows, by duality, that since $p(s)$ and $\tilde{R}(s)$ are relatively left prime, we can find a $\tilde{q}(s)$ and an $\tilde{H}(s)$ such that $\tilde{H}(s)\tilde{q}^{-1}(s)$ is proper with $p(s)\tilde{q}(s) - \tilde{R}(s)\tilde{H}(s)$ equal to desired polynomial $\tilde{p}(s)$ of degree $n+\mu-1$, where μ is the controllability index of the given system. If we therefore choose $\tilde{p}(s)$ to be a stable polynomial with any arbitrary set of $(n+\mu-1)$ zeros, the closed loop transfer matrix of the system depicted in Figure 8.2.25 will be given by the relation:

$$\hat{y}(s) = \tilde{p}^{-1}(s)\tilde{q}(s)\tilde{R}(s)v(s), \tag{8.2.28}$$

with $\tilde{p}^{-1}(s)\tilde{q}(s)\tilde{R}(s)$ a stable, proper transfer matrix. It thus follows that if the control law:

$$u(s) = v(s) + \tilde{H}(s)\tilde{q}^{-1}(s)\tilde{g}\ y(s) \tag{8.2.29}$$

is applied to a system with the cyclic transfer matrix, $T(s)$, then the $(n+\mu-1)$ closed loop poles of the compensated system will equal the zeros of $\tilde{p}(s)$.

We can now summarize all of these single input/output pole placement schemes via the following:

THEOREM 8.2.30: <u>If $T(s)$ is a $(p \times m)$ proper, cyclic transfer matrix with $n = \partial[\Delta]$, then all $n+\nu-1$ $(n+\mu-1)$ closed loop poles of the systems (system) depicted in Figures 8.2.10 or 8.2.21 (Figure 8.2.25) can be arbitrarily assigned</u>.

<u>Proof</u>: The proof of this theorem has already been constructively established.

EXAMPLE 8.2.31: We now illustrate the single input $(k(s) = 0)$ scheme depicted in Figure 8.2.21 using the same transfer matrix as we used in Examples 8.2.6 and 8.2.20. In particular, if

$$T(s) = \begin{bmatrix} \dfrac{1}{s^2+s-1} & , & 0 \\ \dfrac{1}{s^4+s^3-5s^2-4s+4} & , & \dfrac{1}{s^2-4} \end{bmatrix},$$

then as we illustrated in Example 8.2.20, $g = \begin{bmatrix} 1 \\ 0 \end{bmatrix}$ is one vector which "reduces" $T(s)g$ to single input controllable form; i.e. $T(s)g = \dfrac{\begin{bmatrix} s^2-4 \\ -1 \end{bmatrix}}{s^4+s^3-5s^2-4s+4} = \hat{R}(s)p(s)^{-1}$,

where $p(s) = s^4+s^3-5s^2-4s+4 = \Delta(s) = \Delta_m(s)$. We also noted earlier in Example 8.2.20 that $\nu = 2$ and, therefore, that

$$M_e = M_{e2} = \begin{bmatrix} -4 & 0 & 1 & 0 & 0 & 0 \\ -1 & 0 & 0 & 0 & 0 & 0 \\ 0 & -4 & 1 & 0 & 0 & 0 \\ 4 & -4 & -5 & 1 & 1 & 0 \\ 0 & 4 & -4 & -5 & 1 & 1 \end{bmatrix}.$$

We next determine that

$$\hat{M}_e^{-1} = M_e^{-1} = \begin{bmatrix} 0 & -1 & 0 & 0 & 0 & 0 \\ 0 & 0 & 0 & -1 & 0 & 0 \\ 1 & -4 & 0 & 0 & 0 & 0 \\ 0 & 0 & 1 & -4 & 0 & 0 \\ 5 & -16 & -1 & 0 & 1 & 0 \\ -1 & 0 & 6 & -16 & -1 & 1 \end{bmatrix}$$ and hence that $H(s)$ and $q(s)$

can be chosen such that $q^{-1}(s)H(s)$ is proper and $q(s)p(s)-H(s)\hat{R}(s) = \bar{p}(s) = (s+1)^2(s+2)(s+1+j)(s+1-j) = s^5+6s^4+15s^3+20s^2+14s+4$, a stable polynomial of degree $5(=n+\nu-1)$ which we have (arbitrarily) chosen to represent the denominator of the closed loop transfer matrix. Since $\bar{p}(s)$ can be represented as $\beta S_e(s)$, where $\beta = [4,14,20,15,6,1]$ and $S_e(s) = [1,s,s^2,s^3,s^4,s^5]^T$, it follows that $\beta M_e^{-1} = [-49,180,-15,90,-5,-1]$, or, in view of 7.3.34, that $H(s) = [-15s-49, 90s+180]$ and $q(s) = (s+5)$. The reader can now verify that for this particular choice of $H(s)$ and $q(s)$, $q^{-1}(s)H(s)$ is proper and $H(s)\hat{R}(s)+q(s)p(s) = \bar{p}(s) = s^5+6s^4+15s^3+20s^2+14s+4$, the desired denominator polynomial of the closed loop system.

8.3 DYNAMIC DECOUPLING

We now consider the question of compensating a linear multivariable system in order to achieve a "decoupled" design. More specifically, we will consider compensation schemes where each component of the output vector, $y(t)$, is affected by one and only one component of the external input. Such a design obviously implies a large number of zero entries in the transfer matrix of the compensated system and, for convenience, we will choose these zero entries to characterize all of the off-diagonal terms. In view of these requirements, we now precisely define this notion of "decoupling".

DEFINITION 8.3.1: A linear multivariable system is said to be (<u>dy-</u>

namically) decoupled if and only if its transfer matrix is diagonal and nonsingular.

We realize that, in general, a given system will not be decoupled although it might be highly desirable to achieve a decoupled design; e.g. additional compensation could then be applied to each scalar subsystem individually without affecting the other input/output pairs. Also, manual control would generally be facilitated if a decoupled design could be achieved. Whatever the reasons for decoupling a system, the two methods which we will consider in this section for realizing a decoupled design are (i) input dynamics in combination with l.s.v.f. as defined in Section 7.4, and (ii) l.s.v.f. alone.

In view of the results which were presented in Sections 5.5 and 7.4, we can now state and establish an almost obvious result.

THEOREM 8.3.2: A linear multivariable system characterized by a $(p \times m)$ proper transfer matrix, $T(s)$, can be decoupled by input dynamics in combination with l.s.v.f. if and only if it is right invertible.

Proof: If the given system is right invertible, there exists an $(m \times p)$ rational transfer matrix, $T_{RI}(s)$, (not necessarily proper) such that $T(s)T_{RI}(s) = I_p$ (see Section 5.5). If we now divide each column of $T_{RI}(s)$ by a polynomial, $p_i(s)$, of large enough degree to insure that $T_c(s) = T_{RI}(s)\{\text{diag}[p_i(s)]\}^{-1}$ is proper, it follows that $T_c(s)$ can then be implemented by the combined use of input dynamics and l.s.v.f. (see Section 7.4). The resulting compensated system will therefore be characterized by the nonsingular, diagonal transfer matrix, $T_d(s) = T(s)T_c(s) = \{\text{diag}[p_i(s)]\}^{-1}$; i.e. the transfer matrix of a decoupled system.

To establish necessity, we first note that if a system can be decoupled by the combined use of input dynamics and l.s.v.f., then it can be decoupled by input dynamics alone (see Section 7.4.). In particular, if a system can be decoupled by the scheme depicted in Figure 7.4.23, with only p components of $v(s)$ employed, then there exists a proper $(m \times p)$ transfer matrix, $T_c(s)$, such that

$$T(s)T_c(s) = T_d(s), \tag{8.3.3}$$

a nonsingular, diagonal transfer matrix. If we now postmultiply both sides of 8.3.3 by $T_d^{-1}(s)$, it follows that

$$T(s)T_c(s)T_d^{-1}(s) = I_p, \tag{8.3.4}$$

or that $T(s)$ has a right inverse, $T_{RI}(s) = T_c(s)T_d^{-1}(s)$, thus establishing necessity as well as the theorem. In view of Theorems 8.3.2 and 5.5.3, it should now be noted that <u>a system can be decoupled by l.s.v.f. in combination with input dynamics if and only if its transfer matrix, $T(s)$, has rank p</u>, an observation which represents a most direct frequency domain resolution to the decoupling question.

The constructive proof of sufficiency of Theorem 8.3.2 represents a direct but almost "brute force" technique for decoupling and little, if any, insight is provided with respect to a number of important decoupling questions; e.g. nothing has yet been said regarding the stability of the decoupled system or the increase in system order required for decoupling. We realize, however, that it is obviously desirable to minimize the order of any dynamic compensation while simultaneously achieving a stable decoupled design with as much arbitrary pole placement as possible. With these design objectives in mind, we will present an alternative constructive proof of Theorem 8.3.2 which will then enable us to resolve the following decoupling questions:

<u>DQ1</u>: How many poles of the decoupled system can be arbitrarily

assigned?

DQ2: Can the decoupled system be stabilized or, equivalently, what poles of the system are "fixed" or invariant under decoupling compensation?

DQ3: What is the minimum order of any dynamic compensation required for decoupling?

DQ4: Can l.s.v.f. alone decouple the system?

All of these interrelated questions should be considered whenever a decoupling scheme is contemplated, and the remainder of this section will focus on their resolution. In order to achieve this objective, we will first present a general decoupling algorithm for invertible systems (see Section 5.5); i.e. we now outline a three step synthesis procedure which employs the synthesis algorithm of Section 7.4 for decoupling systems which are characterized by a proper, nonsingular transfer matrix. We will later show that the "invertible decoupling algorithm" not only allows us to resolve all of the preceeding decoupling questions in the case of invertible systems, but that it can also be extended and then employed to resolve these same decoupling questions in the more general case as well.

Let us therefore consider any open loop system with a proper, nonsingular $(m \times m)$ transfer matrix:

$$T(s) = R(s)P(s)^{-1}, \qquad 8.3.5$$

where $R(s)$ and $P(s)$ are both nonsingular and relatively right prime, with $P(s)$ column proper. It should be noted here that this factorization of $T(s)$ might follow as the result of applying the structure theorem (Theorem 4.3.3) to a given controllable and observable state-space representation, $\{A,B,C,E\}$, or it might result from factoring a given $T(s)$ in accordance with the constructive proof of Theorem 5.4.1.

<u>The Invertible Decoupling Algorithm</u>:

<u>Step 1</u>: Define $R_d(s)$ as the $(m \times m)$ nonsingular, diagonal polynomial matrix, $\text{diag}[r_{di}(s)]$, with each diagonal element, $r_{di}(s)$, equal to the greatest common (monic) divisor of the corresponding, i-*th* row of $R(s)$; i.e.

$$R(s) = R_d(s)\hat{R}(s), \tag{8.3.6}$$

with $R_d(s)$ a diagonal left divisor of $R(s)$ of maximum row degree.

<u>Step 2</u>: Compute the product, $P(s)\hat{R}(s)^{-1}$, and let

$$P_d(s) = \text{diag}[p_{di}(s)] = \text{diag}[s^{g_i}+a_{i,g_i-1}s^{g_i-1}+\ldots+a_{i1}s+a_{io}], \tag{8.3.7}$$

with each $p_{di}(s)$ an arbitrary (monic) polynomial of degree g_i, where the g_i are chosen to be as small as possible consistent with the requirement that $P(s)\hat{R}(s)^{-1}P_d^{-1}(s)$ be a proper transfer matrix.

<u>Step 3</u>: Let

$$T_c(s) = P(s)\hat{R}(s)^{-1}P_d^{-1}(s) = P(s)[P_d(s)\hat{R}(s)]^{-1}, \tag{8.3.8}$$

a proper transfer matrix, and employ the synthesis algorithm of Section 7.4 to realize this $T_c(s)$ via l.s.v.f. in combination with input dynamics; i.e. since

$$P^{-1}(s)T_c(s) = [P_d(s)\hat{R}(s)]^{-1} = \bar{P}^{-1}(s)L(s), \tag{8.3.9}$$

it is clear that we can choose $L(s) = I_m$ and

$$\bar{P}(s) = P_d(s)\hat{R}(s) = G(s)P(s) - F(s), \tag{8.3.10}$$

which we can now readily solve for $G(s)$ and $F(s)$ (see Theorem 7.4.3). This $G(s)$ and $F(s)$ can then by employed in the compensation scheme depicted in Figure 7.4.14.

Before illustrating the employment of this decoupling algorithm, we will first partially resolve (completely resolve in the case of invertible systems) the four decoupling questions posed earlier via the following observations:

O 1: We first note that $R_d(s)$ and $P_d(s)$ are chosen in order to achieve as much arbitrary pole placement as possible while simultaneously minimizing the order of any input dynamics which may be required for decoupling; i.e. if $\tilde{T}_c(s)$ is any proper dynamic compensator which decouples a given invertible system with transfer matrix $T(s) = R(s)P^{-1}(s)$, then

$$T(s)\tilde{T}_c(s) = R(s)P^{-1}(s)\tilde{T}_c(s) = \tilde{R}_d(s)\tilde{P}_d^{-1}(s), \qquad 8.3.11$$

for some diagonal pair, $\{\tilde{R}_d(s), \tilde{P}_d(s)\}$. In view of 8.3.11, it is clear that

$$P^{-1}(s)\tilde{T}_c(s) = R^{-1}(s)\tilde{R}_d(s)\tilde{P}_d^{-1}(s) = [\tilde{P}_d(s)\tilde{R}_d^{-1}(s)R(s)]^{-1} \qquad 8.3.12$$

or, in view of 7.4.11, that

$$P^{-1}(s)\tilde{T}_c(s) = [\tilde{P}_d(s)\tilde{R}_d^{-1}(s)R(s)]^{-1} = \overline{P}^{-1}(s)L(s), \qquad 8.3.13$$

where $\partial[|\overline{P}(s)|] - \partial[|P(s)|]$ represents the order of an input dynamic compensator required for realizing $T(s)\tilde{T}_c(s)$ via l.s.v.f. and input dynamics. It is thus clear that if $\tilde{R}_d(s) = R_d(s)$, a diagonal left divisor of $R(s)$ of maximum row degree, then $\overline{P}(s) = \tilde{P}_d(s)\hat{R}(s)$ and $L(s) = I_m$. Furthermore, if $\tilde{P}_d(s) = P_d(s)$, an (arbitrary) diagonal polynomial matrix of minimum row degree, then $\partial[|\overline{P}(s)|]$ will be minimized, thus resulting in a minimization of the order of any decoupling dynamic precompensation which may be required.

O 2: In view of Step 3 and 8.3.10 in particular, it is clear that $n_d (= \partial[|P_d(s)|])$ closed loop poles of the decoupled system can be arbitrarily assigned (a partial resolution of DQ1). Furthermore, the zeros of $|\hat{R}(s)|$ represent the only fixed poles of an invertible system decoupled by combined l.s.v.f. and input dynamics (a partial resolution of DQ2), and appear as cancellable pole-zero terms (unobservable modes--see Problem 8-22) in the decoupled transfer matrix $T_d(s) = R_d(s)P_d^{-1}(s)$. It should be noted that these "fixed" poles of the decoupled system correspond to certain of the invariant poles of the inverse of the given system (see Section 5.5), as we will later show (see Theorem 8.3.30). It is thus clear that if one or more of the zeros of $|\hat{R}(s)|$ lie in the unstable half-plane, Re(s) > 0, then the decoupled system will also be unstable. In these latter cases, however, it is still possible to decouple and simultaneously stabilize a given invertible system by employing a "two stage" synthesis procedure which is somewhat more involved than the technique outlined in Section 7.4 (see Problem 8-26).

O 3: In view of 8.3.10 and O1, it is also clear that if $\bar{n} = \partial[|P_d(s)\hat{R}(s)|]$ and $n = \partial[|P(s)|]$, then $\bar{n} - n$ represents the minimun order of any precompensator, $G(s)$, required for decoupling (a partial resolution of DQ3). It therefore follows that a given invertible system can be decoupled via l.s.v.f. alone if and only if $\bar{n} = n$; i.e. if and only if $\partial[|G(s)|] = 0$ (a partial resolution of DQ4).

To illustrate the invertible decoupling algorithm as well as clarify the partial resolution of the four decoupling questions posed earlier, let us now consider the following example which is due to Gilbert [G3].

EXAMPLE 8.3.14: Consider the following controllable and observable state-space system, {A,B,C,E}; i.e.

$$A = \begin{bmatrix} 0 & 0 & 0 & 0 \\ 1 & 0 & 0 & 0 \\ 0 & 0 & 0 & 0 \\ 1 & 0 & 1 & 0 \end{bmatrix}, \quad B = \begin{bmatrix} 1 & 0 \\ 0 & 0 \\ 0 & 1 \\ 0 & 0 \end{bmatrix}, \quad C = \begin{bmatrix} 1 & 1 & 1 & 0 \\ 0 & 0 & 0 & 1 \end{bmatrix}, \text{ and } E = \begin{bmatrix} 0 & 0 \\ 0 & 0 \end{bmatrix}.$$

To determine whether or not this system can be decoupled through the combined use of l.s.v.f. and input dynamics, we will first reduce the system to controllable companion form via the equivalence transformation $Q = \begin{bmatrix} 0 & 1 & 0 & 0 \\ 1 & 0 & 0 & 0 \\ 0 & -1 & 0 & 1 \\ 0 & 0 & 1 & 0 \end{bmatrix}$ (see Section 3.6), and them employ Theorem 4.3.3. In particular,

$$QAQ^{-1} = \hat{A} = \begin{bmatrix} 0 & 1 & 0 & 0 \\ 0 & 0 & 0 & 0 \\ 0 & 0 & 0 & 1 \\ 0 & 0 & 0 & 0 \end{bmatrix}, \quad QB = \hat{B} = \begin{bmatrix} 0 & 0 \\ 1 & 0 \\ 0 & 0 \\ 0 & 1 \end{bmatrix}, \quad CQ^{-1} = \hat{C} = \begin{bmatrix} 1 & 1 & 0 & 1 \\ 1 & 0 & 1 & 0 \end{bmatrix},$$

and $E = \hat{E} = 0$. It is now clear, in view of Theorem 4.3.3, that $T(s) = \hat{C}S(s)\delta(s)^{-1}\hat{B}_m = R(s)P(s)^{-1}$, with $R(s) = \hat{C}S(s) = \begin{bmatrix} s+1, & s \\ 1, & 1 \end{bmatrix}$, and $P(s) = \hat{B}_m^{-1}\delta(s) = \begin{bmatrix} s^2 & 0 \\ 0 & s^2 \end{bmatrix}$. Since $T(s)$ is nonsingular, the system can be clearly decoupled via l.s.v.f. in combination with input dynamics; i.e. we can now employ the invertible decoupling algorithm.

<u>Step 1</u>: Since 1 is the greatest common (monic) divisor of both rows of $R(s)$, $R_d(s) = I_2$ and $\hat{R}(s) = R(s)$.

<u>Step 2</u>: We next determine that $P(s)\hat{R}(s)^{-1} = \begin{bmatrix} s^2, & -s^3 \\ -s^2, & s^3+s^2 \end{bmatrix}$ and, therefore, that $g_1 = 2$ and $g_2 = 3$, the respective row degrees of $P_d(s) = \text{diag}[p_{di}(s)]$. Since only the degrees of the $p_{di}(s)$ are fixed, let us (arbitrarily) choose $P_d(s) = \begin{bmatrix} s^2+3s+2, & 0 \\ 0 & s^3+4s^2+6s+4 \end{bmatrix}$, which will yield a decoupled closed loop system with transfer matrix $T_d(s) = P_d(s)^{-1}$, and $n_d(=5)$ poles at $s = -1, -1\pm j, -2$, and -2.

<u>Step 3</u>: We now determine that

$$\overline{P}(s) = P_d(s)\hat{R}(s) = \begin{bmatrix} s^3+4s^2+5s+2, & s^3+3s^2+2s \\ s^3+4s^2+6s+4, & s^3+4s^2+6s+4 \end{bmatrix},$$ and by utilizing

Theorem 7.4.3, we find that

$$\overline{P}(s)P^{-1}(s) = \underbrace{\begin{bmatrix} s+4, & s+3 \\ s+4, & s+4 \end{bmatrix}}_{G(s)} - \underbrace{\begin{bmatrix} -5s-2, & -2s \\ -6s-4, & -6s-4 \end{bmatrix}}_{F(s)=\hat{F}S(s)} \underbrace{\begin{bmatrix} s^2 & 0 \\ 0 & s^2 \end{bmatrix}^{-1}}_{P^{-1}(s)} .$$

If $G^{-1}(s) = \begin{bmatrix} 1 & \frac{-s-3}{s+4} \\ -1 & 1 \end{bmatrix}$ and $F(s)$, or $Fx(t)$ with

$$F = \hat{F}Q = \begin{bmatrix} -5 & -2 & -2 & 0 \\ -6 & 0 & -6 & -4 \end{bmatrix},$$ are now employed in the compensation scheme

depicted in Figure 7.4.14, the resulting closed loop transfer matrix

will equal $T_d(s) = P_d^{-1}(s) = \begin{bmatrix} \frac{1}{s^2+3s+2}, & 0 \\ 0, & \frac{1}{s^3+4s^2+6s+4} \end{bmatrix}$ as desired.

To resolve the four decoupling questions for this particular example, we note that since $|\hat{R}(s)| = 1$, there are no "fixed" poles associated with this decoupled system; i.e. all $\overline{n} = 5 = \partial[|P_d(s)\hat{R}(s)|]$ closed loop decoupled poles can be arbitrarily assigned. Furthermore, an input compensator, $G^{-1}(s)$, of (minimum) order $1 = \overline{n} - n = 5 - 4$ is required to achieve this design, since l.s.v.f. alone will not decouple this system (a fact we will formally establish via Corollary 8.3.27).

The four decoupling questions which were posed earlier in this section have now been resolved in the case of invertible systems via the invertible decoupling algorithm. It should be noted, however, that this decoupling algorithm does not provide a direct resolution of

(DQ4) the question of whether or not a given system can be decoupled via l.s.v.f. alone; i.e. we must first compute the transfer matrix $P(s)\hat{R}(s)^{-1}$ in order to determine the degree of the determinant of $P_d(s)$. A direct resolution of this question does exist, however, as well as a constructive procedure for decoupling a system via l.s.v.f., as we now show.

To begin, let us consider any system with a proper, right invertible, $(p \times m)$ transfer matrix, $T(s)$. We define the $(p \times p)$ diagonal polynomial matrix, $D(s)$, as $\text{diag}[s^{f_i+1}]$, where the degree of each entry, f_i+1, is chosen to be the least integer, greater than or equal to 0, such that the corresponding (i-th) row of $D(s)T(s)$ is finite but nonzero when evaluated at $s = \infty$. The $(p \times m)$ constant matrix B^* is then defined as follows:

$$B^* = \lim_{s \to \infty} D(s)T(s) \qquad 8.3.15$$

Since B^* is a function of $T(s)$ alone, we will sometimes find it convenient to explicitly display this dependence; i.e. in 8.3.15, $B^* = B^*(T(s))$. The matrix B^* now enables us to directly and completely resolve (DQ4) the question of decoupling via l.s.v.f. alone.

THEOREM 8.3.16: A system with a proper, right invertible, $(p \times m)$ transfer matrix, $T(s)$, can be decoupled via l.s.v.f. alone if and only if there exists some constant $(m \times p)$ matrix G such that $B^*(T(s)G)$ is nonsingular.

Proof: We first establish necessity. In particular, we now recall that $T(s)$ can always be factored as the product: $R(s)P(s)^{-1}$, with $R(s)$ and $P(s)$ relatively right prime, $P(s)$ column proper, and $\partial_c[R(s)] \leq \partial_c[P(s)]$. If we now assume that the system can be decoupled via l.s.v.f. alone, then there must exist a l.s.v.f. pair, $\{F(s),G\}$, with $\partial_c[F(s)] < \partial_c[P(s)]$, such that

$$T_{F,G}(s) = R(s)[P(s) - F(s)]^{-1}G = R(s)P_F^{-1}(s)G = T_d(s), \tag{8.3.17}$$

a nonsingular, diagonal transfer matrix. Since $T_{F,G}(s)$ is both diagonal and nonsingular, it is clear that $B^*(T_{F,G}(s))$ must be nonsingular, i.e. $\lim_{s\to\infty} D(s)T_{F,G}(s)$ must be both finite and nonsingular for the appropriate $D(s)$. Since $\partial_c[F(s)] < \partial_c[P(s)]$, it also follows that

$$B^*(T_{F,G}(s)) = \lim_{s\to\infty} D(s)R(s)P_F^{-1}(s)G = \lim_{s\to\infty} D(s)R(s)P^{-1}(s)G = \lim_{s\to\infty} D(s)T(s)G,$$

or that

$$B^*(T_{F,G}(s)) = B^*(T(s)G), \tag{8.3.18}$$

thus establishing necessity.

In order to (constructively) establish sufficiency, we will assume, without loss of generality, that $T(s) = R(s)P^{-1}(s)$ is invertible; i.e. if a G can be found such that $B^*(T(s)G)$ is nonsingular, then the $(p \times p)$ transfer matrix $T(s)G$ will be invertible. We now factor $R(s)$ as the product: $R_d(s)\hat{R}(s)$, where $R_d(s)$ is the (monic) diagonal left divisor of $R(s)$ of maximum row degree (see 8.3.6), and define $P_d(s)$ as any arbitrary, nonsingular, diagonal polynomial matrix of column degree $f_i + 1 + \partial_{ci}[R_d(s)]$ for all $i = 1,2,\ldots,p$. In view of this particular choice for $P_d(s)$, it now follows that

$$B^*(T(s)) = \lim_{s\to\infty} D(s)R(s)P^{-1}(s) = \lim_{s\to\infty} P_d(s)R_d^{-1}(s)R(s)P^{-1}(s) \tag{8.3.19}$$

or, in view of 8.3.6, that

$$B^*(T(s)) = \lim_{s\to\infty} P_d(s)\hat{R}(s)P^{-1}(s) \tag{8.3.20}$$

We now note that $P_d(s)\hat{R}(s)P^{-1}(s)$ is a proper transfer matrix with its $E = \lim_{s\to\infty} P_d(s)\hat{R}(s)P^{-1}(s) = B^*$, which is nonsingular by assumption. Since $P(s)$ is column proper, it thus follows (see Section 4.3) that

$P_d(s)\hat{R}(s)$ must also be column proper and of the same column degree as $P(s)$; i.e.

$$\partial_c[P_d(s)\hat{R}(s)] = \partial_c[P(s)] \tag{8.3.21}$$

In view of the above, it now follows (see Section 7.3) that $P_d(s)\hat{R}(s)$ can be expressed as the product: $G^{-1}P_F(s) = G^{-1}[P(s) - F(s)]$ for some l.s.v.f. pair $\{F(s),G\}$, with G nonsingular and $\partial_c[F(s)] < \partial_c[P(s)]$. To explicitly determine this pair, we note that in view of 8.3.20,

$$I_m = \lim_{s\to\infty} P_d(s)\hat{R}(s)P(s)^{-1}B^{*-1} = \lim_{s\to\infty} P_d(s)\hat{R}(s)[B^*P(s)]^{-1}, \tag{8.3.22}$$

which immediately implies that

$$\Gamma_c[P_d(s)\hat{R}(s)] = \Gamma_c[G^{-1}P_F(s)] = \Gamma_c[B^*P(s)] \tag{8.3.23}$$

Since $\Gamma_c[G^{-1}P_F(s)] = G^{-1}\Gamma_c[P_F(s)]$, and $\Gamma_c[P_F(s)] = \Gamma_c[P(s)]$, it now follows that

$$G = B^{*-1} \tag{8.3.24}$$

Furthermore, since $P_d(s)\hat{R}(s) = G^{-1}P_F(s) = B^*[P(s) - F(s)]$, it also follows that

$$F(s) = P(s) - B^{*-1}P_d(s)\hat{R}(s) \tag{8.3.25}$$

In summary, we have now constructively established (the sufficiency proof of) Theorem 8.3.16 by showing that if $T(s) = R(s)P(s)^{-1} = R_d(s)\hat{R}(s)P(s)^{-1}$ is invertible, with $B^*(T(s))$ nonsingular, then the l.s.v.f. pair $\{F(s),G\}$, as given by 8.3.25 and 8.3.24, produces the decoupled closed loop transfer matrix:

$$\begin{aligned} T_d(s) = R(s)P_F(s)^{-1}G = R(s)[B^{*-1}P_d(s)\hat{R}(s)]^{-1}B^{*-1} &= R_d(s)\hat{R}(s)\hat{R}^{-1}(s)P_d^{-1}(s) \\ &= R_d(s)P_d^{-1}(s) \end{aligned} \tag{8.3.26}$$

As a direct consequence of Theorem 8.3.16, we now have:

COROLLARY 8.3.27: <u>A system with a proper, invertible transfer matrix,</u>

<u>$T(s)$, can be decoupled via l.s.v.f. alone if and only if $B^*(T(s))$ is nonsingular</u>.

It should be noted at this point that, as in the case of the invertible decoupling algorithm, the "l.s.v.f. decoupling algorithm for invertible systems" (the sufficiency proof of Theorem 8.3.16) also enables us to arbitrarily assign all of the closed loop decoupled poles of the system except those "fixed" poles which correspond to the zeros of $|\hat{R}(s)|$ (see O2). We further note that if the given system is right invertible but not invertible, then a G can always be found such that $T(s)G$ is nonsingular. $T(s)G$ can then be factored as the product:

$$T(s)G = R(s)P^{-1}(s)G = \tilde{R}(s)\tilde{P}^{-1}(s), \qquad 8.3.28$$

with $\tilde{R}(s)$ and $\tilde{P}(s)$ relatively right prime, $\tilde{P}(s)$ column proper, and $\partial_c[\tilde{R}(s)] \leq \partial_c[\tilde{P}(s)]$. If we then decouple this system via either l.s.v.f. alone or l.s.v.f. in combination with input dynamics, it can readily be established that all decoupled poles except those which correspond to the zeros of $|P(s)| \div |\tilde{P}(s)|$ and $|\tilde{R}_d^{-1}(s)\tilde{R}(s)|$, where $\tilde{R}_d(s)$ is any diagonal left divisor of $\tilde{R}(s)$ of maximum row degree, can be arbitrarily assigned (see Problem 8-23). Therefore, care must be exercised when selecting an appropriate G prior to decoupling a system via this "column reduction" procedure.

In view of Theorem 8.3.16, decoupling question DQ4 is completely resolved, and we will now attempt to completely resolve the remaining three decoupling questions in the case of noninvertible systems. To begin, we recall that a proper transfer matrix, $T(s)$, can always be expressed as the product: $R(s)P^{-1}(s)$, with $R(s)$ and $P(s)$ relatively right prime, $P(s)$ column proper, and $\partial_c[R(s)] \leq \partial_c[P(s)]$. Furthermore, if $T(s)$ has a right inverse, $\rho[R(s)] = p$ and we can find a unimodular matrix, $U_R(s)$, which reduces $R(s)$ to lower left triangular form; i.e. in view of R3 of Section 5.5, it follows that

$$T(s) = R(s)P(s)^{-1} = R(s)U_R(s)[P(s)U_R(s)]^{-1} = R_R(s)P_R(s)^{-1}, \quad 8.3.29$$

with $R_R(s)$ in lower left triangular form. If $R_{Rp}(s)$ is defined as the nonsingular matrix composed of the first p (non-zero) columns of $R_R(s)$, it follows (see R3 of Section 5.5) that the zeros of the determinant of $R_{Rp}(s)$ represent the invariant poles of any right inverse system. It is therefore natural to expect that when we decouple a right invertible system, certain of these invariant right inverse poles may represent fixed poles of the decoupled system. As a matter of fact, we have essentially shown this to be the case if the given system is invertible.[†] We will now formalize this observation in the general case.

THEOREM 8.3.30: <u>Consider any system with a proper, right invertible transfer matrix,</u> $T(s) = R(s)P(s)^{-1} = R_R(s)P_R(s)^{-1}$, <u>with</u> $R_R(s)$ <u>in lower left triangular form. If</u> $R_d(s)$ <u>denotes a diagonal left divisor of</u> $R_R(s)$ (or $R(s)$, or $R_{Rp}(s)$) <u>of maximum row degree, then the zeros of the determinant of</u> $\hat{R}_{Rp}(s) \triangleq R_d^{-1}(s)R_{Rp}(s)$ <u>represent the only fixed poles of the system decoupled by l.s.v.f. in combination with input dynamics.</u>

<u>Proof</u>: Since the system is assumed to be right invertible, it can be decoupled by combined l.s.v.f. and input dynamics; i.e. there exists an $(m \times p)$ proper transfer matrix, $T_c(s)$, such that

$$\begin{aligned} T(s)T_c(s) = R_R(s)P_R^{-1}(s)T_c(s) &= T_d(s) \\ &= \tilde{R}_d(s)\tilde{P}_d^{-1}(s) \end{aligned} \quad 8.3.31$$

† In the case of invertible systems, $R_{Rp}(s) = R_R(s)$, and $|R_R(s)| = |R(s)U_R(s)| = \alpha|R(s)| = |R_d(s)|\times|\hat{R}(s)|$ for some scalar α (see 8.3.6). Therefore, the zeros of $|\hat{R}(s)|$, which represent the fixed decoupled poles, correspond to certain of the zeros of $|R_R(s)|$, which represent the invariant poles of any right invertible system.

with $\tilde{R}_d(s)$ and $\tilde{P}_d(s)$ relatively right prime and diagonal. If we now define $\hat{T}(s)$ as the $(p \times p)$ matrix consisting of the first p rows of $P_R^{-1}(s)T_c(s)$, it follows that

$$R_R(s)P_R^{-1}(s)T_c(s) = R_{Rp}(s)\hat{T}(s) = T_d(s), \qquad 8.3.32$$

since the final $(m-p)$ columns of $R_R(s)$ are identically zero. In view of 8.3.32, it is thus clear that

$$T(s) = R_{Rp}^{-1}(s)T_d(s) = R_{Rp}^{-1}(s)\tilde{R}_d(s)\tilde{P}_d^{-1}(s), \qquad 8.3.33$$

where any "uncancelled" zeros of $|R_{Rp}(s)|$ must represent poles of the decoupled system. Since $T_d(s) = \tilde{R}_d(s)\tilde{P}_d^{-1}(s)$ is diagonal however, any common left divisor of $R_{Rp}(s)$ and $\tilde{R}_d(s)$ can be diagonalized. Therefore, only the zeros of $|R_d(s)|$ can be cancelled by the zeros of $T_d(s)$, which directly implies that the zeros of $|\hat{R}_{Rp}(s)|$ must represent fixed poles of the decoupled system.

To show that the zeros of $|\hat{R}_{Rp}(s)|$ represent the <u>only</u> fixed poles, we will now constructively establish that we can decouple the given right invertible system, with transfer matrix, $T(s) = R(s)P(s)^{-1} = R_R(s)P_R^{-1}(s)$, and simultaneously assign all decoupled poles except those which correspond to the zeros of $|\hat{R}_{Rp}(s)|$. Our proof will be based on a direct employment of the invertible decoupling algorithm once the given right invertible system has been "judiciously extended" to invertible form. In particular, let us now define $R_e(s)$ as the $(m \times m)$ nonsingular matrix obtained by appending to $R(s)$, $(m-p)$ linearly independent rows such that the following three conditions are satisfied:

(i) $\partial_c[R_e(s)] \leq \partial_c[P(s)]$,

(ii) $R_e(s)U_R(s)$ is a lower left triangular matrix,

and (iii) the product of each added $(p+i)$-<u>th</u> row of $R_e(s)$ and the corresponding $(p+i)$-<u>th</u> column of $U_R(s)$ is any desired, (stable) polynomial of largest possible degree, consistent with the other two conditions. We merely note here that the added rows of $R_e(s)$ can always be chosen to satisfy these three conditions due to the fact that $U_R(s)$ is a unimodular matrix. Furthermore, given any desired, fixed degree (but otherwise arbitrary) polynomial, $r_{p+i}(s)$, as the $(p+i,p+i)$-th entry of $R_e(s)U_R(s)$, the choice of the added $(p+i)$-th row of $R_e(s)$ which will produce $r_{p+i}(s)$ is generally not unique (see Example 8.3.35 and Problem 8-24).

In view of the above choice for $R_e(s)$, we note that the extended system with transfer matrix, $T_e(s) = R_e^{-1}(s)P(s)$ will be invertible and, therefore, that it can be decoupled via the invertible decoupling algorithm. Furthermore, since

$$\alpha|R_e(s)| = |R_e(s)U_R(s)| = |R_{Rp}(s)| \times \prod_{i=1}^{m-p} r_{p+i}(s)$$

$$= |R_d(s)| \times |\hat{R}_{Rp}(s)| \times \prod_{i=1}^{m-p} r_{p+i}(s), \qquad 8.3.34$$

with $\prod_{i=1}^{m-p} r_{p+i}(s)$ arbitrary, and $\left[\begin{array}{c|c} R_d(s) & 0 \\ \hline 0 & I_{m-p} \end{array}\right]$ a diagonal left divisor of $R_e(s)$, it is clear that the zeros of $|\hat{R}_{Rp}(s)|$ represent the only fixed poles of the extended system when decoupled via the invertible decoupling algorithm. We finally note that since $T(s) = R^{-1}(s)P(s)$, rather than $R_e^{-1}(s)P(s)$, the invertible decoupling algorithm will actually produce a $(p \times m)$ closed loop transfer matrix, $R^{-1}(s)[G(s)P(s) - F(s)]$, with the first p columns decoupled and the final columns identically zero. As a final step in decoupling the right invertible system, we can therefore employ the additional external gain matrix $L(s) = L = \left[\begin{array}{c} I_p \\ \hline 0 \end{array}\right]$ in order to achieve a diagonal decoupled system. Theorem 8.3.30 is thus constructively established.

In view of this theorem, it is now clear that decoupling

questions DQ1 and DQ2 are completely resolved; i.e. we have now shown that it is possible, via the synthesis procedure of Section 7.4, to arbitrarily assign all of the decoupled closed loop poles of a given right invertible system except those fixed poles which correspond to the zeros of $|\hat{R}_{Rp}(s)|$. As we noted earlier, however, even the fixed poles which correspond to the zeros of $|\hat{R}_{Rp}(s)|$ can be altered while decoupling if one is willing to employ a "two stage" synthesis procedure which involves a substantial increase in the order of the compensated system (see Problem 8-26).

The only question which therefore remains partially unresolved is DQ3; i.e. what is the minimum order of any dynamic compensator required for decoupling? Unfortunately, there is no complete resolution of DQ3 at present in the case of right invertible, but noninvertible systems. We should note, however, that the constructive proof of Theorem 8.3.30, which has just been presented, does produce a maximization of the rank of $B^*(T_e(s))$; i.e. in view of condition (iii), each added row of $R_e(s)$ increases $\rho[B^*]$ by one. It therefore follows that if $\rho[B^*(T(s))] = p$, then $\rho[B^*(T_c(s))] = m$, and the given system could then be decoupled via l.s.v.f. alone, with no additional dynamic precompensation (see Example 8.3.35 and Problem 8-24). In any case, the fact that $\rho[B^*(T_e(s))]$ is maximized via the constructive proof of Theorem 8.3.30 tends to minimize the order of any dynamic precompensation which may be required for decoupling, and thus results in a minimal order dynamic compensator in a large number of cases, and certainly in those cases where $\rho[B^*(T(s))] = p$.

EXAMPLE 8.3.35: To illustrate the decoupling algorithm employed to establish sufficiency of Theorem 8.3.30, let us now consider a right invertible system with the transfer matrix

$$T(s) = R(s)P(s)^{-1} = \begin{bmatrix} s+1, & 1, & s \\ s, & s-2, & s-1 \end{bmatrix} \begin{bmatrix} s, & 0, & 0 \\ 0, & s+2, & 0 \\ 0, & 0, & s+1 \end{bmatrix}^{-1}, \text{ with } R(s) \text{ and}$$

$P(s)$ relatively right prime, $P(s)$ column proper, and $\partial_c[R(s)] \leq \partial_c[P(s)]$. We first determine that the unimodular matrix,

$$U_R(s) = \begin{bmatrix} -s+1, & s, & s^2-3s+1 \\ 0, & 0, & -1 \\ s, & -s-1, & -s^2+2s+2 \end{bmatrix}$$ reduces $R(s)$ to the lower left

triangular matrix $R_R(s) = \begin{bmatrix} 1 & 0 & 0 \\ 0 & 1 & 0 \end{bmatrix}$. It is therefore clear that

$R_d(s) = I_2$ and that $R_{Rp}(s) = \hat{R}_{Rp}(s) = I_2$ as well. In view of Theorem 8.3.30 therefore, there are no fixed poles associated with this system when decoupled via l.s.v.f. and input dynamics. We further note that since $B^*(T(s)) = \begin{bmatrix} 1 & 0 & 1 \\ 1 & 1 & 1 \end{bmatrix}$ in this example, it should be possible to decouple this system via l.s.v.f. alone and simultaneously assign all $n(=3)$ decoupled poles. More specifically, if we denote the third row of $R_e(s)$ as: $[as+b, cs+d, es+f]$, so that condition (i) holds, the product of this vector and the third column of $U_R(s)$; i.e. $r_{33}(s)$ is given by: $r_{33}(s) = (a-e)s^3 + (-3a+b+2e-f)s^2 + (a-3b-c+2e+2f)s + b - d + 2f$, a polynomial of (largest possible) degree 3, which clearly can be arbitrarily assigned by appropriate choice of third row elements of $R_e(s)$. For example, if we specify that $r_{33}(s) = (s+1)(s^2+2s+2) = s^3+3s^2+4s+2$, then the particular (nonunique) choice: $a = 2$, $b = 3$, $c = -17$, $d = -7$, $e = 1$, $f = -4$ for the scalars,

or $R_e(s) = \begin{bmatrix} s+1, & 1, & s \\ s, & s-2, & s-1 \\ 2s+3, & -17s-7, & s-4 \end{bmatrix}$, will produce an invertible extended system with transfer matrix $T_e(s) = R_e(s)P(s)^{-1}$, with

$$B^*(T_e(s)) = \begin{bmatrix} 1 & 0 & 1 \\ 1 & 1 & 1 \\ 2 & -17 & 1 \end{bmatrix},$$ a nonsingular matrix. In view of Theorem 8.3.16, therefore, this extended system can now be decoupled via l.s.v.f. alone. In particular, since $R_e(s) = \hat{R}_e(s)$ is column proper and of the same column degree as $P(s)$, $R(s)\hat{R}_e^{-1}(s)$ is a proper transfer matrix and, therefore, $P_{de}(s) = I_3$. In view of 8.3.24 and 8.3.25, it now follows that the l.s.v.f. pair

$$\{F(s),G\} = \left\{\begin{bmatrix} 15 & 59 & 21 \\ 1 & 5 & 1 \\ -16 & -60 & -20 \end{bmatrix}, \begin{bmatrix} -18 & 17 & 1 \\ -1 & 1 & 0 \\ 19 & -17 & 1 \end{bmatrix}\right\} =$$

$= \{P(s)-B^{*^{-1}}P_{de}(s)\hat{R}_e(s),\ B^{*^{-1}}\}$ decouples the extended system, or that

$F(s) = F$ and $\hat{G} = B^{*^{-1}}L = GL = \begin{bmatrix} -18, & 17 \\ -1, & 1 \\ 19, & -17 \end{bmatrix}$ decouples the given right invertible system, and produces the (stable) closed loop transfer matrix: $T_d(s) = R(s)[P(s) - F(s)]^{-1}GL = I_2$, a system of apparent order zero, with three stable, cancelled (unobservable--see Problem 8-22) poles at $s = -1\pm j$ and -1.

8.4 STATIC DECOUPLING

The results presented in the previous section focused on what we termed dynamic decoupling, or simply decoupling; i.e. the attainment of a system characterized by a diagonal, nonsingular transfer matrix. We note that if the initial conditions on the internal state of a dynamically decoupled system are zero, each (i-th) output, $y_i(t)$, is "driven" by the corresponding (i-th) input, $u_i(t)$, and by that input only, regardless of the particular dynamical behavior of $u_i(t)$. We further recall that, in general, linear state variable feedback and often input dynamics in combination with linear state variable feedback must be employed in order to achieve dynamic decoupling.

In certain applications, however, it may be sufficient to attain only "static decoupling", which loosely can be defined at this point as the condition under which a step change in the static, steady-state level of each (i-th) input is reflected by a corresponding change in the steady-state level of the corresponding (i-th) output and only that output.[†] In this section, necessary and sufficient conditions for statically decoupling a given system in either state or differential operator form while arbitrarily altering the (n) closed loop poles are presented, along with constructive procedures for determining appropriate state feedback control laws. Extensions of the theory to include linear output feedback are also discussed.

In order to formalize the somewhat heuristic definition of static decoupling given earlier and establish the main result of this section, we first define the class of systems which will be considered as those whose input/output dynamical behavior can be characterized by a $(p \times m)$ proper rational transfer matrix, $T(s)$; i.e. we will consider only linear, time-invariant, multivariable systems with

$$y(s) = T(s)u(s), \qquad 8.4.1$$

where $y(s)$ and $u(s)$ represent the Laplace transforms of the p-dimensional output and m-dimensional input of the system respectively. It might be recalled here that the relation represented by 8.4.1 assumes zero initial conditions on the internal state of the system which need not be explicitly known, although $T(s)$ can readily be derived from any known state-space representation of the form (see Section 3.2):

---

† In a process control system, for example, it is often desirable for a set point change in each input to "appropriately alter" the steady-state level of each corresponding output without affecting the steady-state levels of the other outputs [B6][S8].

$$\dot{x}(t) = Ax(t) + Bu(t); \; y(t) = Cx(t) + Eu(t), \qquad 8.4.2$$

In particular, if we repeat the procedure outlined in Section 4.2 and solve 8.4.2 for $y(s)$, the Laplace transform of $y(t)$, in terms of $u(t)$, the Laplace transform of $u(t)$, we readily determine that $y(s) = [C(sI-A)^{-1}B + E]u(s)$, or, in view of 8.4.1, that

$$T(s) = C(sI-A)^{-1}B + E \qquad 8.4.3$$

We will assume for convenience that any given state-space system is both controllable and observable and furthermore, that B and C both have full rank (m and p respectively).

As an alternative to the state-space representation, the dynamical behavior of the system may also be specified via a controllable and observable differential operator representation of the form (see Section 5.2):

$$P(D)z(t) = u(t); \quad y(t) = R(D)z(t), \qquad 8.4.4$$

where $P(D)$ and $R(D)$ are relatively right prime polynomial matrices in the differential operator $D = \frac{d}{dt}$ with $P(D)$ nonsingular and column proper and $R(D)$ of column degree less than or equal to the column degree of $P(D)$; i.e.

$$\partial_c[R(D)] \leq \partial_c[P(D)] \qquad 8.4.5$$

As in the state-space case, the transfer matrix of a differential operator representation can also be readily obtained; i.e. if we assume zero initial conditions on $z(t)$ and all of its derivatives, then for all practical purposes D can simply be replaced by the Laplace operator s, and by solving for $y(s)$ in terms of $u(s)$, we determine that $y(s) = R(s)P(s)^{-1}u(s)$ or, in view of 8.4.1, that

$$T(s) = R(s)P(s)^{-1} \qquad 8.4.6$$

We recall at this point that in view of the results presented in Section 4.3, a differential operator representation of the form 8.4.4, which is completely equivalent to a given state-space representation of the form 8.4.2, can directly be determined by first reducing the state-space representation to controllable companion form and then applying the controllable version of the structure theorem. Conversely, as shown in Section 5.2, an equivalent state-space representation of the form 8.4.2 can be obtained from a given differential operator representation by essentially reversing the procedure.

Irrespective of the particular derivation of $T(s)$, we now consider the case where each element of the input is restricted to be a step input of magnitude k_i (a change in the static d.c. level) applied, for convenience, at $t = 0$; i.e. if the Laplace transform of $u_i(t)$, namely $u_i(s)$ is equal to $\frac{k_i}{s}$, and all of the poles of $T(s)$ lie in the stable half-plane $Re(s) < 0$ (if the system is asymptotically stable), then by the well known final value theorem,

$$\lim_{t\to\infty} y(t) = \lim_{s\to 0} sT(s)u(s) = \lim_{s\to 0} T(s)K, \qquad 8.4.7$$

with K equal to the transpose of $[k_1, k_2, \ldots, k_m]$.

DEFINITION 8.4.8: In view of the above, we will say that a system with the proper, rational transfer matrix $T(s)$ is statically decoupled if and only if it is asymptotically stable and the $\lim_{s\to 0} T(s)$ is diagonal, nonsingular, and finite (its elements are finite). The reader can verify that this formal definition of a statically decoupled system is completely consistent with and actually clarifies the somewhat heuristic definition given earlier. Furthermore, static decoupling (thus defined) implies a square transfer matrix,

$T(s) = \left[\frac{n_{ij}(s)}{d_{ij}(s)}\right]$, in which s divides $n_{ij}(s)$ when $i \neq j$ but not when $i = j$.

In general, of course, we would not expect that a given system would satisfy these requirements and therefore, some form of compensation would generally be required in order to achieve a statically decoupled design, provided such a design could be achieved. Since it is perhaps the most common form of linear system compensation, we will initially focus our attention on the employment of linear state variable feedback in order to achieve static decoupling.

In terms of a state-space representation such as 8.4.2, we recall that linear state variable feedback (l.s.v.f.) is defined as the control law (see Section 6.2):

$$u(t) = Fx(t) + Gv(t), \qquad 8.4.9$$

where $v(t)$ is a q-vector called the external input, and F and G are real matrices of the appropriate dimensions. We now recall from the results given in Section 7.3 that for every l.s.v.f. control law 8.4.9 there is a completely equivalent, corresponding feedback control law associated with the differential operator representation 8.4.4 (see Problem 7-21), namely

$$u(t) = F(D)z(t) + Gv(t), \qquad 8.4.10$$

where $F(D)$ is an $(m \times m)$ polynomial matrix in the differential operator D which satisfies the relation: $\partial_c[F(D)] < \partial_c[P(D)]$.

In view of the above, we now note that under l.s.v.f., the closed loop transfer matrix of the state-space system is given by

$$T_{F,G}(s) = (C+EF)(sI-A-BF)^{-1}BG + EG, \qquad 8.4.11$$

while the closed loop transfer matrix of the equivalent differential operator system is given by

$$T_{F,G}(s) = R(s)[P(s) - F(s)]^{-1}G = R(s)P_F^{-1}(s)G, \qquad 8.4.12$$

with $P_F(s) = P(s) - F(s)$. In view of the equivalence of the two representations and the feedback control laws, it also follows that these two transfer matrices are equal, or that

$$T_{F,G}(s) = (C+EF)(sI-A-BF)^{-1}BG + EG = R(s)P_F^{-1}(s)G, \quad 8.4.13$$

with the closed loop poles of the system given by the zeros of either $|sI-A-BF|$ or $|P_F(s)|$; i.e.

$$|sI-A-BF| = \alpha|P_F(s)| \quad 8.4.14$$

for some real scalar α. It is important to note that since we assumed controllability of the given system, the feedback matrix F (or $F(D)$) can be used to completely and arbitrarily assign the (n) zeros of $|sI-A-BF|$ (or $|P_F(s)|$) via the algorithm given in Section 6.3 (or Section 7.3), which represent the closed loop poles of the system. In view of these preliminary observations, we can now state and constructively establish the main result of this section.

THEOREM 8.4.15: <u>The state-space system 8.4.2 (the differential operator system 8.4.4) can be statically decoupled via the l.s.v.f. control law 8.4.9 (8.4.10) if and only if the rank of the composite matrix $\begin{bmatrix} A & B \\ C & E \end{bmatrix}$ is equal to $n+p$ (the rank of $R(0)$ is equal to p)</u>; i.e. <u>static decoupling via l.s.v.f. is possible if and only if</u>

$$\rho\begin{bmatrix} A & B \\ C & E \end{bmatrix} = n+p \quad (\text{or } \rho[R(0)] = p) \quad 8.4.15$$

<u>Proof</u>: We first establish the portion of the theorem statement in parenthesis;

Necessity: We assume that the differential operator system 8.4.4 can be statically decoupled via the l.s.v.f. control law 8.4.10. Therefore, there exists an $F(s)$ such that the zeros of $|P_F(s)|$ lie in the half-plane $Re(s) < 0$. This assumption clearly implies that there can be no pole-zero cancellations between $R(s)$ and $P_F(s)$ at $s = 0$ and therefore that the

$$\lim_{s\to 0} R(s)P_F^{-1}(s)G = R(0)[P(0) - F(0)]^{-1}G = \Lambda, \qquad 8.4.16$$

a nonsingular diagonal matrix of dimension p. Since $R(0)$ is a $p \times m$ matrix, the rank of $R(0)$ must equal p, thus establishing necessity.

Sufficiency: We assume that the rank of $R(0)$ is equal to p and let $F(s)$ be any polynomial matrix which (completely and arbitrarily) positions the (n) zeros of $|P_F(s)|$ in the half-plane $Re(s) < 0$. Since $|P_F(0)|$ is equal to a nonzero scalar which represents the product of the poles, $P_F^{-1}(0)$ exists and G can be chosen so that

$$R(0)P_F^{-1}(0)G = \Lambda, \qquad 8.4.17$$

for any arbitrary $(p \times p)$ matrix, including one which is both diagonal and nonsingular. More specifically, since the rank of $R(0)P_F^{-1}(0)$ is equal to p, an appropriate G can be found by first inverting the matrix formed by the first p linearly independent columns of $R(0)P_F^{-1}(0)$ and then inserting $(m-p)$ zero rows in the resulting inverse corresponding to those numbered (dependent) columns of $R(0)P_F^{-1}(0)$ which were eliminated in the search for its first p independent columns. This choice for G will yield a $\Lambda = I_p$, and any other desired Λ can be obtained by an appropriate postmultiplication.

To establish that part of the theorem statement which is not in parenthesis, we need only establish the equivalence between the rank conditions on $\begin{bmatrix} A & B \\ C & E \end{bmatrix}$ and $R(0)$. However, in view of the structure theorem (Section 4.3), the equivalence between these rank conditions is immediately apparent when the state-space representation is reduced to controllable companion form, and the details associated with the formal establishment of this equivalence relationship will therefore be left as exercise for the reader (see Problem 8-33). It should be noted, however, that if the rank of $\begin{bmatrix} A & B \\ C & E \end{bmatrix}$ does equal $n+p$, then the state-space system 8.4.2 can be statically decoupled via l.s.v.f. by first selecting F to (completely and arbitrarily) position all n closed loop poles in the half-plane $\mathrm{Re}(s) < 0$, and then choosing G in the same way outlined in the differential operator case, namely so that

$$T_{F,G}(0) = [(C+EF)(-A-BF)^{-1}B + E]G = \Lambda, \qquad 8.4.18$$

a nonsingular diagonal matrix; i.e. as in the case of a differential operator system, the $(p \times m)$ matrix $(C+EF)(-A-BF)^{-1}B + E$ will have rank p, thus permitting an appropriate choice for G.

A number of remakrs are now in order.

R 1: The condition that the rank of $R(0)$ equal p (or equivalently, that the rank of $\begin{bmatrix} A & B \\ C & E \end{bmatrix}$ equal $n+p$) implies at least as many inputs as outputs and represents an extension, to the multivariable case, of the condition that the numerator polynomial of a scalar transfer function cannot be divisible by s if the output of the system is to reflect a new steady-state level in response to a step change in the input.

R 2: It should be emphasized that if a given system can be statically decoupled via l.s.v.f., then the state feedback gain matrix F (or $F(D)$) is used solely for arbitrary pole placement in the half-plane $\mathrm{Re}(s) < 0$, while G actually plays the "dominant role" in statically decoupling the system. Since the practical implementation of a l.s.v.f. control law often involves dynamic compensation, in the form of a Luenberger observer, or an excessive number of sensors to measure the entire state, it is often of practical importance to consider alternative static decoupling methods before resorting to linear state feedback. For example, if the given open loop system is asymptotically stable to begin with, static decoupling can be achieved via G (input combination) alone. More realistically, if some degree of pole shifting is desired, linear output feedback (l.o.f.), which has been defined as the control law (see Section 8.2):

$$u(t) = Hy(t) + Gv(t), \qquad 8.4.19$$

can be employed to statically decouple the system, provided the system is asymptotically stabilizable via l.o.f.; i.e. provided H can be chosen so that the (n) zeros of $|P(s) - HR(s)| = |P_H(s)|$ lie in the half-plane $Re(s) < 0$. Although the question of pole placement (or simply stabilization) of a linear multivariable system via l.o.f. has yet to be fully resolved, as noted in Section 8.2, the employment of a differential operator representation often permits a "brute force" solution to the question of pole placement via l.o.f., for purposes such as static decoupling, as the following example illustrates:

EXAMPLE 8.4.20: Let us consider the system in the differential operator form 8.4.4 with $P(D) = \begin{bmatrix} D+1, & 0 \\ -D, & D-2 \end{bmatrix}$ and $R(D) = \begin{bmatrix} 1, & D+3 \\ 1, & D+2 \end{bmatrix}$. Since $R(0) = \begin{bmatrix} 1 & 3 \\ 1 & 2 \end{bmatrix}$ is nonsingular, this system can be statically decoupled via l.s.v.f.; e.g. the feedback pair $F(D) = \begin{bmatrix} 0 & 0 \\ 0 & -4 \end{bmatrix}$ and $G = P_F^{-1}(0)R(0) = \begin{bmatrix} -2 & 3 \\ 2 & -2 \end{bmatrix}$ statically decouples the system and, in view of 8.4.12, yields the closed loop transfer matrix:

$T_{F,G}(s) = \dfrac{\begin{bmatrix} 2 & s(s+4) \\ 0 & (s+1)(s+2) \end{bmatrix}}{(s+1)(s+2)}$, with poles at $s = -1$ and -2. In this example, however, it is not necessary to employ l.s.v.f. since l.o.f. can be used to statically decouple the system while arbitrarily assigning the $(n=2)$ closed loop poles. In particular, if $H = \begin{bmatrix} h_1 & h_2 \\ h_3 & h_4 \end{bmatrix}$ then $|P(s)-HR(s)| = |P_H(s)| = (1-h_1-h_2-h_3-h_4)s^2 + (-1-h_1-h_3)s + (h_2h_3-h_1h_4-5h_3-4h_4-2)$, a polynomial whose roots represent the closed loop poles of the system. Suppose we now require that the closed loop poles of the system lie at $s = -1\pm j$, or, equivalently, that $|P_H(s)| = s^2+2s+2$. In view of the expression given above for $|P_H(s)|$ in terms of the elements of H, this latter requirement clearly implies that (i) $h_1+h_2+h_3+h_4 = 0$, (ii) $h_1+h_3 = -3$, and (iii) $h_2h_3-h_1h_4-5h_3-4h_4-2=4$. In general, of course, neither the existence nor the uniqueness of solutions to such a set of nonlinear algebraic equations is readily apparent. However, in this particular case a "brute force" approach can be employed to achieve a solution. To illustrate, let us arbitrarily set $h_1 = 0$ which, in view of (ii), implies that $h_3 = -3$. Equation (i) then implies that $h_4 = 3-h_2$ and if these values are now substituted into (iii) we readily determine that $h_2 = 1$ and therefore, that $h_4 = 2$. $H = \begin{bmatrix} 0 & 1 \\ -3 & 2 \end{bmatrix}$ thus represents a l.o.f. gain matrix which yields the desired closed loop poles. In view of 8.4.17, G is now given by $P_H(0)^{-1}R(0)\begin{bmatrix} \lambda_1 & 0 \\ 0 & \lambda_2 \end{bmatrix}$ or

$$\begin{bmatrix} 0 & -2 \\ 1 & 3 \end{bmatrix}\begin{bmatrix} -2 & 3 \\ 1 & -1 \end{bmatrix}\begin{bmatrix} \lambda_1 & 0 \\ 0 & \lambda_2 \end{bmatrix} = \begin{bmatrix} -2\lambda_1 & 2\lambda_2 \\ \lambda_1 & 0 \end{bmatrix},$$

resulting in the (l.o.f.) closed loop transfer matrix:

$$T_{H,G}(s) = \frac{\begin{bmatrix} -\lambda_1 s^2-4\lambda_1 s+2\lambda_1, & 2\lambda_2 s^2+8\lambda_2 s \\ -\lambda_1 s^2-3\lambda_1 s \quad , & 2\lambda_2 s^2+6\lambda_2 s+2\lambda_2 \end{bmatrix}}{s^2+2s+2}.$$

R 3: The conditions noted in our main result can be employed to rather easily determine whether or not a given system can be statically decoupled provided the dynamics are expressed in either the state-space form 8.4.2 or the differential operator form 8.4.4. However, if the dynamics are initially expressed via a given proper, rational transfer matrix, $T(s)$, it may also be possible to directly ascertain whether or not the system can be statically decoupled without first deriving a state-space or differential operator realization of $T(s)$. In particular, we now recall that in view of Theorem 5.4.1, $T(s)$ can always be factored as the product of $R(s)$ and $P(s)^{-1}$, with $R(s)$ and $P(s)$ relatively right prime and $P(s)$ column proper. If none of the poles of $T(s)$ lie at $s = 0$, it therefore follows that the

$$\lim_{s\to 0} T(s) = T(0) = R(0)P(0)^{-1}, \qquad 8.4.21$$

and consequently that the rank of $T(0)$ is equal to the rank of $R(0)$. In view of Theorem 8.4.15, it is thus clear that a system described by a rational, proper transfer matrix, $T(s)$, can be statically decoupled (via l.s.v.f. or perhaps some other less ambitious scheme) if $T(0)$ is finite and has rank p.

R 4: The notions of dynamic decoupling and static decoupling are not necessarily compatible; e.g. as the reader can readily verify, the dynamically decoupled system with $T(s) = \begin{bmatrix} \frac{s}{s+1} & 0 \\ 0 & \frac{1}{s+2} \end{bmatrix}$ cannot be statically decoupled via l.s.v.f. while the statically decoupled system with $T(s) = \begin{bmatrix} 1 & 0 \\ \frac{s}{s+1} & \frac{1}{s+1} \end{bmatrix}$ cannot be dynamically decoupled via l.s.v.f. In view of the first transfer matrix given above, it might also be noted at this point that if a system cannot be statically decoupled via l.s.v.f. alone, it still can be statically decoupled via input dynamics in combination with l.s.v.f. (see Section 7.4) provided it has a right inverse (see Section 5.5). We will not pursue this point any further, however, since such a design always requires "exact" pole-zero cancellation at the marginally stable point, $s = 0$.

8.5 EXACT MODEL MATCHING

In this section we will consider another question of considerable importance in the synthesis of linear multivariable systems, namely the question of exact model matching; i.e. given a system with the $(p \times m)$ proper transfer matrix, $T(s)$, and a model (system) with the $(p \times q)$ proper transfer matrix, $T_m(s)$, does there exist a compensation scheme which modifies the given system so that its compensated transfer matrix equals (exactly matches) $T_m(s)$? This question is motivated, in large part, by the notion of a "model following" control system, or a control scheme which employs the error between the output of the given system and that of some model system to modify the dynamical behavior of the given system so that its performance approximates, in some sense, that of the model (see Figure 8.5.1). For example, it may be desirable to design a stability augmentation system for a new aircraft so that its performance, or "handling qualities", is similar to the performance of a model aircraft which possesses certain desirable "handling qualities". One might therefore begin the design process by first asking whether or not an exact match of the transfer matrices of the two systems is possible. It should be realized that, in general, an exact match of the transfer matrices of two arbitrary

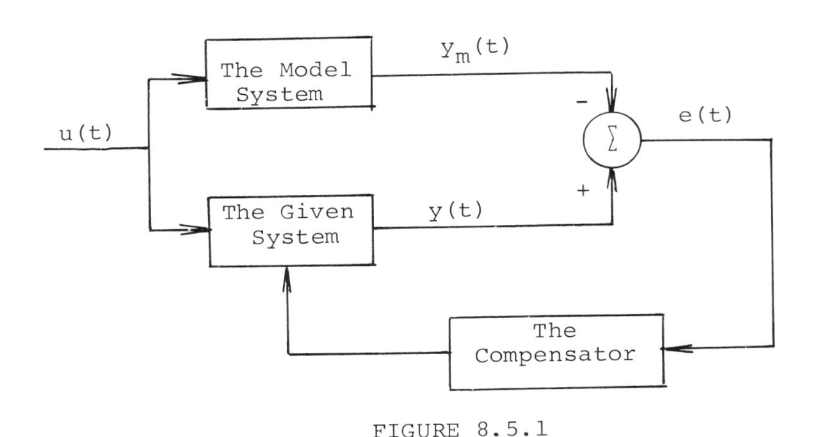

FIGURE 8.5.1

A MODEL FOLLOWING CONTROL SYSTEM

systems is rather difficult to achieve, regardless of the compensator employed, and we will therefore restrict our discussions in this section to exact model matching via the most powerful synthesis technique which we have discussed, namely linear state variable feedback (l.s.v.f.) in combination with input dynamics (see Section 7.4). We further observe that even if an exact match of the two transfer matrices cannot be achieved, an investigation of the question of exact model matching can often provide considerable insight into the structural properties of both the system and the model, and a practical compromise design, based on an appropriate modification of the model, can often result.

In view of the results presented in Section 7.4, we can now readily resolve the question of exact model matching via combined l.s.v.f. and input dynamics.

THEOREM 8.5.2: <u>Consider a given system with the $(p \times m)$ proper, rational transfer matrix, $T(s)$, and a model system with the $(p \times q)$</u>

proper, rational transfer matrix, $T_m(s)$. There exists a compensation scheme which employs input dynamics in combination with l.s.v.f., as defined in Section 7.4, such that the compensated transfer matrix of the given system is equal to $T_m(s)$ if and only if

$$T(s)T_c(s) = T_m(s) \qquad 8.5.3$$

for some proper $(m \times q)$ rational matrix, $T_c(s)$.

Proof: Sufficiency is easily established by noting that if 8.5.3 holds, then $T_c(s)$ can be realized by the (synthesis algorithm of Section 7.4) l.s.v.f./dynamic feedforward compensation scheme depicted in Figure 7.4.14.

In order to establish necessity, we assume that there does exist a compensation scheme, as depicted in Figure 7.4.14, which yields the desired model transfer matrix, $T_m(s)$. Since any proper $(p \times m)$ rational transfer matrix, $T(s)$, can be expressed as the product: $R(s)P^{-1}(s)$, with $R(s)$ and $P(s)$ relatively right prime and $P(s)$ column proper, it follows in light of Figure 7.4.14 that

$$G(s)P(s)z(s) = L(s)v(s) + F(s)z(s); \; y(s) = R(s)z(s), \qquad 8.5.4$$

with $\partial_c[F(s)] < \partial_c[P(s)] \leq \partial_c[G(s)P(s)]$. Since $G^{-1}(s)L(s)$ is a proper transfer matrix by assumption, $P(s)[G(s)P(s)-F(s)]^{-1}L(s)$ is also proper; i.e. since $\partial_c[F(s)] < \partial_c[G(s)P(s)]$,

$$\lim_{s\to\infty} P(s)[G(s)P(s) - F(s)]^{-1}L(s) = \lim_{s\to\infty} G^{-1}(s)L(s), \qquad 8.5.5$$

a constant finite matrix. In view of 8.5.4, it is thus clear that $y(s) = R(s)[G(s)P(s)-F(s)]^{-1}L(s)v(s) = R(s)P^{-1}(s)P(s)[G(s)P(s)-F(s)]^{-1}L(s)v(s)$, or if we define $P(s)[G(s)P(s) - F(s)]^{-1}L(s)$ as $T_c(s)$, that

$$y(s) = T(s)T_c(s)v(s), \qquad 8.5.6$$

thus establishing necessity and completing the proof of the theorem.

It should now be noted that in view of Theorem 8.5.2, _the question of exact model matching via combined l.s.v.f. and input dynamics reduces to a standard problem in linear algebra, namely the determination of all (proper) solutions_, $T_c(s)$, _to the rational matrix equation 8.5.3_. The actual determination of the solutions to 8.5.3 is relatively straightforward as we will now illustrate by first considering the case when the given system is left invertible (see Theorem 5.5.3). In particular, if $\rho[T(s)] = m$ then, provided it exists, a $T_c(s)$ which satisfies 8.5.3 is unique and can readily be found; i.e. if we let $\hat{T}(s)$ denote the $(m \times m)$ nonsingular matrix consisting of the first m linearly independent rows of $T(s)$ and $\hat{T}_m(s)$ denote the corresponding (m) rows of $T_m(s)$, then in view of 8.5.3, it is clear that $T_c(s)$ must satisfy the relation:

$$T_c(s) = \hat{T}^{-1}(s)\hat{T}_m(s) \qquad 8.5.7$$

In view of this observation, we can therefore state:

COROLLARY 8.5.8: _If_ $\rho[T(s)] = m$, _exact model matching via combined l.s.v.f. and input dynamic is possible if and only if_

$$T(s)\hat{T}^{-1}(s)\hat{T}_m(s) = T_m(s), \qquad 8.5.9$$

and $\hat{T}^{-1}(s)\hat{T}_m(s) = T_c(s)$ _is a proper transfer matrix._

We should emphasize at this point that if the conditions of Corollary 8.5.8 are satisfied, then $T_c(s)$ is uniquely specified by 8.5.7, and the problem of realizing $T_c(s)$ via l.s.v.f. in combination with input dynamics can then be resolved via the synthesis algorithm of Section 7.4. In this light, we now recall (see R4 of Section 7.4) that the uncancelled poles of $P(s)T_c^{-1}(s)$ represent the poles of the compensated system. Care should therefore be exercised when defining $T_m(s)$ to insure that all of the uncancelled poles of

$P(s)\hat{T}^{-1}(s)\hat{T}_m^{-1}(s)$ lie in the stable half-plane, $Re(s) < 0$.

EXAMPLE 8.5.10: To illustrate the employment of the synthesis algorithm of Section 7.4 to resolve the question of exact model matching via Corollary 8.5.8, we now consider the following example which is due to Moore and Silverman [M9]. In particular, if $T(s) = \begin{bmatrix} \frac{s+2}{s+1} & \frac{s+3}{s+2} \\ \frac{1}{s+1} & 0 \end{bmatrix}$ represents the open loop transfer matrix of some system, can l.s.v.f. in combination with dynamic compensation be used to achieve the desired (model) transfer matrix, $T_m(s) = \begin{bmatrix} \frac{s+1}{s+4} \\ \frac{-2}{(s+2)(s+4)} \end{bmatrix}$? To resolve this question, we must first determine whether or not there exists a proper $T_c(s)$ such that $T(s)T_c(s) = T_m(s)$. Since $T(s)$ is nonsingular in this example, $T_c(s)$ exists and is unique; i.e. $\hat{T}(s) = T(s)$, $\hat{T}_m(s) = T_m(s)$, and in view of 8.5.7, $T_c(s) = T^{-1}(s)T_m(s) = \begin{bmatrix} \frac{-2(s+1)}{(s+2)(s+4)} \\ \frac{s+2}{s+4} \end{bmatrix}$, a proper transfer matrix. $T(s)$ can now be factored as the product:

$$R(s)P^{-1}(s) = \begin{bmatrix} s+2, & s+3 \\ 1, & 0 \end{bmatrix} \begin{bmatrix} s+1, & 0 \\ 0, & s+2 \end{bmatrix}^{-1}$$

, where $R(s)$ and $P(s)$ are relatively right prime, and $P(s)$ is column proper. We next note that the product: $P^{-1}(s)T_c(s) = \begin{bmatrix} \frac{-2}{(s+2)(s+4)} \\ \frac{1}{s+4} \end{bmatrix}$, and can be expressed as

$$P_c^{-1}(s)L_c(s) = \begin{bmatrix} 0, & s+4 \\ s+2, & 2 \end{bmatrix}^{-1} \begin{bmatrix} 1 \\ 0 \end{bmatrix}$$

with $P_c(s)$ and $L_c(s)$ relatively left prime and $L_c(s)$ column proper and upper right triangular.

Since $P(s)P_c^{-1}(s)$ is a proper transfer matrix, we set $\bar{P}(s) = P_c(s)$ and $L(s) = L_c(s)$ and now employ Theorem 7.4.3 to determine $G(s)$ and $F(s)$. For this example, we readily find that $G(s) = \begin{bmatrix} 0 & 1 \\ 1 & 0 \end{bmatrix}$ and $F(s) = \begin{bmatrix} 0 & -2 \\ -1 & -2 \end{bmatrix}$, which clearly implies that exact model matching is possible via l.s.v.f. alone without the need for any dynamic pre-compensation. Moore and Silverman reach the same conclusion using an alternative approach.

If $\rho[T(s)] = p < m$, the question of exact model matching via combined l.s.v.f. and input dynamics becomes considerably more difficult to resolve,[†] since solutions, $T_c(s)$, to 8.5.3 (if any exist) will not generally be unique. In these cases, however, we can classify all $T_c(s)$ which satisfy 8.5.3 by first reducing $T(s)$ to lower left triangular form via an appropriate unimodular matrix $U_R(s)$. In particular, if we first express $T(s)$ as $\frac{N(s)}{\Delta_m(s)}$, where $N(s)$ is a polynomial matrix and $\Delta_m(s)$ is the least common denominator of all of the elements of $T(s)$ (see Problem 8-14), then in view of Theorem 2.5.11 we can find a unimodular matrix, $U_R(s)$, such that

$T(s)U_R(s) = \frac{N(s)U_R(s)}{\Delta_m(s)}$ is in lower left triangular form. Since $\rho[T(s)] = p$, the first p columns of $T(s)U_R(s)$ will constitute a nonsingular matrix and its remaining columns will be identically zero. Therefore, in light of 8.5.3,

$$T(s)U_R(s)U_R^{-1}(s)T_c(s) = T_m(s), \qquad 8.5.11$$

and the first p rows of $\hat{T}_c(s) = U_R^{-1}(s)T_c(s)$ will be uniquely specified, while the remaining $(m-p)$ rows of $\hat{T}_c(s)$ will be completely

---

[†] The procedure which will be outlined for the case when $\rho[T(s)] = p < m$ can be "appropriately modified" to include the case when $\rho[T(s)]$ is less than both m and p -- see Problem 8-39.

arbitrary. The complete class of $T_c(s)$ which satisfy 8.5.3 is then given by:

$$T_c(s) = U_R(s)\hat{T}_c(s) \qquad 8.5.12$$

Once we have classified all appropriate $T_c(s)$ via this procedure, we can determine $P(s)T_c^{-1}(s) = P(s)U_R(s)\hat{T}_c^{-1}(s)$, with only the final p rows of $\hat{T}_c(s)$ specified, and then attempt to choose the final $(m-p)$ rows of $\hat{T}_c(s)$ so that (a) $T_c^{-1}(s)$ is proper, (b) the uncancelled poles of $P(s)U_R(s)\hat{T}_c(s)$ lie in the stable half-plane, $Re(s) < 0$, and (c) any realization of $P(s)U_R(s)\hat{T}_c^{-1}(s)$ is of "lowest possible" order. These objectives will then directly and correspondingly imply that (a) a solution exists, (b) it is stable, and (c) it involves input dynamics of "relatively low" dynamic order. While we cannot guarantee that this procedure is the most efficient one to use, it does represent a rather direct and systematic method for resolving the model matching question whenever $\rho[T(s)] = p < m$.

EXAMPLE 8.5.13: To illustrate the above procedure for exact model matching when $\rho[T(s)] = p < m$, let us consider a given system with the transfer matrix $T(s) = \begin{bmatrix} \frac{1}{s+1} & 0 & \frac{-s}{s+1} \\ \frac{1}{s} & -2 & -1 \end{bmatrix} = R(s)P^{-1}(s)$, with

$R(s) = \begin{bmatrix} s & 0 & 0 \\ s+1 & -2 & 0 \end{bmatrix}$ and $P(s) = \begin{bmatrix} 0 & 0 & s \\ 0 & 1 & 0 \\ -s-1 & 0 & 1 \end{bmatrix}$. To determine whether or not l.s.v.f. in combination with input dynamics can be employed to compensate this system and obtain the model transfer function:

$T_m(s) = \begin{bmatrix} \frac{s}{s+3} & \frac{-s}{s+3} \\ \frac{s+1}{s+3} & \frac{-3s-7}{s+3} \end{bmatrix}$, we first reduce $T(s)$ to lower left triangular form via the unimodular matrix $U_R(s) = \begin{bmatrix} 1 & 0 & s \\ 0 & 1 & 0 \\ 0 & 0 & 1 \end{bmatrix}$; i.e.

$T(s)U_R(s) = \begin{bmatrix} \frac{1}{s+1} & 0 & 0 \\ \frac{1}{s} & -2 & 0 \end{bmatrix}$. Since $T(s)U_R(s)\hat{T}_c(s) = T_m(s)$, the first two rows of $\hat{T}_c(s)$ are uniquely specified; i.e. $\begin{bmatrix} \frac{1}{s+1} & 0 & 0 \\ \frac{1}{s} & -2 & 0 \end{bmatrix}\hat{T}_c(s) =$ $\begin{bmatrix} \frac{s}{s+3} & \frac{-s}{s+3} \\ \frac{s+1}{s+3} & \frac{-3s-7}{s+3} \end{bmatrix}$, or $\hat{T}_c(s) = \begin{bmatrix} \frac{s^2+s}{s+3} & \frac{-s^2-s}{s+3} \\ 0 & 1 \\ \hat{t}_{c31}(s) & \hat{t}_{c32}(s) \end{bmatrix}$, with $\hat{t}_{c31}(s)$ and $\hat{t}_{c32}(s)$ arbitrary. $T_c(s)$ is then given by 8.5.12; i.e.

$$T_c(s) = U_R(s)\hat{T}_c(s) = \begin{bmatrix} \frac{s^2+s}{s+3} + s\hat{t}_{c31}(s), & \frac{-s^2-s}{s+3} + s\hat{t}_{c32}(s) \\ 0 \quad , & 1 \\ \hat{t}_{c31}(s) \quad , & \hat{t}_{c32}(s) \end{bmatrix}, \text{ and}$$

$$P(s)T_c^{-1}(s) = \begin{bmatrix} \frac{1}{s+3} \quad , & \frac{-1}{s+3} \\ 0 \quad , & 1 \\ \frac{s+1}{s+3} + \hat{t}_{c31}(s), & \frac{-s-1}{s+3} + \hat{t}_{c32}(s) \end{bmatrix}.$$

Since we require (a) that $T_c(s)$ be proper, $\hat{t}_{c31}(s)$ must equal -1 + any arbitrary proper transfer function. Since we also require (b) that the uncancelled poles of $P(s)T_c^{-1}(s)$ be stable and (c) that any realization of $P(s)U_R(s)\hat{T}_c^{-1}(s)$ be of lowest possible order, we will set $\hat{t}_{c31}(s) = -1$ alone; i.e. there is no need to introduce any added dynamics in $T_c(s)$. Using similar arguments, we also set $\hat{t}_{c32}(s) = 1$. $T_c(s)$ is therefore equal to $\begin{bmatrix} \frac{-2s}{s+3} & \frac{2s}{s+3} \\ 0 & 1 \\ -1 & 1 \end{bmatrix}$ and, in view of the synthesis algorithm of Section 7.4,

$$P(s)T_c^{-1}(s) = \begin{bmatrix} \frac{1}{s+3} & \frac{-1}{s+3} \\ 0 & 1 \\ -2 & 2 \end{bmatrix},$$ which we now express as the product: $P_c^{-1}(s)L_c(s)$, with $P_c(s) = \begin{bmatrix} s+3 & 0 & 0 \\ 0 & 1 & 0 \\ 2 & 0 & 1 \end{bmatrix}$ and $L_c(s) = \begin{bmatrix} 1 & -1 \\ 0 & 1 \\ 0 & 0 \end{bmatrix}$.

Since $P(s)P_c^{-1}(s)$ is not proper, we set $G_L(s) = \begin{bmatrix} 1 & 0 & 0 \\ 0 & 1 & 0 \\ 0 & 0 & s+2 \end{bmatrix}$ so that $P(s)P_c^{-1}(s)G_L^{-1}(s) = P(s)[G_L(s)P_c(s)]^{-1} = P(s)\bar{P}^{-1}(s)$ will be proper. If we now complete the steps involved with the general synthesis algorithm, we find that the model can be exactly matched via the l.s.v.f. pair $\{F(s),G\} = \left\{ \begin{bmatrix} 2 & 0 & -2 \\ 0 & 0 & 0 \\ 2 & 0 & 1 \end{bmatrix}, \begin{bmatrix} -2 & 2 \\ 0 & 1 \\ -1 & 1 \end{bmatrix} \right\}$ alone, without the need for any dynamic precompensation. In other words, for this l.s.v.f. pair, $T_{F,G}(s) = R(s)[P(s) - F(s)]^{-1}G = T_m(s)$.

Before we conclude this section, we might again note that the stability of a model matched system is determined by the location of the uncancelled poles of $P(s)T_c^{-1}(s)$, which represent the poles of the compensated system. Furthermore, while there is no direct general resolution to the question of exact model matching via l.s.v.f. alone, as in the case of decoupling, we can readily resolve this question whenever $\rho[T(s)] = \rho[T_m(s)] = m$ (see Problem 8-40).

8.6 CONCLUDING REMARKS AND REFERENCES

We have now presented a number of design algorithms for the dynamical compensation of linear, time-invariant, multivariable systems. The emphasis in this chapter has been on frequency domain methods for two primary reasons, namely (i) the design objectives which we have considered can best be described in the frequency domain in terms of a desired transfer matrix or its determinant and (ii) the general synthesis algorithm developed in Section 7.4 appears to be a most natural method for resolving all of the design objectives considered in this chapter. It should be noted, however, that most of the earlier initial investigations which dealt with multivariable system design goals such as arbitrary pole placement, decoupling, and exact model matching employed state-space methods and, in particular, linear state variable feedback or a combination of input dynamics and l.s.v.f. In view of the fact that the general synthesis algorithm is completely analogous to time domain compensation via l.s.v.f. in combination with input dynamics, it follows that virtually all of these earlier results are subsummed by the synthesis procedures which have been presented in this chapter.

Our initial attention was focused, in Section 8.2, on the problem of complete and arbitrary closed loop pole placement, a question which we considered and partially resolved earlier (in Section 6.3) via the time domain notion of l.s.v.f. We demonstrated one "brute force" method for determining the effect which linear output feedback has on the closed loop poles of a system and indicated why this important question remains largely unresolved, although a result due to Davison [D4] does provide new insight into this problem. We then turned our attention to the development of more systematic pole placement schemes which require "relatively low" order dynamic compensation, when compared to schemes such as the l.s.v.f. pole placement algorithm of Section 6.3. In particular, we first showed that a single input linear system does not require as large an increase in observer dynamics for complete and arbitrary pole placement via the l.s.v.f. scheme of Section 6.3 as a multivariable system does, since only a single linear function of the state need be reconstructed. In view of this observation, we then demonstrated that any system with a cyclic transfer matrix can be reduced to a single input system while retaining complete controllability. This reduction then enabled us to construct a feedback compensator of total order $\nu-1$ (or $\mu-1$ by duality) to achieve any desired closed loop pole configuration. This fact, which was first noted in [B4] and [L5], was later "improved" by Brasch and Pearson [B5] who, employing a time domain approach, showed that l.o.f. can always be used to obtain a cyclic transfer matrix. Chen and Hsu [C5] then gave a frequency domain interpretation of these results similar to the development employed here for the case when $k(s) = 0$ (the schemes depicted in Figures 8.2.21 and 8.2.25).

The question of (dynamic) decoupling was then considered and, for the most part, resolved in Section 8.3. Historically, the question of diagonalizing the closed loop transfer matrix of a linear system via l.s.v.f. alone was first formulated, but only partially resolved, by Morgan [M4] in 1963. It was not until 1967, however, that Falb and Wolovich [F1] discovered (Theorem 8.3.16) that the nonsingularity of the matrix B^*, first defined in the time domain (see Problem 8-20), represents both a necessary and sufficient condition for l.s.v.f. decoupling. Gilbert [G3] later showed that B^* can be directly obtained from the transfer matrix of the system as well (via 8.3.15) and indicated the possibility of decoupling a larger class of system than can be decoupled via l.s.v.f. alone. Morse and Wonham [M5] then defined this larger class of systems and outlined a new and alternative geometric procedure for decoupling this class through combined l.s.v.f. and input dynamics. We have now rather easily established the fact (Theorem 8.3.2) that a system can be decoupled by combined l.s.v.f. and input dynamics if and only if it is right invertible, and we have also outlined a direct, constructive procedure for achieving a stable decoupled design "whenever possible". It might be noted that the decoupling question has intrigued control engineers for more than two decades [M7], and there has recently been a rather substantial resurgent interest in various aspects of the decoupling question; e.g. on triangular decoupling (see Problem 8-29) [M6][S6], decoupling in the presence of plant and measurement noise [S7], decoupling nonlinear systems [N2] and time-varying systems [F3][P6], the observability of decoupled systems [M8][C6], the development of a computer program for decoupling [G4], and static decoupling, which was considered as a separate question in Section 8.4.

In particular, in Section 8.4 we precisely defined a notion of static decoupling and constructively established (via Theorem 8.4.15) that the ability to statically decouple a given linear, time-invariant, multivariable system characterized by a proper transfer matrix depends on certain matrix rank conditions; i.e. in the case of state space or differential operator representation, the rank condition is both necessary and sufficient, while a sufficient condition can readily be applied if the dynamical behavior of the system is initially expressed in terms of a proper transfer matrix. Entire state feedback was used as a means of completely and arbitrarily assigning all (n) closed loop poles of the system, thereby insuring its asymptotic stability, although we constructively illustrated that other less ambitious forms of compensation, such as linear output feedback, can also be employed provided asymptotic stability is assured.

The model matching question which was formulated and, for the most part, resolved here in Section 8.5 was only recently introduced by the author [W2][W10], motivated in part by the decoupling question (which might be viewed as "qualitative model matching") and also by recent model following investigations [E1][W11]. It should be noted that only l.s.v.f. is employed in [W2] and [W10] for exact model matching, although Moore and Silverman [M9] have also resolved this question for the larger class of systems considered here through the combined use of l.s.v.f. and input dynamics. As we illustrated, the question of exact model matching via combined l.s.v.f. and input dynamics reduces, via the synthesis algorithm of Section 7.4, to a standard problem in linear algebra, an observation which formed the basis of our resolution to this question in Section 8.5. Although the astute reader will no doubt realize that exact model matching is rather difficult to achieve, given an arbitrary model to be matched, the investigation of this question via the procedure outlined here can provide the system designer with new insight into the system he is dealing with, hopefully enabling him to make appropriate modifications to either his system or the model in order to arrive at some compromise design.

In view of the results presented in Chapter 8, the reader now has at his disposal a variety of synthesis procedures; e.g. linear output feedback, linear state variable feedback, and linear state variable feedback combined with input dynamics, which can be employed to resolve a variety of synthesis questions; e.g. complete and arbitrary pole placement, dynamic decoupling, static decoupling, and exact model matching. Although these synthesis procedures and questions certainly do not exhaust all possibilities, they do represent a rather thorough compilation of some of the more important and recent techniques which have been devised for resolving a variety of design questions, and future investigations can now hopefully build on the material which has been presented.

PROBLEMS - CHAPTER 8

8-1 Find a l.o.f. gain matrix H in Example 8.2.6 which produces closed loop poles at $s = -1, -2,$ and $-1 \pm j$. Is H unique?

8-2 If $T(s) = \begin{bmatrix} \frac{s+1}{s^2} \\ \frac{-2}{s^2} \end{bmatrix}$, show that l.o.f. can be used to arbitrarily position all $n(=2)$ closed loop poles, and find an H which places the poles at $s = -2 \pm j$. Is H unique?

8-3 Find an H (if one exists) which positions all of the l.o.f. closed loop poles of the system with open loop transfer matrix

$$T(s) = \begin{bmatrix} \frac{s-1}{s}, & \frac{1}{s+1} \\ 2, & \frac{-s}{s+1} \\ \frac{1}{2s}, & 0 \end{bmatrix} \quad \text{at} \quad s = -1.$$

8-4 Determine a triple $\{H(s), k(s), q(s)\}$ such that the single input, pole placement compensator depicted in Figure 8.2.12 places the $n(=4)$ closed loop poles of the system with the open loop transfer matrix $T(s) = \begin{bmatrix} \frac{1}{s(s+1)} \\ \frac{1}{s+2} \\ \frac{2s}{(s+1)(s+3)} \end{bmatrix}$ at $s = -1, -2,$ and $2 \pm j$.

8-5 Consider the controllable and observable state-space system, $\{A,B,C,E\}$, with $A = B = I_2$, $C = \begin{bmatrix} 1, & 2 \\ -1, & 0 \end{bmatrix}$ and $E = \begin{bmatrix} 0 & 0 \\ 0 & 0 \end{bmatrix}$. Show that the system is not single input controllable. Determine $T(s)$ for this system and verify that it is not cyclic; i.e. that an "appropriate transfer vector" cannot be obtained by linearly combining the $m(=2)$ inputs.

8-6 Show that $T(s) = \begin{bmatrix} \frac{1}{s+1} & \frac{2}{s+2} \\ \frac{s}{s-3} & \frac{1}{s-3} \end{bmatrix}$ is cyclic and find an input gain vector g which reduces $T(s)$ to a single input controllable system with the same characteristic polynomial, $p(s)$, as $T(s)$. What is $p(s)$? Using this g, find an $H(s)$ and $q(s)$ which yield a closed loop system, as depicted in Figure 8.2.21, with all $n+\nu-1$ closed loop poles at $s = -1$. What does the integer $n+\nu-1$ equal in this problem?

8-7 Using the same $T(s)$ as in the previous problem (8-6), find a $\tilde{g}$ which produces a single output system having the same characteristic polynomial as $T(s)$ and, using this $\tilde{g}$, find an $\tilde{H}(s)$ and $\tilde{q}(s)$ which yield a closed loop system, as depicted in Figure 8.2.25, with all $n+\mu-1$ closed loop poles at $s = -2$. What does the integer $n+\mu-1$ equal in this problem?

8-8 Show that $\hat{R}(s)$ and $p(s)$ in 8.2.7 are relatively right prime if and only if 1 is a greatest common divisor of $\hat{r}_1(s), \hat{r}_2(s), \ldots, \hat{r}_p(s)$, and $p(s)$.

8-9 Show that if H is chosen so that $\partial[|P_H(s)|] = \partial[|P(s)|] = n$ in 8.2.3, then linear output feedback can be interpreted as a special case of linear state variable feedback. Explicitly determine $F(s)$ and G in terms of $P(s)$ and $P_H(s)$.

8-10 Can a system with the transfer matrix given in Problem 8-3 be decoupled? What about a system with the transfer matrix given in Problem 8-6? Explain your reasoning and, for all affirmative answers, determine whether or not l.s.v.f. alone can be employed for decoupling.

8-11 Show that the state-space system defined in Problem 8-5 can be (dynamically) decoupled by l.o.f. alone and find an output gain matrix H which decouples the system and simultaneously positions as many decoupled poles as possible at $s = -2$. How many decoupled poles can be positioned at $s = -2$?

8-12 Show that the system employed in Example 7.3.43 can be decoupled by l.s.v.f. alone and find a feedback pair, $\{F(s), G\}$, which places the decoupled $n(=3)$ closed loop poles at $s = -2$ and $-1 \pm j$.

8-13 It can be shown [Y1][C5] that if $\{A, B, C, E\}$ is a minimal realization of some $(p \times m)$ proper transfer matrix $T(s)$, with all (rational) elements in prime form, then the minimal polynomial of A, $\Delta_m(s)$, (as defined in Section 2.4) is equal to the least common denominator of all of the elements of $T(s)$, while the characteristic polynomial of A, $\Delta(s)$, (as defined in Section 2.4) is equal to the least common denominator of all of the minors of $T(s)$. Verify this fact in the following cases:

(a) $T(s) = \begin{bmatrix} \frac{1}{s} & \frac{2}{s} \\ 0 & -\frac{1}{s} \end{bmatrix}$ (b) $T(s) = \begin{bmatrix} \frac{1}{s+1} & \frac{2}{s+1} \\ \frac{-1}{s+1} & \frac{-2}{s+1} \end{bmatrix}$

(c) $T(s) = \begin{bmatrix} \frac{1}{s+1} & 3 \\ \frac{2}{s} & \frac{-2}{s+1} \\ 0 & \frac{1}{s} \end{bmatrix}$ (d) $A = \begin{bmatrix} 0 & 1 \\ 0 & 0 \end{bmatrix}$, $B = C = I_2$, and $E = 0$

(e) $A = B = C = E = I$.

8-14 In view of Problem 8-13, we will define $\Delta_m(s)$, the least common denominator of all of the elements of $T(s)$, as the <u>minimal</u>

polynomial of $T(s)$ and $\Delta(s)$, the least common denominator of all of the minors of $T(s)$, as the characteristic polynomial of $T(s)$. In view of these definitions, verify that $T(s)$ is cyclic if and only if its minimal and characteristic polynomials are equal.

8-15 Show that A is cyclic if all of its (n) eigenvalues are distinct. Similarly, show that a proper transfer matrix, $T(s) = R(s)P^{-1}(s)$, is cyclic if all of the zeros of $|P(s)|$ are distinct. Why don't we require that $R(s)$ and $P(s)$ be relatively right prime?

8-16 It can be shown [B5] that if the proper $(p \times m)$ transfer matrix $T(s) = R(s)P^{-1}(s)$, with $R(s)$ and $P(s)$ relatively right prime, is not cyclic it can always be made cyclic via l.o.f.; i.e. there always exists an H such that $T_H(s) = R(s)[P(s)-HR(s)]^{-1} = R(s)P_H^{-1}(s)$, with $\partial[|P_H(s)|] = [|P(s)|] = n$, is cyclic and furthermore, "almost any" H will work. Show that this result allows us to remove the restriction that $T(s)$ be cyclic in virtually all of the scalar pole placement schemes presented in Section 8.2 and therefore, in view of Theorem 8.2.30, that all $(n+q)$ closed loop poles of any system characterized by a rational proper transfer matrix can be completely and arbitrarily assigned via a compensator of total order $q = \min(\nu-1, \mu-1)$.

8-17 Show that the observability (controllability) index of a system characterized by a cyclic transfer matrix, $T(s)$, is identical to the observability (controllability) index of a single input (output) controllable (observable) system with the transfer matrix $T(s)g$ ($\tilde{g}T(s)$).

8-18 By using the single input controllable companion form, show that if A is cyclic, then A^T is also cyclic which, in turn, implies that whenever A is cyclic, the observable pair, $\{A,C\}$ is <u>single output observable</u>; i.e. there exists a vector $\tilde{g}$ such that the single output pair $\{A,\tilde{g}C\}$ is observable.

8-19 Consider a system with the $(m \times m)$ proper, nonsingular transfer matrix $T(s) = R(s)P(s)^{-1}$, where $R(s)$ and $P(s)$ are relatively right prime and $P(s)$ is column proper. Discuss the implications of compensating this open loop system by the feedback scheme depicted below if $|R(s)|$ is a Hurwitz polynomial and

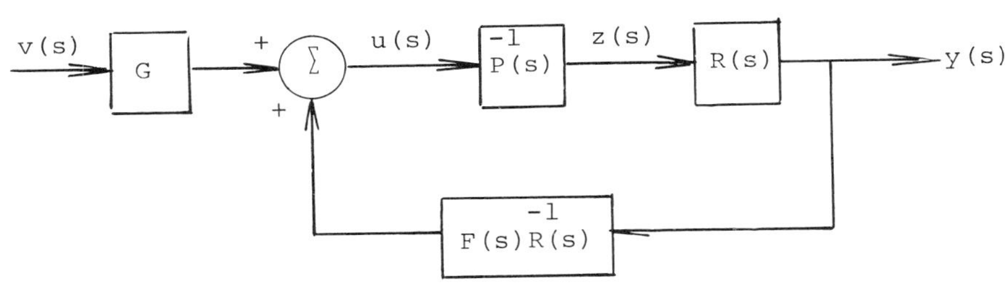

$F(s)R(s)^{-1}$ is proper. In particular, what is the closed loop transfer matrix (between $y(s)$ and $v(s)$) of this system? What is the actual order of the closed loop system? the apparent order? Under what conditions (if any) will the system be stable? Discuss the controllability and observability aspects of the closed loop system. How does this compensation scheme compare to the l.s.v.f. equivalent scheme outlined in Section 7.3? (Hint: Let $\hat{z}(s)$ represent the partial state of the system defined by the feedback transfer matrix $F(s)R(s)^{-1}$ and complete the following frequency domain description:

$$\left[\begin{array}{c|c} \quad & \quad \\ \hline \quad & \quad \end{array}\right] \left[\begin{array}{c} z(s) \\ \hline \hat{z}(s) \end{array}\right] = \left[\begin{array}{c} \quad \\ \hline \quad \end{array}\right] v(s); \quad y(s) = \left[\begin{array}{c|c} \quad & \quad \end{array}\right] \left[\begin{array}{c} z(s) \\ \hline \hat{z}(s) \end{array}\right]).$$

8-20 By using Faddeev's algorithm (Problem 4-16), it can be shown that the constant matrix B^*, defined in the frequency domain via 8.3.15, admits a direct time domain interpretation. In particular, show that if we consider any state-space representation $\{A,B,C,E\}$, and let E_i and C_i represent the i-th rows of E and C respectively, then each (i-th) row of B^*, B_i^*, is given by either E_i (if $E_i \neq 0$) or $C_iA^{f_i}B$ (if $E_i = 0$), where f_i is the least integer, greater than or equal to zero, for which $C_iA^{f_i}B \neq 0$; i.e. show that B^*, as defined by 8.3.15, is also given by the following:

$$B^* = \begin{bmatrix} E_1 \text{ or } C_1A^{f_1}B \text{ (if } E_1 = 0) \\ E_2 \text{ or } C_2A^{f_2}B \text{ (if } E_2 = 0) \\ \vdots \\ E_p \text{ or } C_pA^{f_p}B \text{ (if } E_p = 0) \end{bmatrix}$$

Note: The matrix B^* was first defined in the time domain and employed to resolve the question of decoupling via l.s.v.f. alone. The interested reader is referred to [F1] for additional details.

8-21 Show that if $\rho[B^*(T(s))] = p$, then $T(s)$ is right invertible. Verify, by example, that the converse does not hold.

8-22 Show that the fixed poles (Theorem 8.3.30) of a system decoupled by l.s.v.f. and input dynamics represent unobservable modes of the decoupled system.

8-23 In view of 8.3.28, show that all of the decoupled poles of an invertible system with the open loop transfer matrix $T(s)G = R(s)P(s)^{-1}G = \tilde{R}(s)\tilde{P}(s)^{-1}$ can be arbitrarily assigned via the synthesis procedure of Section 7.4 except those which correspond to the zeros of $|P(s)| \div |\tilde{P}(s)|$ (and represent uncontrollable modes of the decoupled system) and those fixed poles which correspond to the zeros of $|\tilde{R}_d^{-1}(s)\tilde{R}(s)|$ (and represent unobservable modes of the decoupled system--see Problem 8-22), where $\tilde{R}_d(s)$ is any diagonal left divisor of $\tilde{R}(s)$ of maximum row degree.

8-24 Show that a system with the transfer matrix

$$T(s) = \begin{bmatrix} \frac{1}{s+1}, & \frac{-s-1}{s+2}, & \frac{s+2}{s}, & \frac{s^2+2s+2}{(s-1)^2} \\ 0 & \frac{1}{s+2}, & \frac{-2}{s}, & \frac{-2s-1}{(s-1)^2} \end{bmatrix} = R(s)P^{-1}(s), \text{ with}$$

$$R(s) = \begin{bmatrix} 1, & -s-1, & s+2, & s^2+2s+2 \\ 0, & 1, & -2, & -2s-1 \end{bmatrix} \quad \text{and}$$

$$P(s) = \begin{bmatrix} s+1 & 0 & 0 & 0 \\ 0 & s+2 & 0 & 0 \\ 0 & 0 & s & 0 \\ 0 & 0 & 0 & (s-1)^2 \end{bmatrix}$$

can be decoupled via l.s.v.f. alone and, furthermore, that all five decoupled poles can be arbitrarily assigned. (Hint: Use the constructive proof of Theorem 8.3.30 to show that $R(s)$ can be extended such that $B^*(R_e(s)P^{-1}(s))$ is nonsingular and $|R_e(s)|$ is any arbitrary polynomial of degree four.)

8-25 Consider a system with the transfer matrix:

$$T(s) = \begin{bmatrix} \frac{s+1}{s} & \frac{1}{s+1} & \frac{1}{s+2} \\ \frac{1}{s} & 0 & 0 \end{bmatrix} = \underbrace{\begin{bmatrix} s+1, & 1, & 1 \\ 1, & 0, & 0 \end{bmatrix}}_{R(s)} \underbrace{\begin{bmatrix} s & 0 & 0 \\ 0 & s+1 & 0 \\ 0 & 0 & s+2 \end{bmatrix}^{-1}}_{P^{-1}(s)}$$

Verify that this system cannot be decoupled via l.s.v.f. alone, although it can be decoupled via l.s.v.f. in combination with an input dynamic compensator of (minimum) order one. Furthermore, show that all four decoupled poles can be arbitrarily assigned. If $R(s) = \begin{bmatrix} 1 & 0 & s \\ 0 & 1 & s \end{bmatrix}$ instead of $\begin{bmatrix} s+1, & 1, & 1 \\ 1, & 0, & 0 \end{bmatrix}$ in the above, with $P(s)$ unaltered, show that the resulting system can be decoupled and simultaneously stabilized via l.s.v.f. alone.

8-26 Consider an invertible system with the $(m \times m)$ proper transfer matrix $T(s) = R(s)P^{-1}(s)$, with $P(s)$ column proper and $R(s)$ and $P(s)$ relatively right prime. If $\partial_{ci}[P(s)] = \mu$ for all $i = 1,2,\ldots,m$, a condition which can easily be achieved via input dynamics, it is clear that $G^{-1}P_F(s) = \text{diag}[(s+\lambda)^{\mu}]$ can be obtained via l.s.v.f. alone and, therefore, that (for this choice of $\{F(s),G\}$) $T_{F,G}(s) = R(s)P_F^{-1}(s)G = \dfrac{R(s)}{(s+\lambda)^{\mu}}$. If we now let $D(s)$ represent a stable, diagonal polynomial matrix which satisfies the condition: $\partial_c[D(s)] \geq \partial_c[R^+(s)]$, and place a system with the proper transfer matrix $T_c(s) = R^+(s)D(s)^{-1}$ in series with the given system, it follows that the transfer matrix of the series compensated system will be decoupled and given by $\dfrac{|R(s)|}{(s+\lambda)^{\mu}} D(s)^{-1}$. In view of the above, show that if $T(s)$ is invertible, then complete and arbitrary pole placement is possible while decoupling provided one is willing to use this "two stage" synthesis procedure which involves a "substantial increase" in the order of the closed loop system. Extend this observation to include systems which are right invertible.

8-27 Consider the system with the open loop transfer matrix

$$T(s) = \begin{bmatrix} \frac{s+1}{s^2}, & \frac{1}{s-1} \\ \frac{s-1}{s^2}, & \frac{1}{s-1} \end{bmatrix}.$$ Show that this system can be decoupled.

Can this system be decoupled by l.s.v.f. alone? Can the system be stabilized under decoupling compensation; i.e. what are the fixed poles of the decoupled system. Design a compensator which decouples this system and arbitrarily positions all of those decoupled poles which are not fixed at $s = -1$.

8-28 Show that if it is possible to obtain $\mathrm{diag}\left[\frac{r_i(s)}{p_i(s)}\right]$ as the proper, decoupled, closed (under l.s.v.f.) loop transfer matrix of some given system then $\mathrm{diag}\left[\frac{1}{\hat{p}_i(s)}\right]$ can also be achieved via l.s.v.f., where only the degree of each $\hat{p}_i(s)$ is fixed.

8-29 A linear dynamical system will be called <u>triangularly decoupled</u> [M6] if and only if its transfer matrix is lower left triangular. In view of the constructive proof of Theorem 8.3.30, show that any right invertible system with a $(p \times m)$ proper transfer matrix, $T(s) = R(s)P(s)^{-1} = R(s)U_R(s)[P(s)U_R(s)]^{-1} = R_R(s)P_R(s)^{-1}$ (see 8.3.29) can be triangularly decoupled via l.s.v.f. alone and, furthermore, that all $n = \partial[|P(s)|] = \partial[|P_F(s)|]$ triangularly decoupled closed loop poles can be arbitrarily assigned; i.e. in view of 8.3.29 and Theorem 8.3.30, show that if it is possible to choose $G\,P_F(s)^{-1}$ such that:

(i) $\partial_c[G\,P_F(s)^{-1}] = \partial_c[P(s)]$,

(ii) $G\,P_F(s)^{-1}U_R(s)$ is lower left triangular,

and (iii) the product of each (i-<u>th</u>) row of $G\,P_F(s)^{-1}$ and the corresponding column of $U_R(s)$ is any desired (stable) polynomial of "largest possible" degree, consistent with the other two conditions, then $T_{F,G}(s) = R(s)P_F^{-1}(s)G$ and the closed loop system will be triangularly decoupled.

If $T(s) = R(s)P^{-1}(s)$, with $R(s) = \begin{bmatrix} 1 & s^2 \\ 1 & s^2+2s \end{bmatrix}$ and

$P(s) = \begin{bmatrix} s & s^2 \\ s & 1 \end{bmatrix}$, verify that $U_R(s) = \begin{bmatrix} 1 & -s^2 \\ 0 & 1 \end{bmatrix}$ "reduces" $R(s)$ to lower left triangular form, and that $G^{-1}P_F(s) = \begin{bmatrix} 1 & s^2 \\ -s & as^2+bs+c \end{bmatrix}$ satisfies all of the (3) conditions noted above, with $|G^{-1}P_F(s)| = s^3+as^2+bs+c$. Explicitly determine G and $F(s)$ and verify that $T_{F,G}(s)$ represents the transfer matrix of a triangularly decoupled system.

8-30 If $R(s)$ and $P(s)$ are relatively right prime, show that for any constant matrix H, $R(s)$ and $P_H(s) = P(s) - HR(s)$ are also relatively right prime.

8-31 Given a system with the $(p \times m)$ proper rational transfer matrix, $T(s)$, verify by example that $T(0)$ need not be finite in order to statically decouple the system.

8-32 Directly establish the fact that an invertible system with the proper transfer matrix, $T(s)$, can be statically decoupled via l.s.v.f. if and only if the $\lim_{s\to 0} T^{-1}(s)$ is finite.

8-33 By assuming a state space representation of the form 8.4.2 in controllable companion form and an equivalent differential operator representation of the form 8.4.4 with $R(D) = CS(D) + EB_m^{-1}\delta(D)$ and $P(D) = B_m^{-1}\delta(D)$ (see 4.3.8 and 4.3.9), establish the equivalence between $\rho\begin{bmatrix} A & B \\ C & E \end{bmatrix} = n+p$ and $\rho[R(0)] = p$.

(Hint: First consider the case when $E \equiv 0$).

8-34 Consider a system with the proper invertible transfer matrix: $T(s) = R(s)P(s)^{-1}$, with $R(s)$ and $P(s)$ relatively right prime and $P(s)$ column proper. Show that the condition: $|R(0)| = 0$ implies that certain of the zeros of $R(s)$ are located at $s = 0$ and would have to be cancelled by any form of compensation discussed in this text in order to achieve a (marginally stable) statically decoupled design.

8-35 If the proper transfer matrix $T(s) = R(s)P(s)^{-1}$ of a given system is invertible, with $|R(s)| = a_q s^q + \ldots + a_1 s + a_o$, prove that the system can be statically decoupled if and only if $a_o \neq 0$.

8-36 Consider the following scalar state-space system:

$$A = \begin{bmatrix} 0 & 1 & 0 \\ 0 & 0 & 1 \\ 1 & 0 & -1 \end{bmatrix},\ b = \begin{bmatrix} 0 \\ 0 \\ 1 \end{bmatrix},\ c = [1,\ -2,\ 0],\ \text{and}\ e = 0,\ \text{and}$$

the model transfer function $t_m(s) = \dfrac{1}{s^2+3s+2}$. Show that there does exist a l.s.v.f. pair $\{f,g\}$ such that the given model is "matched". Determine an appropriate feedback pair. Is the pair unique? Is the compensated state-space system controllable? observable? stable? Explain your answers.

8-37 Consider the state-space system: $A = \begin{bmatrix} 0 & 0 & 0 \\ 1 & 0 & 0 \\ 1 & 0 & 0 \end{bmatrix}$,

$B = \begin{bmatrix} 1 & 0 \\ 0 & 0 \\ 0 & -1 \end{bmatrix}$, $C = [1,\ -1,\ 2]$, and $E = [0\ \ 1]$. Show that it is possible to exactly match the model transfer matrix $T_m(s) = \left[\frac{3}{s^2+3s+2}, \frac{s-2}{s+2}\right]$ via l.s.v.f. and find a l.s.v.f. pair $\{F,G\}$ which works. Show that the feedback pair is not unique.

3-38 Consider the system with the open loop transfer matrix

$$T(s) = \begin{bmatrix} \frac{s+1}{s^2}, & \frac{s}{s-1} \end{bmatrix} = [s+1, \quad s]\begin{bmatrix} s^2, & 0 \\ 0, & s-1 \end{bmatrix}^{-1} = R(s)P^{-1}(s).$$

Show that it is possible to exactly match the model transfer matrix, $T_m(s) = [0, \ 1]$, via l.s.v.f. while arbitrarily specifying all (3) closed loop poles. Determine the eliminant matrix, M_e, of the pair $\{R(s),P(s)\}$ and find an $H(s)$ and a $K(s)$ such that $H(s)R(s) + K(s)P(s) = Q(s)F(s)$ if $P_{F,G}(s) = \begin{bmatrix} s^2+6s+5, & -6 \\ s+1, & s \end{bmatrix}$ and $Q(s) = \begin{bmatrix} s^2, & 6s+2 \\ -1, & s^2+4s+7 \end{bmatrix}$. If this triple $\{H(s),K(s),Q(s)\}$ is now employed in the feedback compensation scheme depicted in Figure 7.3.25, what will the resulting closed loop transfer matrix be? What can you conclude regarding the controllability, observability, and stability of the closed loop system?

3-39 Show that by initially considering only the first $\hat{p}$ linearly independent rows of $T(s)$, whenever the rank of $T(s)$ is less than both p and m, the question of exact model matching via combined l.s.v.f. and input dynamics can be resolved in essentially the same way as it is resolved when $\rho[T(s)] = p < m$.

3-40 In view of 8.5.3, show that whenever $\rho[T(s)] = \rho[T_m(s)] = m$, then $\rho[T_c(s)] = m$ and $T_c(s)$ will be unique. In view of this observation and R3 of Section 7.4, completely resolve the question of exact model matching via l.s.v.f. alone in this case. (Hint: Use the result obtained in Problem 7-15.)

8-41 Consider a system which has a proper $(p \times m)$ transfer matrix $T(s)$, precompensated (as shown in Figure 8-41a below) by a system which has a proper $(m \times p)$ transfer matrix, $G(s)$. Show that

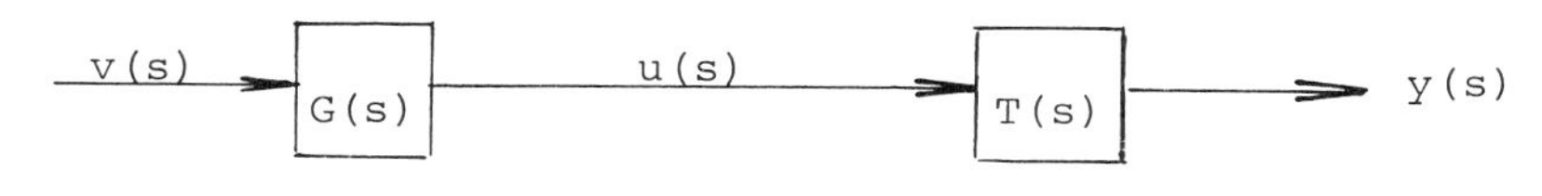

FIGURE 8-41a

the unity feedback design depicted in Figure 8-41b results in an equivalent closed loop transfer matrix, $T(s)G(s)$, provided that (i) $H(s) = G(s)[I_p - T(s)G(s)]^{-1} = [I_m - G(s)T(s)]^{-1}G(s)$ and (ii) either of the indicated inverses exists. Show that if $T(s)G(s)$ is strictly proper then $H(s)$ will be proper.

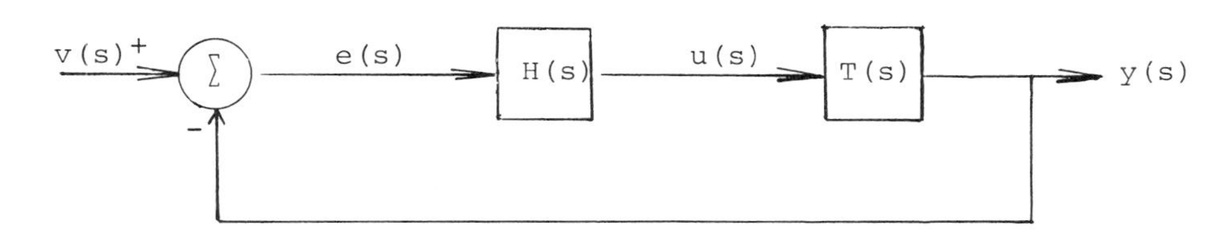

FIGURE 8-41b

8-42 Consider the composite system consisting of the series connection of $G(s) = P_G^{-1}(s)Q_G(s)$ and $T(s) = R_T(s)P_T^{-1}(s)$ (as depicted in Figure 8-41a) with $R_T(s)$ and $P_T(s)$ ($P_G(s)$ and $Q_G(s)$) relatively right (left) prime. Show that any state-space system which is equivalent to the series connected composite system is controllable (observable) if and only if $P_c(s) = P_G(s)P_T(s)$ and $Q_G(s)$ are relatively left prime ($R_T(s)$ and $P_c(s)$ are relatively right prime).

8-43 Consider a given scalar system with transfer function,

$t(s) = \frac{r(s)}{p(s)}$, and a model transfer function $t_m(s) = \frac{r_m(s)}{p_m(s)}$.
Show that the given system can be compensated via the scheme outlined in Section 7.4 to produce an exact model match if and only if $\{\partial[p(s)] - \partial[r(s)]\} \le \{\partial[p_m(s)] - \partial[r_m(s)]\}$.

8-44 Show that l.s.v.f. can sometimes be used to completely alter the zeros of the transfer matrix elements of a multivariable system while leaving its poles unchanged. In particular, given the open loop system with transfer matrix $T(s) = \left[\frac{s+1}{s^2}, \frac{s}{s^2-1}\right] =$

$R(s)P(s)^{-1}$, with $R(s) = [s+1, \; s]$ and $P(s) = \begin{bmatrix} s^2 & 0 \\ 0 & s^2-1 \end{bmatrix}$,

verify that the l.s.v.f. pair $\{F(s),G\} = \left\{ \begin{bmatrix} \frac{1}{5} & -\frac{2}{5} \\ -\frac{2}{5} & -\frac{1}{5} \end{bmatrix}, \begin{bmatrix} 2 & -1 \\ 1 & 2 \end{bmatrix} \right\}$

yields the closed loop transfer matrix,

$$T_{F,G}(s) = R(s)[P(s) - F(s)]^{-1}G = \left[\frac{3s+2}{s^2}, \; \frac{s-1}{s^2-1}\right].$$

REFERENCES

[A1] Athans, M. S., and Falb, P. L., Optimal Control, McGraw Hill, Inc., 1966.

[A2] Anderson, B. D. O., and Moore, J. B., "State Estimation via the Whitening Filter," 1968 JACC Preprints (Ann Arbor, Michigan), pp. 123-129.

[B1] Brockett, R. W., Finite Dimensional Linear Systems, John Wiley and Sons, Inc., 1970.

[B2] Brockett, R. W., and Mesarovic, M. D., "The Reproducibility of Multivariable Systems," J. Math. Analysis and Applications, Vol. 11, 1965, pp. 548-563.

[B3] Brockett, R. W., "Poles, Zeros, and Feedback: State Space Interpretation," IEEE Trans. Auto. Cont., Vol. AC-10, April, 1965, pp. 129-135.

[B4] Bass, R. W., and Gura, I., "High Order System Design via State-Space Considerations," Preprints 1965 JACC, Rensselaer Polytechnic Institute, Troy, New York, pp. 311-318.

[B5] Brasch, F. M., and Pearson, J. B., "Pole Placement Using Dynamic Compensators," IEEE Trans. on Auto. Control, Vol. AC-15, No. 1, Feb., 1970, pp. 34-43.

[B6] Bristol, E. H., "On a New Measure of Interaction for Multivariable Process Control," IEEE Trans. Auto. Cont., Vol. AC-11 (1), Jan., 1966, pp. 133-134.

[C1] Chen, C. T., Introduction to Linear System Theory, Holt, Rinehart and Winston, Inc., 1970.

[C2] Coddington, E. A., and Levison, N., Theory of Ordinary Differential Equations, McGraw Hill, 1955.

[C3] Chen, C. T., "Design of Feedback Control Systems," Proceedings of the National Electronics Conference, Vol. 57, 1969, pp. 46-51.

[C4] Chen, C. T., "A New look at Transfer Function Design," Proceedings of the IEEE, Vol. 59, No. 11, Novenber, 1971, pp. 1580-1585.

[C5] Chen, C. T., and Hsu, C. H., "Design of Dynamic Compensators for Multivariable Systems," Paper No. 8-E1, 1971 JACC Conference Preprints, Washington University, St. Louis, Missouri, August 11-13, 1971, pp. 893-900.

[C6] Chandrasekharan, P. C., "Observability and Decoupling", IEEE Trans. Auto. Control, Vol. AC-16 (5), October, 1971, pp. 482-484.

[D1] Desoer, C. A., Notes for a Second Course on Linear Systems, Van Nostrand Reinhold Company, 1970.

[D2] Dorato, P., "On the Inverse of Linear Dynamical Systems," IEEE Trans. Systems Science & Cybernetics, Vol. 5, No. 1, January, 1969, pp. 43-48.

[D3] Davison, E. J., "On Pole Assignment in Multivariable Linear Systems," IEEE Trans. Auto. Control, Vol. AC-13, No. 6, December, 1968, pp. 747-748.

[D4] Davison, E. J., "On Pole Assignment in Linear Systems with Incomplete State Feedback," IEEE Trans. Auto. Control, Vol. AC-15 (3), June, 1970, pp. 348-351.

[E1] Erzberger, H., "Analysis and Design of Model Following Control Systems by State Space Techniques," Proc. 1968 JACC (Ann Arbor, Michigan), pp. 572-581.

[E2] Evans, W. R., "Control System Synthesis by Root Locus Method," Trans. AIEE, Vol. 69, pp. 66-69, 1950.

[F1] Falb, P. L., and Wolovich, W. A., "Decoupling in the Design and Synthesis of Multivariable Control Systems," IEEE Trans. Auto. Cont., Vol. AC-12, December, 1967, pp. 651-659.

[F2] Faddeeva, D. K., and Sominskii, I. S., Problems in Higher Algebra, 2nd. ed., Moscow, 1949; 5th ed., Moscow: Gostekhizdat, 1954.

[F3] Freund, D., "Design of Time-Variable Multivariable Systems by Decoupling and by the Inverse," IEEE Trans. Auto. Control, Vol. AC-16 (2), April, 1971, pp. 183-185.

[G1] Gantmacher, F. R., Theory of Matrices, Chelsea, 1959, Vol. 1.

[G2] Gilbert, E. G., "Controllability and Observability in Multivariable Control Systems," SIAM J. on Cont., Vol. 1 (1963), pp. 128-151.

[G3] Gilbert, E. G., "The Decoupling of Multivariable Systems by State Feedback," SIAM J. Control, Vol. 7, Feb., 1969, pp. 50-64.

[G4] Gilbert, E. G., and Pivnichny, J. R., "A Computer Program for the Synthesis of Decoupled Multivariable Feedback Systems," IEEE Trans. Auto. Control, Vol. AC-14 (6), Dec., 1969, pp. 652-659.

[H1] Hoffman, K., and Kunze, R., Linear Algebra, Prentice Hall, Inc., 1961.

[H2] Ho, B. L., and Kalman, R. E., "Effective Construction of Linear, State-variable Models from Input/Output Functions," Proc. Third Allerton Conference (1965), pp. 449-459.

[H3] Ho, Y. C., and Behn, R. D., "On a Class of Linear Stochastic Differential Games," IEEE Trans. Auto. Cont., Vol. AC-13, June, 1968, pp. 227-240.

[H4] Heymann, M., "Comments on Pole Assignment in Multi-Input Controllable Linear Systems," IEEE Trans. Auto. Control, Vol. AC-13, No. 6, Dec., 1968, pp. 748-749.

[K1] Kalman, R. E., "Contributions to the Theory of Optimal Control," Boletin de la Sociedad Matematica Mexicana, 1960.

[K2] Kalman, R. E., "Canonical Structure of Linear Dynamical Systems," Proc. Nat. Acad. Sci., U. S., Vol. 48, pp. 596-600, 1962.

[K3] Kalman, R. E., "Mathematical Description of Linear Dynamical Systems," SIAM J. Cont., Vol. 1, No. 2, 1963, pp. 152-192.

[K4] Kreindler, E., and Sarachick, P. E., "On the Concepts of Controllability and Observability of Linear Systems," IEEE Trans. Auto. Control, Vol. AC-9, April, 1964, pp. 129-136.

[K5] Kalman, R. E., "On the Structural Properties of Linear, Constant, Multivariable Systems," Proc. Third IFAC Congress, June, 1966.

[L1] Liapunov, A. M., "Probleme General de la Stabilite du Mouvement," Ann. Fac. Sci. Toulouse, Vol. 9 (1907), pp. 203-474.

[L2] LaSalle, J. P., "The Time-Optimal Control Problem," Contributions to Differential Equations, Vol. V, pp. 1-24, Princeton University Press, Princeton, New Jersey, 1960.

[L3] Luenberger, D. G., "Canonical Forms for Linear Multivariable Systems," IEEE Trans. Auto. Control, AC-12 (1967), pp. 290-293.

[L4] Luenberger, D. G., "Observing the State of a Linear System," IEEE Trans. on Military Electronics, Vol. MIL-8, pp. 74-80, April, 1964.

[L5] Luenberger, D. G., "Observers for Multivariable Systems," IEEE Trans. Auto. Control, Vol. AC-11, No. 2, April, 1966, pp. 190-197.

[M1] MacDuffee, C. C., _The Theory of Matrices_, Chelsea, 1956.

[M2] Mayne, D. Q., "Computational Procedure for the Minimal Realization of Transfer Function Matrices," Proc. IEEE, 115 (1968), pp. 1363-1368.

[M3] Mehra, R. K., "Inversion of Multivariable Linear Dynamic Systems Using Optimum Smoothing," IEEE Trans. Auto. Cont., April, 1970, p. 252.

[M4] Morgan, B. S., "The Synthesis of Single Variable Systems by State Variable Feedback," Proc. Allerton Conf. on Circuit and Systems Theory, Urbana, Illinois, November, 1963.

[M5] Morse, A. S., and Wonham, W. M., "Decoupling and Pole Assignment by Dynamic Compensation," SIAM J. Control, Vol. 8, 1970, pp. 317-337.

[M6] Morse, A. S., and Wonham, W. M., "Triangular Decoupling of Linear Multivariable Systems," IEEE Trans. Auto. Cont., Vol. AC-15 (4), August, 1970, pp. 447-449.

[M7] Morse, A. S., and Wonham, W. M., "Status of Noninteracting Control," IEEE Trans. Auto. Control, Vol. AC-16 (6), Dec., 1971, pp. 568-581.

[M8] Mufti, I. H., "On the Observability of Decoupled Systems," IEEE Trans. Auto. Cont., Vol. AC-14 (1), Feb., 1969, pp. 75-77.

[M9] Moore, B. C., and Silverman, L. M., "Exact Model Matching by State Feedback and Dynamic Compensation," Proc. 1971 IEEE Conference on Decision and Control, Miami Beach, Florida, Dec., 1971, pp. 114-120.

[N1] Newton, G. C., Gould, L. A., and Kaiser, J. F., Analytical Design of Linear Feedback Controls, Wiley Publishing Co., New York, New York, 1957.

[N2] Nazar, S., and Rekasius, A. V., "Decoupling of a Class of Non-linear Systems," IEEE Trans. Auto. Cont., Vol. AC-16 (3), June, 1971, pp. 257-260,

[O1] Orner, P. A., "Construction of Inverse Systems," IEEE Trans. Auto. Control, Vol. AC-17, No. 1, February, 1972, pp. 151-153.

[P1] Pontryagin, L. S., V. Boltyanskii, R. Gamkrelidze, and E. Mishchenko, The Mathematical Theory of Optimal Processes, Interscience Publishers, Inc., New York, 1962.

[P2] Polak, E., "An Algorithm for Reducing a Linear, Time-Invariant Differential System to State Form," IEEE Trans. Auto. Control, AC-11 (1966), pp. 577-579.

[P3] Porter, W. A., "Decoupling and Inverses for Time-Varying Linear Systems," IEEE Trans. Auto. Cont., August, 1969, pp. 378-380.

[P4] Pearson, J. B., and Ding, C. Y., "Compensator Design for Multivariable Linear Systems," IEEE Trans. on Auto. Control, Vol. AC-14, No. 2, April, 1968, pp. 130-134.

[P5] Popov, V. M., Lecture notes in Mathematics, 144, Seminar on Differential Equations and Dynamical Systems II, Springler-Verlag, (1969), pp. 169-180.

[P6] Porter, W. A., "Decoupling of and Inverses for Time-varying Linear Systems," IEEE Trans. Auto. Cont., Vol. AC-14 (4), August, 1969, pp. 378-380.

[R1] Rosenbrock, H. H., State-Space and Multivariable Theory, John Wiley & Sons, Inc., 1970.

[R2] Rosenbrock, H. H., "On Linear System Theory," Proc. IEEE, Vol. 114, No. 9, September, 1967, pp. 1353-1359.

[R3] Rosenbrock, H. H., "Least Order of System Matrices," Electronics Letters, Vol. 3, No. 2, February, 1967, pp. 58-59.

[R4] Rosenbrock, H. H., "Generalised Resultant," Electronics Letters, Vol. 4, No. 12, June, 1968, pp. 250-251.

[S1] Silverman, L. M., "Inversion of Multivariable Linear Systems," IEEE Transactions on Auto. Cont., Vol. AC-14 (3), June, 1969, pp. 270-276.

[S2] Singh, S. P., "A Note on Inversion of Linear Systems," IEEE Trans. Auto. Cont., August, 1970, pp. 492-493.

[S3] Sain, M. K., and Massey, J. L., "Invertibility of Linear Time-Invariant Dynamical Systems," IEEE Trans. Auto. Cont., Vol. AC-14 (2), April, 1969, pp. 141-149.

[S4] Sain, M. K., "Functional Reproducibility and the Existence of Classical Sensitivity Matrices," IEEE Trans. Auto. Cont., Vol. AC-12, August, 1967, p. 458.

[S5] Silverman, L. M., and Payne, H. J., "Input-Output Structure of Linear Systems with Application to the Decoupling Problem," SIAM J. Control, Vol. 9, 1971, pp. 199-233.

[S6] Sato, S. M., and Lopresti, P. V., "On Partial Decoupling in Multivariable Control Systems," Proc. 1970 JACC, Atlanta, Ga., June, 1970, pp. 814-819.

[S7] Sankaran, V., and Srinath, M. D., "Decoupling of Systems with Plant and Measurement Noise," IEEE Trans. Auto. Control, Vol. AC-16 (2), April, 1971, pp. 202-203.

[S8] Shinskey, F. G., Process Control Systems, McGraw Hill Book Company, 1967.

[T1] Thomas, J. M., Theory of Equations, McGraw Hill, 1938, pp. 169-181.

[W1] Wolovich, W. A., "Equivalent Representations and Realizations of Linear Multivariable Systems," Brown Engineering Report NSF GK-2788/2, October, 1970.

[W2] Wolovich, W. A., "The Application of State Feedback Invariants to Exact Model Matching," Proceedings of the Fifth Annual Princeton Conference on Information Sciences and Systems, 1971, pp. 387-392.

[W3] Wolovich, W. A., "A Direct Frequency Domain Approach to State Feedback and Estimation," 1971 IEEE Decision and Control Conference, Miami, Florida, Dec., 1971.

[W4] Wolovich, W. A., and Falb, P. L., "On the Structure of Multivariable Systems," SIAM J. Control, Vol. 7, No. 3, August, 1969, pp. 437-451.

[W5] Wolovich, W. A., "The Determination of State-Space Representations for Linear Multivariable Systems," 2nd IFAC Symposium on Multivariable Technical Control Systems, Duesseldorf, Germany, October, 1971.

[W6] Wonham, W. M., "On Pole Assignment in Multi-Input Controllable Linear Systems," IEEE Trans. Auto. Control, Vol. AC-12, No. 6, Dec., 1967.

[W7] Wolovich, W. A., "A Frequency Domain Approach to the Design and Analysis of Linear Multivariable Systems," NASA TN D-5743, May, 1970.

[W8] Wiener, N., The Extrapolation, Interpolation, and Smoothing of Stationary Time Series, Wiley Publishing, Inc., 1949.

[W9] Wonham, W. M., and Morse, A. S., "Decoupling and Pole-Assignment in Linear Multivariable Systems: A Geometric Approach," SIAM J. Control, Vol. 8, Feb., 1970, pp. 1-18.

[W10] Wolovich, W. A., "The Use of State Feedback for Exact Model Matching," SIAM J. on Control, Vol. 10, No. 3, August, 1972, pp. 512-523.

[W11] Winsor, C. A., and Roy, R. J., "The Application of Specific Optimal Control to the Design of Desensitized Model Following Control Systems," IEEE Trans. Auto. Cont., AC-15 (3), June, 1970, pp. 326-333.

[W12] Wolovich, W. A., "Static Decoupling," Preprints of the 1973 JACC, Ohio State University, Columbus, Ohio, June 20-23, 1973.

[W13] Wolovich, W. A., "On the Synthesis of Multivariable Systems," Preprints of the 1972 JACC, Stanford University, August, 1972, pp. 158-165.

[Y1] Youla, D. C., and Tissi, P., "n-Port Synthesis via Reactance Extraction," IEEE International Convention Record, 1966, pp. 183-205.

[Z1] Zadeh, L., and Desoer, C. A., Linear System Theory, McGraw-Hill, 1963.

INDEX